A Primer of Molecular Population Genetics

Timeline of molecular population genetic history

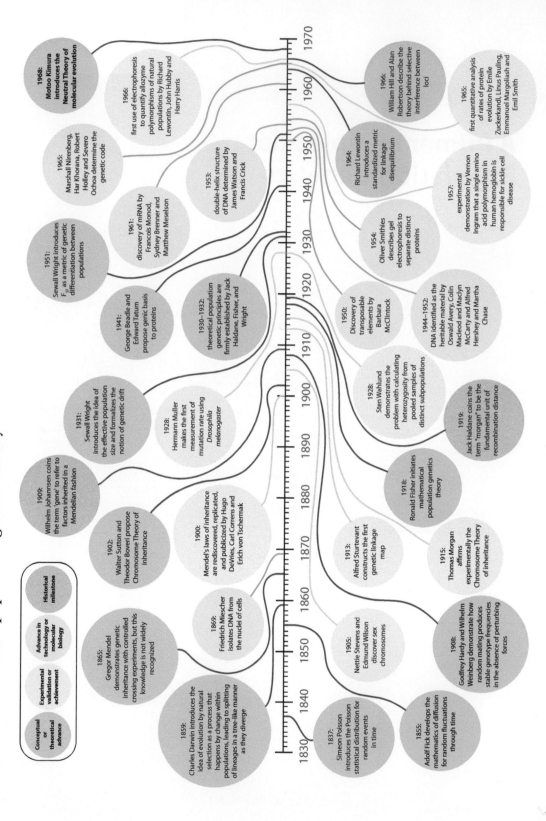

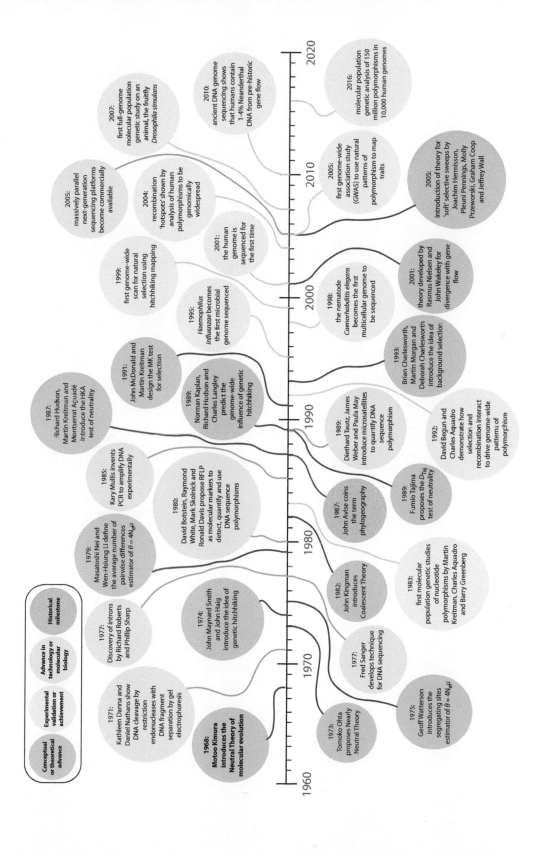

1960   1970   1980   1990   2000   2010   2020

1966: Motoo Kimura introduces the Neutral Theory of molecular evolution

1973: Tomoko Ohta proposes Nearly Neutral Theory

1971: Kathleen Danna and Daniel Nathans show DNA cleavage by restriction endonucleases with DNA fragment separation by gel electrophoresis

1977: Fred Sanger develops technique for DNA sequencing

1975: Geoff Watterson introduces the segregating sites estimator of $\theta = 4N_e\mu$

1974: John Maynard Smith and John Haig introduce the idea of genetic hitchhiking

1977: Discovery of introns by Richard Roberts and Phillip Sharp

1979: Masatoshi Nei and Wen-Hsiung Li define the average number of pairwise differences estimator of $\theta = 4N_e\mu$

1982: John Kingman introduces Coalescent Theory

1983: first molecular population genetic studies of nucleotide polymorphisms by Martin Kreitman, Charles Aquadro and Barry Greenberg

1980: David Botstein, Raymond White, Mark Skolnick and Ronald Davis propose RFLP as molecular markers to detect, quantify and use DNA sequence polymorphisms

1985: Kary Mullis invents PCR to amplify DNA experimentally

1987: John Avise coins the term phylogeography

1989: Fumio Tajima proposes the D_{Taj} test of neutrality

1992: David Begun and Charles Aquadro demonstrate how selection and recombination interact to drive genome-wide patterns of polymorphism

1987: Richard Hudson, Martin Kreitman and Montserrat Aguadé introduce the HKA test of neutrality

1991: John McDonald and Martin Kreitman design the MK test for selection

1989: Norman Kaplan, Richard Hudson and Charles Langley predict the genome-wide influence of genetic hitchhiking

1989: Diethard Tautz, James Weber and Paula May introduce microsatellites to quantify DNA sequence polymorphism

1993: Brian Charlesworth, Martin Morgan and Deborah Charlesworth introduce the idea of background selection

1995: *Haemophilus influenzae* becomes the first microbial genome sequenced

1998: the nematode *Caenorhabditis elegans* becomes the first multicellular genome to be sequenced

2001: theory developed by Rasmus Nielsen and John Wakeley for divergence with gene flow

1999: first genome-wide scan for natural selection using hitchhiking mapping

2001: the human genome is sequenced for the first time

2004: recombination 'hotspots' shown by analysis of human polymorphisms to be genomically widespread

2005: massively parallel next-generation sequencing platforms become commercially available

2005: first genome-wide association study (GWAS) to use natural patterns of polymorphism to map traits

2005: introduction of theory for 'soft' selective sweeps by Joachim Hermisson, Pleuni Pennings, Molly Przeworski, Graham Coop and Jeffrey Wall

2007: first full-genome molecular population genetic study on an animal, the fruitfly *Drosophila simulans*

2010: ancient DNA genome sequencing shows that humans contain 1–4% Neanderthal DNA from pre-historic gene flow

2016: molecular population genetic analysis of 150 million polymorphisms in 10,000 human genomes

Conceptual or theoretical advance

Experimental validation or achievement

Advance in technology or molecular biology

Historical milestone

A Primer of Molecular Population Genetics

ASHER D. CUTTER

Department of Ecology & Evolutionary Biology
University of Toronto

OXFORD
UNIVERSITY PRESS

Great Clarendon Street, Oxford, OX2 6DP,
United Kingdom

Oxford University Press is a department of the University of Oxford.
It furthers the University's objective of excellence in research, scholarship,
and education by publishing worldwide. Oxford is a registered trade mark of
Oxford University Press in the UK and in certain other countries

First Edition published in 2019

Published in the United States of America by Oxford University Press
198 Madison Avenue, New York, NY 10016, United States of America

British Library Cataloguing in Publication Data
Data available

Library of Congress Control Number: 2019934003

ISBN 978–0–19–883894–4 (hbk.)
ISBN 978–0–19–883895–1 (pbk.)

DOI: 10.1093/oso/9780198838944.001.0001

Printed and bound by
CPI Group (UK) Ltd, Croydon, CR0 4YY

For my descendants and for my ancestors,
starting with Beatrix and Oona and with Bradley and Lillian

Preface
Why you should read this book

My mind's-eye title for this book is *A Fireside Chat about Molecular Population Genetics*. Winters are cold here in Canada, and I like to sit back by a fire and warm up to the alluring collection of ideas in molecular population genetics. I'd like you to, as well. If you live somewhere hot, then substitute the fireplace and cushions with a shady tree and hammock. Let this book be your hot chocolate or your lemonade. That sums up my outlook for this *Primer*.

So, what are these alluring ideas? Here are a few savory questions that molecular population genetics aims to answer. What are the genomic inner-workings of adaptations and how do we see them in DNA? How much of evolution is actually driven by natural selection versus something else? How does the ebb and flow in the abundance of individuals over time get marked onto chromosomes to record this history? Molecular population genetics is the main way that researchers apply theory to data to answer questions like these. It provides the way to learn about how evolution works and how it shapes species by looking at DNA. It lets us understand the logic of how mutations originate to then change in abundance in populations to potentially get locked-in as DNA sequence divergence between species. This crucial role in modern science stems in no small part from the mainstreaming of population genomic sequencing technologies that reinforce the ever-growing relevance of molecular population genetics to diverse problems in biology. All of this makes it important for you to start on your way to learning about molecular population genetics.

I presume that you are a novice to studying population genetics from a molecular angle. Welcome! I also presume that you would like your first foray into mol-pop-gen to be friendly and unintimidating, while nevertheless enlightening you about how we can think about evolution in a genetic way. That is what I aim to do with this book. This *Primer* is not, however, an all-encompassing text in population genetics. I provide only limited detail of classic population genetics theory. Fortunately, many excellent resources for classic population genetics theory are readily available, from the introductory to the advanced. I *do* expect that you have basic knowledge about genetics, molecular biology, and evolutionary principles. If you are rusty, take heart, as I provide quick reminders about critical basic background at the relevant points. My reason for minimizing coverage of both basic biology and classic population genetics is so that we can jump right in to the portion of population genetics that deals explicitly with molecular data and ideas. I take DNA sequence as our starting point, and then I bring in key concepts from classic theory as we need them.

There are many ways that I could have chosen to organize the material in this book. I settled on a way that makes sense to me and that I hope will make sense to you. I start with the tangible and build toward the more abstract concepts, then bring it all home with integrative case studies. I start by jumping in with the nuts and bolts of molecular population genetics: what is genetic variation at the molecular level and how do we measure it? With these concrete metrics in hand (Chapters 1–3), I then take a pause for a moment to step back and deal with the inevitable questions: why are we doing this? What is the context? What does it mean to see new mutations and polymorphisms and DNA differences between species? What should we *expect* to see? This gives us a chance to connect ideas of neutral molecular evolution (Chapter 4), the pistons and motors in this tenuous mechanical analogy, to those nuts and bolts of mutation and genetic variation. With this new firm footing in both practical and conceptual topics, I then drive into some more challenging but important concepts relating to genealogies in genome evolution (Chapter 5), including the basics of coalescent theory and phylogenies, and to recombination and linkage disequilibrium (Chapter 6). That wraps up the groundwork.

The next phase of the book explores the theme of confronting neutral models with data (Chapters 7–8). What happens when the real world does not match the convenient predictions of the standard neutral model? How should we anticipate selection and demography to perturb molecular signatures in genomes? How can we detect it? As a capstone to having worked through and built up all these views toward thinking about evolution in populations with molecular data, I end with a series of case studies that integrate many of the ideas and approaches from throughout the book (Chapter 9). I hope that this flow works and helps you to get the ideas. I also hope that it keeps you engaged to see why so many people have been enchanted by the genetics of evolution for over 100 years and why molecular population genetics, in particular, remains such a flourishing discipline today.

The biochemical details of DNA mixed together with mathematical models of evolutionary change, spiced up with the lifestyle realities of wild organisms, makes for a rich and complex scientific soup. For our purposes here, we focus on the essentials of molecular population genetics. We will zoom in and out to think about single nucleotides and individual gene loci and whole genomes; I will take for granted that "genomics" is simply an integrated part of modern molecular population genetics.

My intention with this book is to provide a concise and readable introduction to the important ideas and caveats for understanding evolution at the molecular level. Whether your interest is the adaptive process or the historical demography of populations, there is something to chew on in these pages. To give you a peek at some of the complexities and the emphasis of current research in this field, I also include some intriguing teasers for topical advanced issues. However, this book does not aim to detail thoroughly the nuances of advanced topics in molecular population genetics. I also mostly avoid details about specific analysis software and sequencing technologies, given how rapidly these techniques turn over; I take it for granted that there are robust ways to identify and quantify differences in DNA sequences. Instead, I hope that undergraduates, junior graduate students, and practicing scientists from cognate fields hoping to gain a basic understanding and appreciation for molecular population genetics will all find this *Primer* valuable for that simple goal stated in the top sentence of this paragraph.

Different traditions of study emphasize to greater degrees different portions of population genetic phenomena. In the immortal words of Christopher Wallace, "It's all good." For example, you may be familiar with the focus on the history and timing of demographic dynamics of species by phylogeographers and molecular anthropologists to learn about the history of migration and population size change over time. Or perhaps the molecular phylogeneticist's emphasis on delineating species trees in evolutionary history to understand the relationships among taxa. Or the selectionist's aim to localize recent molecular targets of adaptation in genomes. In the chapters that follow, we will consider all these views, acknowledging each as an important facet of the dynamics of DNA change in populations from mutational origin to fixed divergence between species. They are all good ways to help us appreciate from the inside what Charles Darwin appreciated from the outside, how "from so simple a beginning endless forms most beautiful and most wonderful have been, and are being, evolved."

Acknowledgments

I owe special thanks to all the researchers who made their data publicly available or who so graciously provided me their data to graph as examples throughout the text (Philip Awadalla, Thomas Bataillon, David Begun, Vincent Castric, Frank Chan, William Cresko, Matthew Dean, Hans Ellegren, Nicolas Galtier, Nandita Garud, Matthew Hahn, Daniel Halligan, Felicity Jones, Peter Keightley, Theresa Lamagni, Donna Lehman, Luke Mahler, Michael Nachman, Rob Ness, Sarah Ng, Friso Palstra, John Parsch, Dmitri Petrov, John Pool, Qiang Qiu, Noah Rosenberg, Alisa Sedghifar, Jay Shendure, Alexander Suh, Paul Verdu, Nagarjun Vijay, John Welch, Robert Williamson, Jochen Wolf). Similarly, I thank Brian Baer, Spencer C.H. Barrett, Julian Bermudez, Rob Colautti, Nicholas Ellis, Eddie Ho, Hopi Hoekstra, Lucia Kwan, Luke Mahler, Craig Miller, Rob Ness, Rebecca Schalkowski, Janice Ting, and Holly Warland for enhancing the organismal context of examples with their photographic contributions. I also must express my gratitude to the chronology of mentors I've had over the years in Arizona and Edinburgh who did more than they might have realized in sparking my interest in population genetics and molecular evolution: Leticia Avilés, Brian Charlesworth, Deborah Charlesworth, Wayne Maddison, Rick Michod, Nancy Moran, Michael Nachman, Howard Ochman, Bruce Walsh, Sam Ward. My exceptional colleagues in the evolutionary genetics group at the University of Toronto helped me to hone my thinking in writing this book and by providing me invaluable feedback on the content. I am indebted to the advice and discussions with Aneil Agrawal, Nicole Mideo, Alan Moses, and Stephen Wright, to Joanna Bundus, Amardeep Singh, Jasmina Uzunovic, and to Caressa Tsai. I am also grateful for the careful and constructive suggestions and corrections from Charles Baer, Andrea Betancourt, Yaniv Brandvain, Ian Dworkin, Andrew Kern, Bret Payseur, Todd Schlenke, Jon Seger, and Nadia Singh. But the buck stops here, and any errors are my own. Special thanks to Yee-Fan Sun for holding down the fort and for all her continual support. Finally, it can't go without saying how appreciative I am for the patience of the hundreds of undergraduate students who have graced the seats of the lecture halls over the years as I aimed to bring molecular population genetics into your lives; a heap of rough-hewn course notes evolved into this book as you taught me how to explain myself.

Contents

CHAPTER 1

Introduction

What is molecular population genetics?

Do you want to travel through time? Well, you're in luck. Because molecular population genetics is a time machine and, for many purposes, it may be the best we'll ever have. It transports us backward through time, forward through time, takes snapshots of points in time. When we look out of the windows of *this* time machine, we don't see wormholes and ethereal gases spinning in spirals—instead we see the evolution of the spiral helix of DNA. We see mutations arise and spread like ripples through populations, producing adaptations for all of life, the baleen of the whale, the eye of the squid, the horn of the rhinoceros, the light of the firefly (Figure 1.1). We see our ancestors winking at us at the far ends of gene trees. We see what we share with chimpanzees, with kangaroos, with clams and algae and *Salmonella*. We see some populations explode with growth to expand across continents and others shrink to petite refuges and still others spring like rubber bands, their numbers bigger and smaller and bigger and smaller, over the passage of time. We see animals shuttle back and forth as migrants between different groups, and we see them stop their shuttling, changing those groups forever. We see it all in our genes, in the simple differences of those four nucleotide letters of DNA that share parts of their history with everyone else. The time machine of molecular population genetics lets us get inside evolution to see what is going on and to really understand how it unfolds.

1.1 On the origins of molecular population genetics

Everyone now takes it for granted that changes to DNA are inextricably woven into thinking about evolution (Box 1.1). But, of course, this was not always the case. Evolution, as Charles Darwin wrote about it publicly from 1859 until his death in 1882, describes the process of heritable trait change across generations. The mechanism of the "heritable" piece of this process, however, was unfortunately and famously unknown to Darwin. This missing link stymied a quantitative view of evolutionary change for a while. The logic of genetics only became appreciated in 1900, after the rediscovery of Gregor Mendel's cross-breeding experiments that he had performed four decades earlier. Mendel demonstrated that genes are "particulate," inherited intact from parent to offspring, and that their transmission from parent to offspring follows certain mathematical rules. With this simple insight, progress was set to accelerate.

Now, with more than a century of genetic research in hand, we can define evolution in the most basic of ways: evolution is the change across generations in the relative abundance of different forms of genes in a population (Box 1.2). The clarity of genetic

A Primer of Molecular Population Genetics. Asher D. Cutter, Oxford University Press (2019).
© Asher D. Cutter 2019. DOI: 10.1093/oso/9780198838944.001.0001

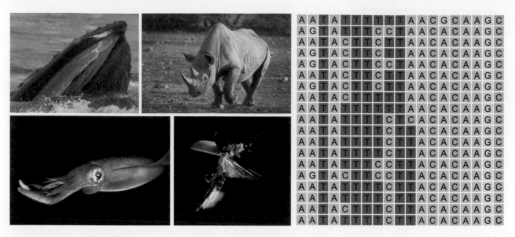

Figure 1.1 Adaptations.

Adaptations abound in nature. The dramatic and charismatic traits of organisms most easily capture our imaginations. You see them everywhere: the filter-feeding baleen of toothless whales that allow them to extract and eat small invertebrates from ocean water, or the keratin horns of black rhinoceros used in social interactions and defense against would-be predators, or the magnificent compound eyes of invertebrates like squid that evolved independently of similar structures in vertebrates, or the almost-magical glow of luciferase from the light-emitting organs of fireflies. As Charles Darwin put it, "we see beautiful adaptations everywhere and in every part of the organic world." All of these features evolved from mutation to and selection on DNA sequences that got recorded in those species' genomes. An alignment of homologous DNA, with each row a different species of *Anolis* lizard (see Figure 4.2), shows the similarities and differences among them as a result of mutational changes getting fixed by natural selection and genetic drift over the course of evolution. Humpback whale feeding photo by NOAA Photo Library, reproduced under the CC-BY 2.0 license (cropped from original). Rhino © 2630ben/Shutterstock.com. Squid © uatari/Shutterstock.com. Firefly by Terry Priest and reproduced under the CC BY-SA 2.0 license.

thinking spurred mathematical summaries of evolutionary change that reinforced and elaborated on Darwin's foundation for how to conceive of the living world (Boxes 1.3 and 1.4). The mathematical theory and empirical analysis of changes to gene frequencies over time is the branch of science that we call **population genetics**.

The basic ingredients to population genetic theory about evolution were firmly established in the first half of the last century by Ronald Fisher, Jack Haldane, and Sewall Wright, among other luminaries of biology (see Timeline of molecular population genetic history). The five major forces of evolution—mutation, genetic drift, migration (gene flow), recombination, natural selection—secured their place in evolutionary thinking nearly 100 years ago. All this discovery took place without any knowledge of what is the principal heritable material that transmits from parent to offspring: DNA. This classic and exceptionally general population genetics theory forms the backbone of modern evolutionary understanding (Box 1.2), including how we think about the evolution of DNA sequences. Michael Lynch (2007) emphasized the importance of understanding population genetics in saying "nothing in evolution makes sense except in the light of population genetics" in his saucy rephrasing of Theodosius Dobzhansky's famous quote about the light that evolution brings to biology.

There are two key ways of thinking when thinking about population genetics, and both are foreign to most of us in our everyday individual experience: population thinking and tree thinking. We are used to how cause and effect interact for an individual, like how when you take a bite of an artisanal chocolate, you sense a delicious taste in your mouth. Population effects are a bit different. The consequences for the population result from the collective effects on individuals: you tell your friends about the new chocolatier in town, and pretty soon the general vibe of the neighborhood is much happier from all the endorphin stimulation from the increased chocolate consumption. In population genetics, we are concerned with the genetic outcomes for a population as a consequence of the survival and reproduction of its individual members. How many individuals have this gene copy or that one, and how well overall does a given copy propagate to the next generation? Another component of population thinking is that we generally do not have information about every individual; we just have data for a sample of them that we must use as a representative group, presuming that they are a random subset of individuals.

But individuals in populations are not static or fully independent of one another, their composition changes over time and their gene copies are related to one another: these features lead us to tree thinking. The idea is that we can use graphical branching diagrams to give a concrete representation of relationships between gene copies that are present in different individuals, whether those individuals are members of a single species or even from different species (Figure 1.2). DNA sequences give us a natural basis for quantifying homologous features found in different individuals (see section 5.3), features that share a common ancestor but that have changed due to some or all of the factors that influence how common is a given gene copy in a population. A virtue of both population thinking and tree thinking is that we can use mathematics and statistics to integrate ideas and data to describe evolution in genetic terms.

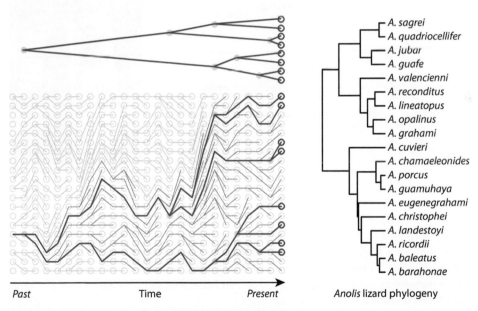

Past *Time* *Present*	*Anolis* lizard phylogeny

Figure 1.2 Evolutionary trees for tree thinking in evolution.
Using tree thinking about evolution lets us visualize the relationships between homologous gene copies found in different individuals in a population (left, see section 5.1) as well as gene copies found in different species (right, see section 5.3; for details on *Anolis* see Figure 4.2). Each circle in the left diagram represents a different copy of a gene found in different individuals in a population. The purple lines trace the genealogical relationships between a subset of gene copies to the ancestral gene copy that represents their most recent common ancestor (leftmost filled green circle), with the gene tree on top having been pruned of extraneous information. For the species gene tree in the right diagram, a single representative gene copy was sampled from each species to determine the branching pattern that defines shared ancestry among species.

Box 1.1 DNA

DNA is a beautiful thing. It is beautiful to visualize and it is beautiful to conceptualize. DNA's double-helix chemical structure, built from just four nucleotides (A = adenine, T = thymine, G = guanine, C = cytosine) and a single kind of sugar, lays the groundwork for the elegant biochemical mechanisms of replication and transcription. As James Watson and Francis Crick noted in their seminal article on the structure of DNA in 1953, with a healthy dose of cheek, "It has not escaped our notice that the specific pairing that we have postulated immediately suggests a possible copying mechanism for the genetic material." DNA comprising the entire genome copies itself before cells divide (replication), whereas individual genes copy themselves into RNA (transcription), with the help of different kinds of polymerase enzymes. But it was Ruth Padel, Charles Darwin's great-great-granddaughter, who wrote most poetically of the genetic material as "that other / secret scripture DNA: a hidden barcode / invisible as a string of fireflies / sleeping on a leaf-edge in the pre-dusk blue of day." Four lines from "Giant Bugs from the Pampas," first published in *Darwin: A Life in Poems* (2009) with permission from United Agents LLP on behalf of Ruth Padel.

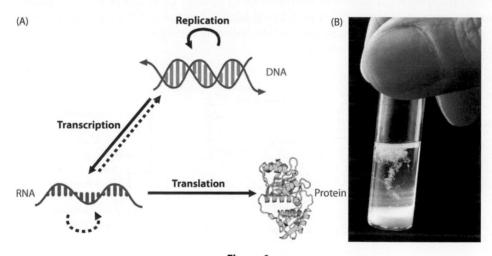

Figure 1

Figure 1 (A) The Central Dogma of Molecular Biology holds that biological information flows from DNA to RNA to protein within cells, and replication of DNA ensures its transmission to daughter cells (solid lines). Cellular machinery also can allow some RNA to beget more RNA, or for RNA to be reverse-transcribed back into DNA (dotted lines). (B) The white stringy material in the test tube is DNA extracted from plant material. Photo by CSIRO and reproduced under the CC-By 3.0 license.

The second half of the last century saw the dawn of molecular biology, after Alfred Hershey, Martha Chase, Francis Crick, and James Watson pinned down the chemical structure of DNA as the heritable material in the early 1950s (Box 1.1). The Central Dogma of Molecular Biology soon followed, delineating how transcription of RNA from DNA is followed by translation of protein from RNA.

Box 1.2 EVOLUTION IN POPULATIONS: CHANGE IN ALLELE FREQUENCY

At its simplest, evolution is the change in the population frequency (p) of alleles across generations. To illustrate the idea of how selection can drive those changes, we can ask what happens to a beneficial allele that is initially rare in a population. For the time being, we won't worry about how populations having a finite number of individuals will lead to some degree of stochastic fluctuation in allele frequencies (genetic drift, see Boxes 1.3 and 3.5). For an allele i that is rare in a population but confers a survival or reproductive advantage to individuals that have a copy (selection coefficient s; dominance coefficient h), how quickly does it rise in abundance to outcompete another allele j for that gene locus?

We can predict allele frequency in the next generation (p_i') as: $p_i' = (p_i^2 \cdot w_{ii} + p_i \cdot (1-p_i) \cdot w_{ij}) / \overline{w}$, where the fitness ($w$) of homozygous genotypes is $w_{ii} = 1 + s$, the fitness of heterozygous genotypes is $w_{ij} = 1 + hs$, the population mean fitness is $\overline{w} = p_i^2 \cdot w_{ii} + 2 \cdot p_i \cdot (1-p_i) \cdot w_{ij} + (1-p_i)^2 \cdot w_{jj}$, and individuals homozygous for the initially more common j allele have fitness $w_{jj} = 1$. With this setup, beneficial effects of allele i give $s > 0$; for example, $s = 0.02$ means that individuals homozygous for the ii genotype have 2% higher fitness than individuals with the homozygous jj genotype, on average. The change in frequency from generation to generation is then $\Delta p_i = p_i' - p_i$. The graphs of allele frequency change (Figure 2) show the effects of different amounts of dominance (for example, purely recessive $h = 0$, purely dominant $h = 1$, perfectly additive $h = \frac{1}{2}$) and different strengths of selection on changes in allele frequency as evolution proceeds from an initial 2% frequency of the beneficial allele in generation 0. In this simple model, the rare beneficial allele will spread and eventually become **fixed** in the population to reach a population frequency of 100%.

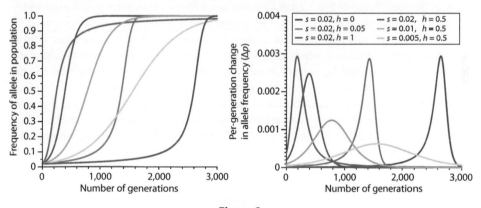

Figure 2

From your day-to-day life, you are already familiar with many of the slight heritable variations between individuals in what traits they have: height, hair color, earlobe attachment. But even with this phenotypic variation, people still look like people, mostly having traits that they share and that make humans quite distinct from other species. What about DNA? How polymorphic would molecules be within a species? Richard Lewontin and Jack Hubby demonstrated in 1966 that protein differences among individuals were extremely common, and Martin Kreitman in 1983 showed that the DNA sequences for the gene alcohol dehydrogenase from a collection of *Drosophila melanogaster* fruit flies differed between every single copy that he looked at. Human molecular variation told the same story, as Harry Harris first saw for protein differences and Charles Aquadro and Barry Greenberg found for DNA. Everyone already knew that genetic variability existed, but molecules turned out to be rife with variation, with variability so much more pervasive than anyone could have guessed until they looked. Such molecular genetic variation in DNA represents *the* most fundamental kind of genetic variation.

Biologists realized that DNA sequence variation holds important clues to the past, about the history of populations and the evolutionary forces that shaped them, if only there were a way to wrangle it. As Jack King and Thomas Jukes wrote in 1969, "Patterns of evolutionary change that have been observed at the phenotypic level do not necessarily apply at the genotypic and molecular levels. We need new rules in order to understand the patterns and dynamics of molecular evolution." What to do?

This connection between the concept of **genes** and **alleles** with their physical and chemical basis spawned a new series of mathematical models of evolution. These models built on classic population genetics theory (Boxes 1.2, 1.3, and 1.4), but extended it to incorporate realistic details from molecular biology. Importantly, it is that same set of five key evolutionary forces from the pre-molecular age of evolutionary thinking that also influences changes at the molecular level. The most important among these theoretical developments arrived in 1968: the Neutral Theory of Molecular Evolution, introduced by Motoo Kimura.

Box 1.3 EVOLUTION AS THE PROBABILITY OF FIXATION FOR NEW MUTATIONS

Classic population genetics theory provides mathematical equations that describe the change in frequency of mutations in a finite population. This adds a realistic complication to the predictions shown in Box 1.2, because stochastic fluctuations enter into the dynamics and mean that the allele with the most beneficial effects is not guaranteed to spread to fixation. If there are just two alleles of a gene, then one of them could get lost from a population (allelic extinction), which, stated the other way around, means that the alternate allele would get fixed and be present in every individual. This loss versus fixation can be caused by genetic drift (see Box 3.5) or by natural selection (see Chapter 7). We can think of this happening to new mutations that are so rare that there is only a single copy of that new allele, or we can think of mutations that have been around for a while and are still rare but not so extremely rare as are single-copy alleles. Two common features of this evolutionary dynamic deserve a lot of attention: the *probability* of fixation and the *time* to fixation (Box 1.4).

Positive selection is a directional evolutionary force, allowing us to predict how likely it is that a beneficial new mutation will eventually become fixed in the population. To eventually get fixed, a new mutation first has to escape loss by drift when rare, because even beneficial alleles are susceptible to the stochastic effects of genetic drift when they are very rare or in a small population. Population size holds a fundamental place in understanding evolution because it influences how effective the force of positive selection is in actually producing evolutionary change.

In general, the probability of fixation depends positively on the strength of selection favoring the beneficial allele (selection coefficient, s), the degree to which dominance masks the effect of the alternate allele (h), the frequency of the allele (p), and the effective population size (N_e; see Box 3.1). Combined together, the expected probability of fixation is approximately $P_{fix} = \frac{1-e^{-4N_e hsp}}{1-e^{-4N_e hs}}$ (see Figure 4.5). A good rule of thumb is that this simplifies down to $P_{fix} \approx 2hs$ for new mutations. When selection is negligible, then $P_{fix} \approx p$. For simplicity, we often presume additive dominance effects ($h = \frac{1}{2}$), which is usually what is implied in population genetics equations that show s but not h.

What was so inspired in Kimura's theory that set it as a key milestone in evolutionary biology? His research introduced elegant and deceptively simple predictions about how changes in DNA ought to work. The Neutral Theory provides a null model that we can compare to observed patterns of genetic variation from the real world. This comparison gives us a "**test of neutrality**" to take those evolutionary forces that depend on chance events, the neutral forces, and see how well they can do on their own in explaining changes to DNA. The "standard neutral model" is the simplest null model based on the Neutral Theory, and it uses many of the same assumptions as other classic population genetic models (random mating, stable population size, mutation-drift equilibrium). Like all models, the standard neutral model is thus a simplification of the natural world. Scientists must look at models with their eyes wide open, taking the same outlook that statistician George Box so aptly invoked: "All models are wrong, but some are useful." The oversimplification of models, including the Neutral Theory, is intentional and we can use it to help us understand the additional complexities of nature.

Box 1.4 THE TIME TO FIXATION FOR NEW MUTATIONS

The probability of fixation tells us how likely it is that a given allele will get fixed *eventually*, given an infinite amount of time (Box 1.3). But how long should it take on average? After T generations, a beneficial allele is expected to reach a frequency of $p_T = \left[1 + \left(\frac{1}{p_0} - 1\right) \cdot e^{-sT}\right]^{-1}$. Consequently, it will take approximately $T_{fix} = \frac{2 \cdot \ln(2N_e)}{s}$ generations to go from an initial frequency of $p_0 = 1/(2N)$ to become essentially fixed in the population ($p_{fix} = 1 - 1/(2N)$). For an allele with a 1% fitness advantage ($s = 0.01$), T_{fix} will range from about 300 to 150 generations for population sizes (N_e) between 1000 and 1,000,000. By contrast, for a selectively neutral allele, it will take on average $4N_e$ generations to fix (from 4000 to 4,000,000 generations). This tells us that natural selection will fix mutations in a tiny fraction of the time that it takes for them to get fixed by genetic drift. Figure 3 shows how the time to fixation (T_{fix}) scales with population size for alleles subjected to different strengths of selection; note the log scale of the axes.

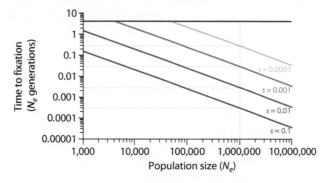

Figure 3

We can, of course, make this molecule-inspired-but-simple model of evolution even more realistic. We can incorporate added layers of biological complexity. One of the aims of this book is to point out how well and how poorly the oversimplified models perform, when it matters, and how to modify them appropriately. The most obvious oversimplified piece of Neutral Theory is its lack of integration with the process of adaptive evolution by natural selection. Many biologists focus on natural selection as the main non-neutral force of interest, as our innate inquisitiveness about all the life around us often leads us to ask about how organisms adapt to their world (Figure 1.1). Ironically, we can use *neutrality* to learn about selective *non*-neutrality. But another common goal that Neutral Theory helps us out with is in detecting demographic changes in a population's history using genetic data, as for understanding our own human past through "molecular anthropology."

How much evolution at the molecular level can be explained just by chance evolutionary forces, by mutation and genetic drift? Tests of neutrality once were limited to small individual cases, to evaluate how single genes are affected by natural selection. But modern molecular population genetics has scaled up to let us scan across entire genomes. Think for a moment about what it means to do such "molecular population genomics." If you were to analyze the DNA for just 50 humans, you would be dealing with 300 billion nucleotides and over 10 million nucleotide differences. And nowadays, many thousands of human genomes are sequenced on a regular basis. How to make sense of all that? The ideas and techniques of molecular population genetics show us the way.

1.2 What is the use of molecular population genetics?

The remainder of this book aims to introduce, in an accessible way, the bare essentials of the theory and practice of molecular population genetics. You can use this book to develop an understanding of the origins and implications of molecular diversity within populations, molecular divergence between populations and species, and how diversity and divergence connect with the five major forces of evolution. Sometimes referred to as **microevolution**, we will focus on the evolution that happens within species and that contributes to divergence between species that are very closely related. As opposed to studying deep-time **macroevolution** of distantly related organisms (Figure 1.2), we will primarily have in mind a relatively short timescale for understanding evolutionary change, usually spanning from a few generations to a few million generations. This timeframe of DNA sequence evolution is the purview of molecular population genetics. We will filter these ideas through "genome thinking" to learn about the powerful and general forces that control the evolutionary process.

Molecular population genetics has its own abstract parts—for example, coalescent theory (see section 5.2), statistical models of sequence evolution (see section 5.3), the implications of selection interacting with genetic linkage (see section 7.1)—but, by and large, molecular population genetics is embedded in empirical patterns seen in real-world observations (see Chapters 2 and 3). We will take advantage of these concrete features of DNA to make sure we also grasp the more ethereal concepts. That is, how can we interpret patterns of DNA sequence differences within populations, between populations, and between species? Why does a given species have as much genetic variation as it does? What do those molecular signals tell us about natural selection and the **demographic history** of organisms? What are the roles of mutational input and genetic drift and recombination in genome evolution?

This list of questions might sound esoteric, but molecular population genetics also helps us deal with important and interesting problems in applied biology. What is the likelihood that this blood sample from a crime scene belongs to the accused individual? How much of human genetic material derives from extinct human species like Neanderthals? What genetic changes are associated with inherited diseases? How does HIV drug resistance evolve within an infected person? Where has this invasive crop pest species colonized from? Are populations of this rare organism actually genetically distinct enough to warrant protection under endangered species legislation?

Thus, molecular population genetics provides the tools and framework to address a wide spectrum of inquiry, with the emphasis of analysis depending on the goals or interests of the scientist. The topics span from purely academic problems that often extend over prehistoric timescales all the way to applied problems that operate on the most contemporary of timescales. At one end of the spectrum, we can think of the selectionist's research program. From this view, one aims to identify the targets of natural selection in genomes, to infer the prevalence of selection across the genome, and to estimate the magnitude of selective differences at the molecular targets of selection. For these purposes, other evolutionary pressures, including demographic perturbations, represent "nuisance" processes that must be accounted for simply as part of the baseline null model to extract signal from noise.

At the other end of the spectrum, for the molecular anthropologist or phylogeographer or landscape geneticist, it is exactly these details of demographic history that formulate the key questions of interest. Think of applied problems that operate in the present day, such as conservation genetics of human-impacted populations and for human populations themselves. In fact, selection in the genome may be the "noise" if we want to know how the geography and distribution and abundance of individuals over time and space have left their marks in genomes for us to read. And still many other perspectives and goals abound for wanting to distill insights from molecular population genetic data, from forensic analysis, to epidemiological inference, to mapping of disease alleles, genetic dissection of animal and plant domestication, conservation biology, and molecular ecology.

In practice, these disciplinary subdivisions are not hard boundaries. Most practitioners of molecular population genetics are interested in each of these facets to differing degrees. The priority of your emphasis depends on the interests of *you*, the investigator, and the questions piqued by the particular biology of your study system. But you don't want tunnel vision to blind you to alternative explanations. Consequently, it is important to gain a full appreciation for the influences of both selection and demography in molecular population genetics in order to address properly any given problem.

Because our aim here is to focus on *molecular* evolutionary change, I will leave to the side most things about the phenotypes of organisms and the details of how genotypes map to phenotypes (Figure 1.3). In Chapter 9, however, I will walk you through a series of evolutionary vignettes that do feature exciting phenotypic change and also feature all of the molecular population genetic tools from the preceding chapters. In the interim, when using phenotypic examples and analogies, I will primarily use obvious and direct connections between genotype, phenotype, and fitness. This, of course, is a big simplification. Some caveats to this simplicity will crop up as signposts for more advanced study of molecular population genetics beyond the scope of this book. What I want to do mainly is to distill down to its essence the logic of the population genetic process of evolution at the level of DNA as the physical basis of evolution. That essence will be

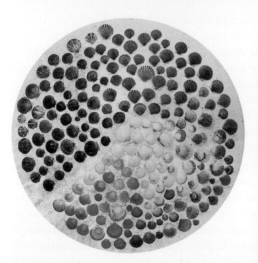 

Figure 1.3 Phenotypic variation with a heritable basis.
Atlantic bay scallop shells (*Argopecten irradians*) that I collected with my daughters from a beach on Cape Cod during a week in May 2016. The white versus yellow or orange background coloration has a simple genetic basis comprised of two genes (Adamkewicz and Castagna 1988). Striping of the shell coloration also varies consistently, though deposition of pigment often changes over ontogenetic growth, as seen from the radial arcs of pigmentation. (photos by A. Cutter)

the fuel that drives our molecular population genetic time machine to let us relish the splendor of evolution from the inside out.

Further reading

Adamkewicz, L. and Castagna, M. (1988). Genetics of shell color and pattern in the bay scallop *Argopecten irradians*. *Journal of Heredity* 79, 14–17.

Casillas, S. and Barbadilla, A. (2017). Molecular population genetics. *Genetics* 205, 1003–35.

Charlesworth, B. and Charlesworth, D. (2017). Population genetics from 1966 to 2016. *Heredity* 118, 2–9.

Darwin, C. R. (1859). *On the Origin of Species by Means of Natural Selection*. John Murray: London.

Hartl, D. (2000). *A Primer of Population Genetics*. Sinauer Associates: Sunderland, MA.

Hubby, J. L. and Lewontin, R. C. (1966). A molecular approach to the study of genic heterozygosity in natural populations. I. The number of alleles at different loci in *Drosophila pseudoobscura*. *Genetics* 54, 577–94.

Kimura, M. (1968). Evolutionary rate at molecular level. *Nature* 217, 624–6.

King, J. L. and Jukes, T. H. (1969). Non-Darwinian evolution. *Science* 164, 788–98.

Lynch, M. (2007). The frailty of adaptive hypotheses for the origins of organismal complexity. *Proceedings of the National Academy of Sciences USA* 104 Suppl 1, 8597–604.

Padel, R. (2010). "Giant Bugs from the Pampas." In: *Darwin: A Life in Poems*. Vintage Classics: London.

Watson, J. D. and Crick, F. H. C. (1953). Molecular structure of nucleic acids—a structure for deoxyribose nucleic acid. *Nature* 171, 737–8.

CHAPTER 2

The origins of molecular diversity

At its simplest, evolution is change in the relative abundance of alternative alleles, from one generation to the next. In describing this evolutionary change, we often place a great emphasis on how natural selection and genetic drift drive changes of allelic differences in populations. But where do these different alleles come from? What causes new alleles to enter a population, and what are the features of these novel genetic variants? In molecular population genetics, we aim to bring the reality of DNA into evolutionary thinking, tying theory and data together especially tightly. The input of new mutations into a population forms the critical first step for incorporating biological detail into how we conceive of genome evolution. This extra bit of biological realism that we add to otherwise abstract and simplifying mathematical models of evolution thus gives us a closer connection to the mechanisms of evolution and shows us the way toward a deeper analysis of genomes.

New alleles can enter a population by two main routes: **migration** and **mutation**. It is simple to envision new alleles entering a population by migration, as it just requires that one population lacks a particular allele until a migrant from another population arrives and contributes to reproduction. Migration can be an important source of new genetic variants for a given population, as we will consider in more detail in Chapter 3. Sometimes the gene flow through "migration" can even come from **hybridization** between different species. But learning that new alleles arrived by migration still does not answer the more fundamental question of *where did the genetic novelty arise in the first place*? The answer is simple: mutation.

Mutations are the ultimate origin of all genetic novelty. Despite the facts that DNA is a very stable chemical and that cells replicate DNA with incredibly high fidelity, DNA does break and the fixing and copying are never error-free. Mutations happen. However, there are many mechanisms of mutation that occur at different rates and with different potential effects on gene function and organism fitness. It is also important to remember that the abstract allele designations like "big *A*" and "little *a*" that we might use in classic two-allele genetic models can refer to mutations that arose by *any* mutational process. In this chapter, we will start at the start by making clear all the different kinds of mutations that create molecular diversity.

2.1 Kinds of mutations

2.1.1 *Single nucleotide mutation*

A **point mutation** (Figure 2.1) involves one DNA base nucleotide (A, T, G, or C) being replaced accidentally by another during DNA replication in germ cells and that manages to evade perfect repair by cellular DNA repair machinery (Box 2.1). Point mutations

A Primer of Molecular Population Genetics. Asher D. Cutter, Oxford University Press (2019).
© Asher D. Cutter 2019. DOI: 10.1093/oso/9780198838944.001.0001

Point mutation in a non-coding sequence

```
...AACTCG... --C→T--> ...AATTCG...
```

Synonymous-site point mutation to a protein-coding sequence

```
...CCG TGC CAG TTC ATG CAT AAC...  --G→A-->  ...CCG TGC CAA TTC ATG CAT AAC...
...Pro Cys Gln Phe Met His Asn...            ...Pro Cys Gln Phe Met His Asn...
```

Replacement-site point mutation to a protein-coding sequence

```
...CCG TGC CAG TTC ATG CAT AAC...  --T→C-->  ...CCG TGC CAG CTC ATG CAT AAC...
...Pro Cys Gln Phe Met His Asn...            ...Pro Cys Gln Leu Met His Asn...
```

Nonsense point mutation to a protein-coding sequence

```
...CCG TGC CAG TTC ATG CAT AAC...  --C→A-->  ...CCG TGA CAG CTC ATG CAT AAC...
...Pro Cys Gln Phe Met His Asn...            ...Pro
```

Figure 2.1 Consequences of point mutations.
Single nucleotide mutations, point mutations, change one nucleotide to another one of the three alternative possibilities. When point mutations occur in a coding sequence, then three qualitatively distinct functional consequences for the encoded protein sequence can result. First, mutation to a synonymous site will not alter the amino acid that is encoded within that codon, leaving the corresponding peptide sequence of the protein product unaltered. Second, a mutation to a non-synonymous site, also known as a replacement site, will alter the encoded amino acid and will therefore change the peptide sequence of the protein. Third, a nonsense mutation can result from point mutation of a nucleotide that changes an amino acid-encoding codon into a termination signal codon, leading to a shorter, prematurely truncated version of the original protein product. In non-coding DNA sequence, point mutations also can have important functional consequences, as when they alter promoter motifs or RNA genes. For these non-coding changes, however, there is no simple mapping like the genetic code offers.

also are called **single nucleotide mutations** or base substitutions. When we identify such mutations so that both the original ancestral allele and the new derived allele are present, yielding a polymorphic site, we call it a **single nucleotide polymorphism** (**SNP**, pronounced "snip"). If the new mutation gets fixed in a population (see Box 1.2), whether due to selection or genetic drift, we then call it a **substitution** that forms the basis of **fixed differences** in the **divergence** between populations or species. Such "evolutionary substitutions" can involve any kind of DNA sequence difference that becomes fixed in a population and that then represents the sole new allele, not just the fixation of one SNP over another. By contrast, sometimes the term "substitution mutation" is used for the new mutant variant itself, which applies to point mutations only. To avoid this ambiguity, here we will try to restrict our usage of "substitution" to refer only to fixations. SNPs will be one of the most important players in molecular population genetic analysis, so let's now spend some time going into more detail about the point mutations that lead to SNPs in genomes.

A **locus**, such as a gene that encodes a protein (Box 2.2), can be made up of hundreds or thousands of nucleotides, any one of which could potentially mutate and become polymorphic. In fact, *many* of those sites could be polymorphic simultaneously. We could use the generic term "allele" to refer to different forms of the same locus, but when we talk about differences at the DNA sequence level, the word allele can often get clumsy: does it refer to the different nucleotide states at a single site? Or the combination of polymorphic nucleotides of the entire gene? Or only the functionally distinct forms of the gene? To be more specific, we will use two new words: variant and haplotype. We will refer to the individual nucleotide differences as **variants**. So, for example, the SNP at position 6 in the locus shown in Figure 2.2 has C and A variants in the population, and position 6 is one of the three variant sites in that 21 base pair (bp) long locus.

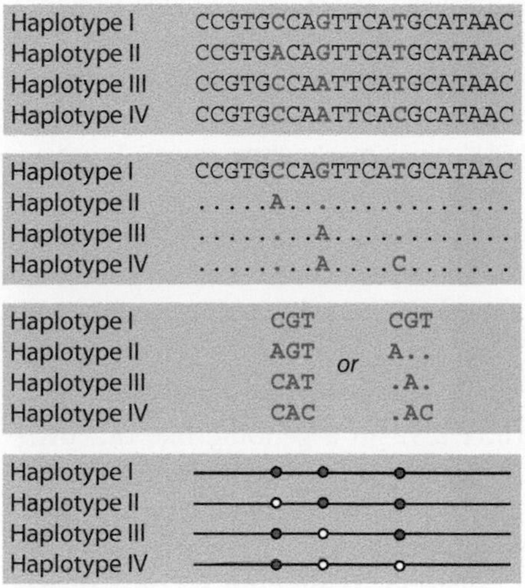

Figure 2.2 The distinction between polymorphic sites and haplotypes.

Haplotypes represent unique strings of DNA sequence, made unique by their particular combination of polymorphic sites (bold red letters). When we look at the sequence of distinct haplotypes in a population, most nucleotide positions will be identical (black letters). In this example, the first five positions in the sequence in the alignment are monomorphic, whereas the sixth position is polymorphic. Haplotypes I, III, and IV have the C variant at the first polymorphic site (occurring at position 6 in the sequence), whereas haplotype II has an A variant. Each of the haplotypes is unique because of their distinct combination of variants of the three polymorphic sites, which occur at positions 6, 9, and 14 in the sequence shown. Rather than showing all the nucleotides that are identical, we can simplify their representation diagrammatically by representing variants equivalent to the top "reference" sequence with a " . " Even more simply, if we only need to focus on the polymorphic sites themselves, we can remove the monomorphic sites from our representation of the alternative haplotypes. An abstraction is to use a horizontal line to represent a string of DNA with polymorphic sites indicated by circles and the absence of circles implying monomorphic sites; different colors of the circles represent distinct variant states.

Haplotype I	CCGTGCCAGTTCATGCATAAC
Haplotype II	CCGTGACAGTTCATGCATAAC
Haplotype III	CCGTGCCAATTCATGCATAAC
Haplotype IV	CCGTGCCAATTCACGCATAAC

Haplotype I	CCGTGCCAGTTCATGCATAAC
Haplotype II	.....A...............
Haplotype III	........A............
Haplotype IV	........A....C.......

Haplotype I	CGT		CGT
Haplotype II	AGT	or	A..
Haplotype III	CAT		.A.
Haplotype IV	CAC		.AC

Haplotype I
Haplotype II
Haplotype III
Haplotype IV

By contrast, a **haplotype** refers to any unique sequence of linked genetic markers like SNP variants, but also may include genetic differences arising from other mutational processes (e.g. indels, microsatellites). The word "haplotype" is what the French would term a portmanteau: an abbreviated fusion of "haploid genotype," indicating that a haplotype is made of a particular combination of variants along a single copy of DNA. In this sense, it is the haplotypes that define different "alleles" of a gene between individuals. However, the term "variant" is especially useful for DNA sequence polymorphisms because a given gene typically will have multiple haplotypes that are each made up of different combinations of SNP variants at many distinct **variant sites** (Figure 2.2).

It is important to understand the distinctions between haplotypes and variants (Figure 2.2), as these ideas about allelic variation in a population are intrinsic to quantifying, modeling, and analyzing molecular population genetic data. Keep in mind that the words "allele" and "locus" are general and flexible terms and, depending on the context, can refer to "variants" or "haplotypes" or "genes."

Box 2.1 ERRORS IN DNA REPAIR IN THE FORMATION OF MUTATIONS

Breaks in DNA can be induced by environmental mutagens, damaging cell metabolic byproducts, errors in replication, and from intentional cleavage of the double helix that is required for completion of meiosis. Germ cells have diverse repair mechanisms to mend double-strand DNA breaks to maintain the integrity of the genome and cell survival, with

(*Continued*)

Box 2.1 CONTINUED

the fixes sometimes leaving mutational scars in the genome. Non-homologous end-joining (NHEJ) is an "emergency response" by the cell that uses DNA ligase enzymes to glue together the broken ends of DNA, but removes some of the DNA from each end, thus leading to deletion mutations. Homologous recombination, sometimes termed gene conversion, uses the cell's recombination machinery to provide another mechanism of repairing double-strand breaks that is less error-prone than NHEJ. DNA mismatch repair (MMR) occurs when the cell recognizes a base-pairing error during DNA replication and then removes, or excises, the offending region and resynthesizes the DNA. Nucleotide excision repair fixes damage caused by ultraviolet radiation, important in somatic cell DNA repair as well as germline DNA repair in some organisms. Cells use several other DNA repair mechanisms as well. These different DNA repair pathways use slightly different combinations of proteins to recognize and fix DNA damage, which lead to distinct likelihoods of introducing new mutations and distinct kinds of mutations that tend to get introduced.

Point mutations arise in eukaryotes at a rate of about one in every 10^8–10^{10} base pairs each generation, depending on the species (Box 2.3). In a genome like our own human genome, this corresponds to about 30 point mutations each generation in every copy of our genome that gets put into a gamete. In a simple sense, we could think of all possible base substitutions being equally likely. In reality, this is not the case. For example, **transition mutations** between purines (adenine, A; guanine, G) or between pyrimidines (cytosine, C; thymine, T) occur more frequently than do **transversion mutations** from a purine to a pyrimidine (or from a pyrimidine to a purine). We will discuss the issue of models of base substitution in more detail in Chapter 5. Point mutations can have a very broad range of possible effects on gene function, depending on (1) where the mutation falls in the genome and (2) what kind of change in protein structure, RNA stability, or gene regulation it might induce (Figure 2.3). As Motoo Kimura and James Crow (1964) put it, "a single nucleotide substitution can have the most drastic consequences, but there are also mutations with very minute effects."

Single nucleotide mutations within a coding exon of a gene can cause three main types of change, as a consequence of the structure of the **genetic code** (Box 2.2). **Synonymous** (sometimes called neutral or silent) mutations change one codon to a different codon of the same amino acid, for example, glutamine's CAG to CAA. In contrast, **non-synonymous** mutations, a.k.a. **replacement** or **missense mutations**, change a codon of one amino acid to a codon of a different amino acid, for example, phenylalanine TTC to leucine CTC (Box 2.2).

In contrast to a missense change, a **nonsense mutation** changes the information content of a codon from an amino acid into a termination signal codon, for example, cystine TGC to stop codon TGA. Nonsense mutations, also termed **premature stops**, are a special type of mutation to a replacement site that leads to premature truncation of the encoded peptide by disrupting proper translation. The methionine codon ATG has a dual purpose: it both encodes an amino acid and also encodes information to trigger the start of translation. As a result, mutation of the non-degenerate **start codon** could result in failure of, or improper, translation.

Box 2.2   GENE STRUCTURE AND THE GENETIC CODE

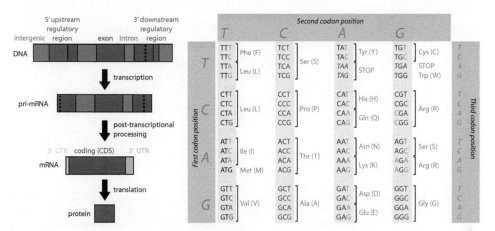

Figure 1

Figure 1 The expression and function of protein-coding genes depend on regulatory elements in noncoding DNA sequences, usually in the 5'-DNA located upstream of the transcription start codon ATG but also in downstream 3'-DNA and in introns. Genes in eukaryotes usually have alternating exons and introns, with introns spliced out of the primary RNA transcript to yield the mature mRNA used as the template for translation of the **coding sequence** of exons only. The **untranslated regions** (UTRs) of the mRNA are important in the post-transcriptional stability and regulation of gene expression. The portion of genes that ultimately gets translated into amino acid peptides encodes 64 possible nucleotide triplets, **codons**, for the corresponding amino acids. The standard genetic code comprises 20 amino acids, 18 of which have **degenerate** alternative synonymous codons usually due to "wobble" of the third codon position, as well as a translation stop signal encoded by one of three codons. Amino acid names often are shown with their standardized three-letter and one-letter abbreviations.

Conveniently, the so-called universal genetic code or standard genetic code that shows us the mapping of DNA onto protein is nearly identical for all organisms. But the genetic code is not quite universal, as some bacteria have alternative codes and the mitochondrial genomes of eukaryotes use a distinct code from their nuclear genomes. For brevity, we will use the standard genetic code (Box 2.2), keeping in mind that a given organism may need appropriate customization to define its synonymous and non-synonymous sites.

It is critical to understand the differences between synonymous and non-synonymous mutations, because we will use this as the basis of much more detailed analyses of molecular evolution of gene sequences. As we will discuss further with respect to fitness effects of mutations (see section 2.2), non-synonymous and nonsense changes alter the peptide sequence of the encoded protein. Because proteins are a key unit of biological function in the cell, such changes are often subject to selective pressure.

Box 2.3 MUTATION RATES

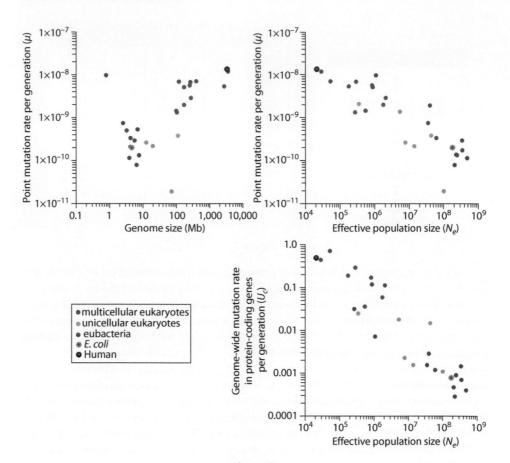

Figure 2

Figure 2 The point mutation rate per generation (μ, the Greek letter for "m," pronounced "mew") tends to be higher for eukaryotes than prokaryotes, higher for eukaryotes with large genomes and prokaryotes with small genomes, and higher for species with small effective population sizes (N_e; Box 3.1). Note that we usually denote μ in units per site and per generation, though sometimes it is given per gene and per year. Integrating the neutral input of mutations with their potential to reduce fitness when they hit protein-coding genes across the entire genome leads to the genome-wide deleterious mutation rate (U). Species with smaller N_e tend to have a larger total input of deleterious mutations into their genomes each generation. Michael Lynch details the evolutionary theory to describe how mutations to the DNA repair process itself get eliminated more efficiently in species with large N_e, leading those species to have lower mutation rates (data replotted from Lynch et al. 2016).

Mutations that occur in non-genic and intronic portions of genomes are usually considered to be selectively neutral ("**silent**"), although mutation of some non-coding sequence elements can, in fact, drastically alter gene function. For example, regulatory

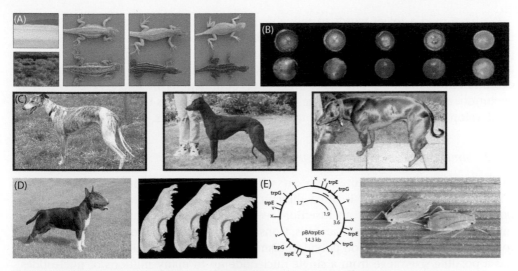

Figure 2.3 Diverse mutation types can cause striking phenotypic effects.

(A) Single nucleotide mutations to replacement sites in the *Mc1r* gene of fence lizards are responsible for differences in scale color. Lizard populations found in white sandy habitats have blanched coloration, conferring crypsis in that environment. Mutations to this same gene in different species also affect lizard coloration, providing an example of convergent evolution at the level of the phenotype and gene despite distinct mutational origins (lizard photos reprinted from Rosenblum et al. 2010, Molecular and functional basis of phenotypic convergence in white lizards at White Sands. *Proceedings of the National Academy of Sciences USA* 107, 2113–17). (B) Retroelement insertion into the promoter region of the *VvmybA1* gene of the wine grape genome is responsible for producing white grapes, and DNA alterations left in the promoter after the element excised itself yield red grapes, In contrast to black grapes that have *VvmytA1* alleles unaffected by the transposable element (photo reprinted by permission from Springer Nature: This et al. (2007). Wine grape (*Vitis vinifera* L.) color associates with allelic variation in the domestication gene *VvmybA1*. *Theoretical and Applied Genetics* 114 (4), 723–30.). Insertion of a transposable element into a regulatory intron of the *cortex* gene of the peppered moth leads to upregulation of its expression and the melanic phenotype that became common in industrial areas in the United Kingdom (Box 8.1) (van't Hof et al. 2016). (C) A 2-bp indel into the myostatin gene of dogs disrupts the protein and is responsible for the "double-muscle" phenotype characteristic of bull whippets compared to standard whippets, with homozygotes having more extreme muscling than heterozygotes (Photo published under the CC-BY license. Taken from Mosher et al. (2007). A mutation in the myostatin gene increases muscle mass and enhances racing performance in heterozygote dogs. *PLoS Genetics* 3, 779–86. https://doi.org/10.1371/journal.pgen.0030079). (D) Distinct tri-nucleotide microsatellite repeat counts within the *runx-2* transcription factor protein-coding sequence, selected for by breeders, have led to the characteristic downturned snout shape of bull terriers (Fondon and Garner 2004, photo © 2004 National Academy of Sciences). (E) The genomes of *Buchnera*, bacterial endosymbionts of *Schizaphis* aphids, have duplicated the *trpEG* gene repeatedly to produce 14 copies of the gene, which allows the bacteria to produce vast quantities of the essential amino acid tryptophan for its aphid host, a limiting resource in aphid nutrition (Lai et al. 1994, photo © 1994 National Academy of Sciences). Miniature Bull Terrier © Capture Light/Shutterstock.com.

elements or intron-exon splice recognition sites represent important functional elements of genomes that experience selection when changes to them affect the fitness of the organism (Figure 2.3). RNA genes also form a functionally important class of non-coding DNA that includes ribosomal-RNA genes, transfer-RNA genes, and micro-RNA genes among others. The lack of a simple genetic code in non-coding DNA complicates modeling selection in non-coding DNA, compared to coding exons. It is therefore harder to know which nucleotide changes will be functional or not within RNA genes or regulatory motifs.

We have used a lot of words to talk about what seems to be such a simple type of mutation. But it is exactly this simplicity that makes single nucleotide mutations so powerful. Because point mutations can be well-described in a mathematical sense and are easy to study in real data, we will mostly focus on them as we consider molecular evolution in populations. The many other kinds of mutations are also very important in evolution, and we will consider some specific cases as examples throughout this book as we develop our understanding of molecular population genetics.

2.1.2 *Simple insertions and deletions*

Another common type of mutation occurs when one or more bases are deleted or inserted (Figure 2.4). In practice, it is often difficult to determine whether such variants in a population were caused by an insertion event or a deletion event, or the distinction might be tangential to a given analysis. Consequently, such mutations are often referred to jointly as insertion-deletion polymorphisms or "**indels**." Generally speaking, an indel has a possible size range from a single nucleotide up to many millions of nucleotides, though small indels of tens of nucleotides or fewer are most common (Figure 2.4). In most cases, indel polymorphisms that occur in non-coding DNA will have no obvious effect on traits or fitness, despite some striking examples of how indel mutations can influence phenotypes (Figure 2.3). If an indel with a length that is a multiple of three nucleotides occurs in a coding sequence, then this will result in the addition or removal (and possibly change) of one or more amino acids in the resulting protein. Indels with a length that is not a multiple of three nucleotides that occur in a coding sequence will cause a **frame-shift**, totally changing the downstream amino-acid sequence and usually leading to a premature transcription termination signal further along in the coding sequence (Figure 2.4). Even a three-bp indel can disrupt multiple codons, however, if it does not occur in "phase 0," occurring cleanly between two intact codons. Very large deletions could remove an entire gene or even multiple genes. Indels in non-coding sequences can potentially affect regulatory elements and RNA genes.

2.1.3 *Tandem repeat mutations*

A special class of indels are defined by changes in the number of repeats in stretches of repeated nucleotides. For example, a **CG** dinucleotide repeat locus might have a 7-repeat motif allele (**CG**$_7$) and an 8-repeat allele (**CG**$_8$) present in the population:

... **gtcattaCGCGCGCGCGCGCGttc** ... *and*
... **gtcattaCGCGCGCGCGCGCGCGttc** ...

Loci of this highly mutable class have many names. They are called **VNTRs** (variable number of tandem repeats), minisatellites, or **microsatellites**, with microsatellites in turn also termed short tandem repeats (STRs) or simple sequence repeats (SSRs). The mutation rate at microsatellite loci is generally in the range of 10^{-3}–10^{-6} per locus per generation, meaning that a given allele can mutate into a different **length variant** in roughly 1 out of every 1,000 gametes. This is quite a high rate of DNA mutational change. It is this kind of mutation that historically has been used in **forensics** for DNA fingerprinting and paternity/maternity testing. To be considered a microsatellite, there must generally be at least five repeat units, whether it be a **mononucleotide repeat** (for example AAAAAAAAA, an A$_9$ allele), **dinucleotide repeat** (CACACACACA, a CA$_5$ allele), trinucleotide repeat (ATTATTATTATTATT, an ATT$_6$ allele), tetranucleotide repeat,

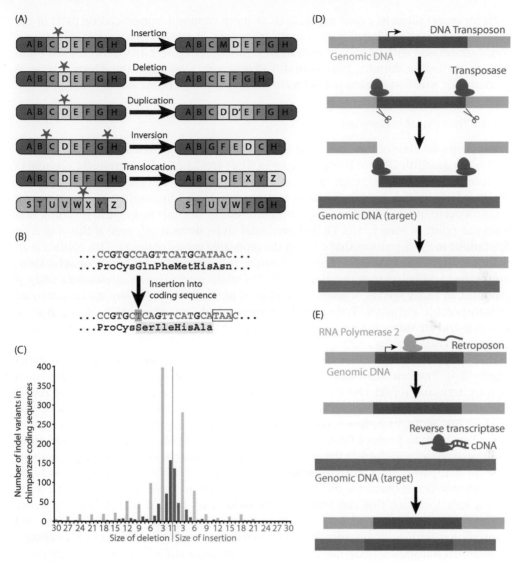

Figure 2.4 Other mutational examples.

(A) Diagrammatic representation of how distinct mutational types can influence gene identity and order along a chromosome. (B) Example of how even a single nucleotide insertion can disrupt the reading frame of a coding sequence to lead to a premature stop codon from the frame-shift. (C) Most insertion and deletion alleles found in populations are short, as in this example for coding sequences of chimpanzees, in which 80% of indels are less than seven bp long (data replotted from Bataillon et al. 2015). Notice how indel sizes in a multiple of three (light-colored bars) are much more abundant than other sizes (dark bars), because they do not cause a frame-shift and so have less drastic effects on the encoded protein. Transposable elements insert themselves through one of two general pathways: (D) "cut-and-paste" DNA transposons and (E) "copy-and-paste" retroelements that pass through a transcribed RNA intermediate phase.

etc. Loci with more repeat units often experience a higher mutation rate than do shorter loci (e.g. an A_{10} allele is more likely to mutate to A_9 or A_{11} than an A_5 allele is to mutate to A_4 or A_6), making biologically realistic modeling of the mutation process fairly complicated. Minisatellites have a more complex repeat unit structure with a sequence motif that itself is made up of approximately 15–50 nucleotides that is then repeated

2–20 (or more) times in a row. Repeat loci are more common in non-coding parts of the genome, although some genes do contain trinucleotide repeats that encode strings of a single amino acid with important roles in gene function (Figure 2.3). Coding repeats often include glutamine, forming poly-Q motifs in a protein such as in the *Ataxin-1* gene in humans, for which expanded poly-Q tracts lead to spinocerebellar ataxia.

2.1.4 *Transposable elements*

Another special class of indels are caused by **transposable elements** (TEs), sometimes referred to as selfish DNA or **jumping genes**. They are called **selfish genetic elements** because they control their own insertion and/or excision in the genome independently of when DNA replicates during mitosis and meiosis. More importantly, TE insertion and excision occur independently of the fitness interests of the host individual, meaning that from the point of view of the TE it is beneficial to replicate itself, even if this might be deleterious to the organisms that contain the replicated copies of the TE. This conflict is an example of how the evolution of genomes can be subject to different **levels of selection**, from gene to organism to population. While TEs might sound like an interesting oddity of genomes, in many species, it turns out that most of the DNA in the genome is comprised of transposable elements. Transposable elements make up about 45% of the human genome, and an astonishing 85% of the maize genome, whereas "just" 17% of *C. elegans'* genome is TEs.

There are many kinds of TEs, defined by their different molecular properties (Figure 2.4). The first main class of TE includes "copy-and-paste" **retroelements**, such as long terminal repeat retroposons (LTRs) and human *Alu* short interspersed repeat elements (SINEs), also termed "class I" elements. Retroelements bypass the Central Dogma of Molecular Biology (see Box 1.1) by making reverse transcriptase and integrase enzymes that allow them to produce DNA copies of their transcribed RNA and to then insert these additional copies elsewhere in the genome. The other major class of transposable elements are "cut-and-paste" **DNA transposons**, like *Drosophila's* P-elements, also termed "class II" elements. Transposons use transposase enzymes to hop out of one location and insert into a new location; this can leave an "excision scar" behind in the DNA sequence of the old location. Both classes of jumping genes hijack the cell's transcription and translation machinery to produce their specialized enzymes. Some TEs do not even encode a functional transposase gene, like the nearly 180,000 copies of MITEs (miniature inverted-repeat transposable elements) in the genome of rice. Instead, these **non-autonomous elements** utilize transposases expressed by *other* selfish elements to facilitate their own transposition—such elements are parasites of genetic parasites! Curiously, most DNA transposons in the genomes of rice and other grasses are made up of such "ultra-selfish" non-autonomous elements.

Insertion of a TE into a coding sequence will seriously disrupt gene function, and insertion nearby to a gene can drastically alter the gene's expression by perturbing regulatory elements, to either good or ill effect on the organism's fitness (Figure 2.3). When TEs expand into a family of nearly identical sequence copies in the genome, the sequence identity between the copies might create problems during meiosis. Specifically, it might lead to incorrect pairing of non-homologous regions of the genome with any resulting crossover causing large-scale chromosomal rearrangements. Such **ectopic recombination** typically leads to dire fitness consequences, so this negative feedback on TE abundance in the genome potentially limits further proliferation of TEs.

2.1.5   *Duplications and large-scale mutations*

Duplications can be thought of as a type of insertion whereby an additional copy of some DNA gets added to the genome (Figure 2.4). This can result from unequal crossing over during synapsis, the phase of chromosome pairing in meiosis. **Segmental duplications** refer to long stretches of sequence getting replicated, often many megabases (Mb) in length. More localized duplications of a few genes or of part of a gene also occur, for example, potentially resulting in a duplicated protein domain within a given gene. Paralogous genes oftentimes occur as **tandem duplicates** in close physical proximity to one another, in which a duplicate copy of a gene lies immediately downstream or upstream of its progenitor copy. The creation and expansion of **multi-gene families** within genomes result from repeated rounds of duplication mutations.

Initially following a duplication mutation, the duplicate copy will represent a **copy number variant** (CNV) in the population such that some genomic haplotypes will have both copies and other haplotypes will have just the single ancestral copy of that DNA segment (Figure 2.5). Even if the haplotype with both copies eventually becomes fixed among all members in the population, over the long term, the retention of functionally identical replicate copies is evolutionarily unstable. In that polymorphic interim, however, data for humans imply that the total amount of DNA sequence that differs among individuals is due more to CNVs than to SNPs. Cases of **polyploidization** represent extreme examples of duplication, specifically of whole genome duplication. Gene and genome duplication is a very important process in evolution over the long term because it can set the stage for the resulting duplicate gene copies to evolve specialized or novel functions.

DNA double-strand breaks are inevitable. In fact, breakage of DNA is an essential part of meiosis. Double-strand breakage of DNA can result in the cell trying to repair its genome in any way possible. Sometimes this results in large-scale and seemingly improbable genomic changes. Mutational **rearrangements** involve a segment of DNA that moves from one spot on a chromosome to another location on the same chromosome. **Translocation** refers to the movement of a DNA sequence from one chromosome to a different, non-homologous chromosome (Figure 2.4). **Reciprocal translocations** occur when segments of DNA from different chromosomes swap places; for example, part of the long arm of human chromosome 5 and part of the long arm of chromosome 17 changed places sometime in the history of divergence between humans and gorillas. **Inversion** mutations involve a DNA segment reversing its order with respect to the

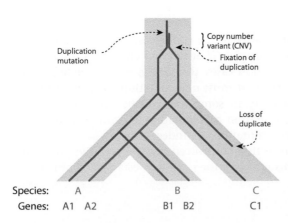

Figure 2.5 Evolution by duplication: copy number variants and paralogs.
Genes are homologs if they descend from a common ancestor, with orthologs being the special case of gene copies from different species that represent each other's closest relative (e.g. genes A1 and B1; Box 5.2). Duplication mutations will initially be polymorphic within a species, termed **copy number variants** (CNV). **Paralogs** represent sets of genes that are related to one another by duplication (e.g. genes A1 and B2 from species A and B in the diagram). Paralog copies can be lost in some lineages, leaving different counts of paralogs in different species.

ancestral state (Figure 2.4). As a result, individuals heterozygous for the inversion and the ancestral "standard" chromosome copies will experience atypical recombination profiles because the mis-oriented sequence in the inverted stretch of DNA generally precludes recombination. This lack of recombination is exploited by experimental geneticists when they create "**balancer chromosomes**" to maintain lab alleles with a defined genotype. These kinds of large-scale mutations are relatively rare within populations compared to other types of mutation, but can nonetheless be very important in evolution over the long term. We will primarily focus on smaller-scale mutational changes, which are more abundant and replicated within genomes.

2.2 Fitness effects of new mutations

Now that we have completed our march through the diverse types of mutations that one might encounter when analyzing DNA sequences in a population, we will next think more explicitly about what effects those mutations can exert on organismal fitness. Qualitatively, fitness effects come in three flavors: deleterious, neutral, and beneficial.

The magnitude of the effect of a mutation on fitness affects directly its likelihood of increasing or decreasing in frequency. We quantify this fitness effect as the **selection coefficient**, s. We usually define s as the difference in average (mean) fitness between individuals that are homozygous for one allele and individuals homozygous for the allele that confers highest fitness, so s is a measure of its **relative fitness** (Box 2.4). We can also define s in terms of a new mutation relative to the homozygous genotype of a pre-existing allele that already is established in the population. Specifically, we may set the fitness of individuals that carry the ancestral allele as equal to 1 and that of individuals containing the new mutation as $1 - s$ for homozygotes and $1 - hs$ for heterozygotes, where h is the **dominance coefficient**. In this way, $s > 0$ implies a *deleterious* new mutation and $s < 0$ implies a *beneficial* new mutation. Sometimes it may be convenient to define deleterious mutations as having negative values instead (see Box 1.2), so it is important to pay attention to what the sign of s actually means in a given situation. We know that some mutations can have large effects and others can have very small effects: neutral mutations have $s = 0$, a lethal mutation has $s = 1$, a beneficial mutation like one for antibiotic resistance in a bacterium would have $s << 0$ in an environment containing the antibiotic (note that "$<<$" means "*much* less than").

2.2.1   Deleterious mutations

Imagine a mutation that causes a random change in the coding sequence of a gene that already produces an efficiently functioning protein. That random change in the gene is unlikely to improve the protein function. In most cases, the change will be deleterious or have no effect. Therefore, we should expect that most new mutations *that affect fitness* by influencing gene function or gene activity in some way will be deleterious. Does this prediction jive with reality? In fact, mutation accumulation experiments in many systems have demonstrated that deleterious mutations do indeed prevail (Box 2.5). Peter Keightley and Michael Lynch (2003) wrote this point even more forcefully: "the vast majority of mutations are deleterious. This is one of the most well-established principles of evolutionary genetics."

Box 2.4 FITNESS

We measure the survival and reproduction of individuals with raw counts and probabilities, meaning that we can associate genotypes with the **absolute fitness** of the individuals that have a given genotype or trait value. Absolute fitness integrates survival from birth to adulthood and the total number of offspring produced. In practice, biologists usually measure several **fitness components** that together approximate total fitness. The "intrinsic" growth rate for the number of individuals of a given genotype also is a measure of absolute fitness, as often denoted with the variable R_0 in population dynamic models with discrete generations (r or λ for continuous time or overlapping generations; Figure 3 assumes asexual reproduction from an initial set of 100 individuals of each genotype).

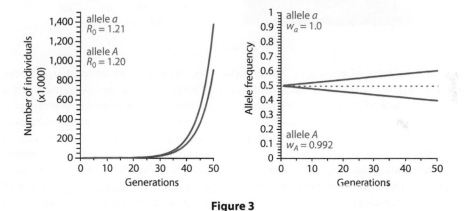

Figure 3

It is how these metrics compare to one another among individuals of different genotype, however, that is most important for natural selection. The relative magnitude of these measures of survival and reproduction defines the relative fitness of different genotypes (w; must always be ≥ 0). It is just shorthand to say that selection acts on genotypes or alleles, because selection does not actually act on genotypes directly. The net reproductive output of individuals gets mediated by phenotypes, with evolutionary responses depending in part on the details of the genetic encoding of those phenotypes. But here we are interested primarily in those genotypic responses that get recorded in the genome, so we will skip over the complexities of explicit mapping of fitness onto phenotype onto genotype. Instead, we'll proceed by focusing on the average survival and reproductive advantage or disadvantage of individuals that have a given DNA variant.

The standard notation for describing fitness in this way is that the genotype with greatest fitness, say diploid genotype aa, has relative fitness $w_{aa} = 1$. The diploid homozygote for the alternate allele will then have fitness $w_{AA} = 1 - s$. This relative fitness difference s between individuals that have one or the other homozygous genotype summarizes the strength of selection, the selection coefficient (s). I've presumed two alleles at the locus for simplicity, though you could define fitness values for any arbitrary number of homozygous genotypes. For heterozygotes, their fitness will depend on the dominance coefficient h relating the alternative alleles, so $w_{Aa} = 1 - hs$. Selection acts to increase the **population mean fitness** ($\overline{w}$) every generation, meaning that in the short term, from generation to generation, selection will deterministically push changes in allele frequencies toward $\overline{w} = 1$.

Box 2.5   MUTATION ACCUMULATION EXPERIMENTS

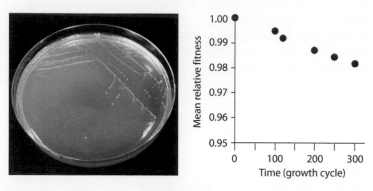

Figure 4

To study the input of mutations into a population, the experimenter must bypass the ability of natural selection to favor or disfavor alternative alleles. By transferring just a single individual offspring (or bacterial colony) each generation for many generations, new mutations effectively hide themselves from selection so that they can accumulate over time. This is the essence of a **mutation accumulation (MA)** experiment. By measuring the fitness of the genotypes that are the descendants of this process and then comparing them to the ancestral genotype, we can quantify what is the average effect of mutational input on fitness. An MA experiment with the bacterium *E. coli* (Figure 4) gives one of many such examples that demonstrate how new mutations generally act to reduce fitness: descendant genotypes have lower and lower relative fitness across MA generations (data redrawn from Kibota and Lynch 1996). Similar experiments in animals, plants, and viruses show the same pattern. The rate of decline in fitness across generations, in combination with the increasing variability in fitness among replicated lineages subjected to MA, also allows the experimenter to quantify the rate of mutational input and the average effect size of mutations. These days, you can also sequence the genomes of the evolved genotypes to identify all of the mutational changes in the DNA. In addition, Michael Lynch and other evolutionary biologists have used population genetic models to study the evolution of mutation rates themselves and why selection does not drive them to zero. (photo of *E. coli* courtesy R. Schalkowski)

Consequently, most of the time that a new mutation is affected by natural selection, selection will be acting to eliminate deleterious mutations. This process is termed **purifying selection** and is responsible for the conservation of DNA and protein sequences over time, despite the constant input of new mutations. Purifying selection, sometimes also referred to as **negative selection**, is subject to the same limitations as any form of natural selection: its effectiveness is limited by the mutation rate (μ), the magnitude of the mutational effects on fitness (the strength of selection or selection coefficient, s), and the effective population size (N_e; see Box 3.1).

Recall the term **stabilizing selection** from introductory biology: it describes the variation-reducing force of natural selection that favors an intermediate phenotype. It may be helpful to think of purifying selection as the molecular, sequence-based analog of what we call stabilizing selection at the phenotypic level (see section 7.1.4). This line of thinking presumes that most mutations that affect fitness are deleterious. In fact, it turns out that the molecular evolutionary signature will tend to look like purifying selection for those genes underlying a polygenic trait that is subject to stabilizing selection (see section 7.1.4).

For some kinds of questions, it is important to think about the input of *all* deleterious mutations into a genome each generation, not just the expected input at a given nucleotide site. This genome-wide deleterious mutation rate, usually indicated with the variable *U*, integrates both per-site mutational input and genome size, which influences the "mutational target" size of fitness-affecting elements in a species.

2.2.2 *Neutral and nearly neutral mutations*

Despite the fact that most new mutations that affect fitness are deleterious, many mutations do *not* affect fitness. These mutations are selectively **neutral**. The proportion of new mutations that are visible to selection in the genome of an organism depends on its effective population size and on the fraction of its genome that is functional (see Chapter 4). For example, species with very compact genomes that contain a high density of genes will have a greater proportion of new mutations being deleterious than will species with a lot of intervening non-coding sequence, given that both species have similar effective population sizes (N_e). Another way to state this point is that species with a high density of genes in their genome will have a greater **mutational target size** for fitness-affecting changes.

Some mutations might influence the fitness of an organism by only a very small amount, making them **nearly neutral**. We will explore this idea in more detail in Chapter 4. Examples of nearly neutral mutations are changes to synonymous sites and to noncoding nucleotides that can affect translational efficiency, RNA stability, and other factors in subtle ways. Mutations with selection coefficients $s < 1/(4N_eh)$, or rearranged mathematically to $4N_ehs < 1$, are **effectively neutral** (see Figure 4.5). In Chapter 4, we will go into why this is the case. So, mutations causing equivalent effects on fitness could be effectively neutral in smaller populations but non-neutral in larger populations. For example, the TTC codon for phenylalanine might have an additive ($h = \frac{1}{2}$) fitness benefit of 0.000002 relative to a synonymous variant (phenylalanine codon TTT). In a species with $N_e = 10,000$, the C and T variants would be effectively neutral with respect to each other because of the low $N_ehs = 0.01$. This effectively neutral mutational effect would mean that changes in their frequency over time will be caused primarily by genetic drift and not by natural selection. Genetic drift is the evolutionary force that controls the changes in the frequencies of mutations that are neutral or effectively neutral within populations (see Boxes 3.1 and 3.5).

In molecular population genetics, we often presume changes to be effectively neutral if they occur in synonymous sites or non-coding sites in the genome. This is a good assumption in many situations, especially for species with small population sizes. It is indeed easiest to think about mutations to selectively unconstrained segments of DNA creating selectively neutral alleles. But can mutation to non-synonymous sites be selectively neutral? Can a mutation that induces a large phenotypic effect be selectively neutral? The answer is "yes" to both of these questions. A mutation that changes one amino acid into another can be functionally equivalent and not induce a fitness difference, thus making the alternative alleles selectively neutral relative to one another.

2.2.3 *Advantageous mutations*

The final, and least common, class of new mutations actually increase the fitness of the host organism: advantageous or **beneficial mutations**. Such mutant alleles experience **positive selection** that favors their increase in frequency, also sometimes termed directional selection. Much of our interest in biology and evolution is focused on understanding adaptation by natural selection, which, ultimately, depends on advantageous alleles arising by mutation. As a consequence, much of molecular population genetics is devoted to trying to distinguish these adaptive mutational changes from those that are deleterious or neutral, using the information content of polymorphisms and fixed differences in the DNA of populations.

Beneficial mutations could confer a very weak benefit, which could be acted upon by selection only if the population size is sufficiently large (that is, $N_e > 1/(4sh)$). Resurrecting our phenylalanine example from section 2.2.2 ($h = \frac{1}{2}, s = 0.000002$), if we found the same polymorphism and selection coefficient in a species with a larger $N_e = 10,000,000$, then selection would be able to effectively discriminate the tiny difference in fitness ($N_esh = 10$): we would expect evolution eventually to fix the favored C variant that corresponds to the TTC codon. This example helps us to see that a beneficial mutation does not necessarily imply a dramatic adaptive change in phenotype, although some of the most clear-cut examples of adaptation in nature and domestication do involve large effect mutations (Figure 2.3).

Despite advantageous mutations arising only rarely, they have (1) a much higher probability of fixation than neutral or deleterious mutations and (2) a shorter time to fixation than neutral mutations (see Box 1.4 and Figure 4.5). As a consequence, advantageous mutations may account for a surprisingly high fraction of the sequence differences between species. For example, analyses of the genome for the fruit fly *D. melanogaster* indicate that positive selection fixed 50% or more of the replacement-site differences between species (see Chapter 8). In humans, however, our much smaller N_e appears to have led to at most 15% of replacement-site differences from other primates being the result of positive selection on beneficial mutations.

Advantageous mutations need not always experience positive *directional* selection, however. Sometimes selection can favor the maintenance of multiple alleles in a species, reflecting one of the many forms of **balancing selection** (see section 7.1.2). Even when this happens, however, what ultimately manifests as balancing selection to maintain alleles at intermediate frequencies in a population will "look like" positive directional selection from the point of view of a new, rare beneficial mutation. When rare, both balancing selection and positive selection will favor an allele's increase in frequency.

It is important to keep in mind that it is relative fitness that is important for understanding the selective changes in frequency of alternative alleles. Consequently, what constitutes an advantageous mutation depends on the context. For example, **reversion** and **compensatory** mutations can be considered beneficial mutations by allowing recovery of a function in the context of a genetic background that had previously lost that function.

2.2.4 *Magnitude and distribution of effects*

We can easily picture the *qualitative* effects of mutations: deleterious, neutral, beneficial. But the magnitude of selection on a new mutation can vary *quantitatively*, with s taking on a continuous possible range of values. So, what does the **distribution of fitness**

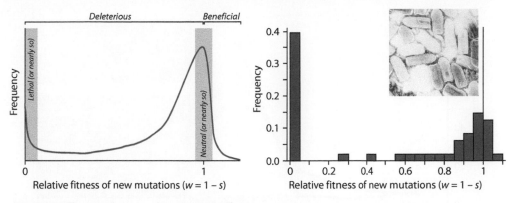

Figure 2.6 The distribution of mutational effects.
Many new mutations have strongly detrimental effects and are lethal or nearly so, as we might predict for a hypothetical distribution of mutational effects (left). Experimental data from the vesicular stomatitis virus show a qualitatively similar distribution (right; data redrawn from Sanjuan et al. 2004). However, many mutations also have no detectable effect and so are neutral or nearly so. Smaller proportions of mutations have intermediate deleterious effects or beneficial effects on fitness.

effects (DFE) look like? We might expect that the DFE for new mutations would have: (1) lots of strongly deleterious mutations with lethal or nearly lethal effects on fitness, (2) lots of neutral or nearly neutral mutations with little detectable effect on fitness ($N_e s \cong 0$), and (3) relatively few beneficial mutations (Figure 2.6). Does this verbal hypothesis hold up to experimental scrutiny? The answer is "Yes," although experiments to validate this idea are challenging to conduct in most organisms and so the best data come from microbes and viruses (Figure 2.6).

It is too bad that logistics limit to only a few kinds of organisms our ability to perform experiments to measure fitness effects of new mutations. But available experiments *do* validate our intuition about the general shape of the DFE. So, can we use molecular data to learn about the DFE in other organisms? In fact, we can (Figure 2.7). We can draw inferences about the DFE from molecular information about polymorphic nucleotides within a population and fixed differences between species, analyzed separately for nucleotide sites that selection influences either directly or not at all. We will explore the logic behind this approach in detail in the coming chapters and in Chapter 8 in particular.

2.3 Models of mutation

Now that we have an idea of the *kinds* of mutations that can arise in a genome and what kinds of *effects* they can have on gene function and on an organism's fitness, we can think about how to describe their input into populations quantitatively. That is, we can **model** the mutation process. Remember that in science we use models as tractable simplifications of reality. We use models because they provide a way to understand the most important features about how nature works and because they are useful in prediction, so long as the simplifying assumptions that they depend on are not too strongly violated in the real world. Most mutational models focus on those classes of mutation that are more practical for quantifying allele variation within populations, such as point mutations and short tandem repeat mutations.

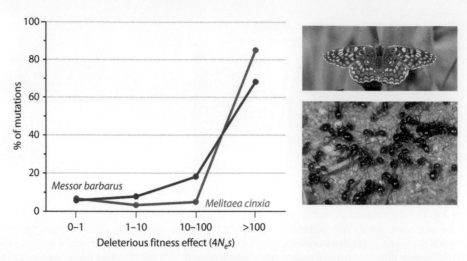

Figure 2.7 The molecular distribution of deleterious mutational effects.
We can characterize the relative abundance of weakly and strongly detrimental changes to proteins by quantifying the changes to replacement sites for protein-coding genes and comparing those changes to neutral synonymous-site changes. We also have to compare sequence changes within versus between species. The resulting pattern is the distribution of deleterious fitness effects. As in the Glanville fritillary butterfly (*Melitaea cinxia*) and harvester ant (*Messor barbarus*) shown here, most mutations to replacement sites are strongly selected against ($4N_es > 100$; data redrawn from Romiguier et al. 2014). Because the selection coefficients here refer explicitly to deleterious mutations only (strongly deleterious and lethal to the right, weakly deleterious and neutral to the left), larger values of s imply more detrimental alleles (in contrast to Figure 2.6). Glanville Fritillary photo by Steve Childs, reproduced under the CC-BY 2.0 license (cropped from original). European Harvester Ants © WildPictures/Alamy Stock Photo.

All of the models we will consider allow an equilibrium to be reached, usually a **mutation-drift equilibrium**, even in the absence of natural selection. This ability to achieve equilibrium is important, as we would like to be able to predict what is the balance between mutation increasing the number of alleles and genetic drift decreasing the number of alleles (see Box 3.1). Mathematical models of the mutation process specify how the input of mutation will generate polymorphism within populations and divergence between populations and species. We will explore these explicit connections as the crux of Chapters 3–5. Here, we will simply introduce the conceptual logic of several of the most common and important models of mutation in molecular population genetics.

2.3.1 Recurrent mutation model

With its origin in the notion of a "wildtype" mutating to a "mutant" type at a low but constant rate, the simple recurrent mutation model is a two-allele model of mutation that gives rise to the classical genetic notation a and A (for wildtype and mutant alleles, respectively). One can also imagine backward or **reverse mutation** (from A to a with rate v) in addition to forward mutation (from a to A with rate u), which yields a simple equilibrium frequency for allele A of $f_A = u/(u + v)$. Mutation models of this sort have a long and useful history in population genetics for understanding how diverse evolutionary forces influence allele frequency changes. While convenient and general, this mutation model does not match well to nucleotide change, because of the four possible allelic states at a given site in DNA and because of the fact that a DNA sequence locus may contain thousands of sites.

2.3.2 *Infinite alleles model*

In terms of DNA sequence, the two-allele recurrent mutation model is unrealistic: it seems implausible that the same change in DNA sequence would simply toggle back-and-forth by mutation. Because mutation at any nucleotide in a gene would generate a new haplotype for that gene, Motoo Kimura and James Crow in 1964 proposed the infinite alleles model of mutation, in which it is assumed that *each new mutation creates a unique allele*. Conceptually, you can think of each new mutant allele as representing a distinct, unique haplotype. However, this mutational model ignores the facts that, in reality, some haplotypes are genetically more similar than others and that recombination can create new unique haplotypes. Alternative alleles are different, but we don't know how different under the infinite alleles mutation model. This model makes the approximation that there are infinitely many sites in a gene, which of course is a simplification, because each gene must be finite in length even if a gene may be very long and made up of thousands of nucleotides. Nevertheless, this model often describes well the change at microsatellite loci and the mitochondrial genome. The infinite alleles model of mutation is a useful tool for understanding molecular evolution, in spite of the simplifications, particularly for loci like these that experience little to no recombination. But it is used less and less in practice, as DNA sequences let us look at all the variant sites along a locus.

2.3.3 *Infinite sites model*

The infinite sites model of mutation is the most important mutational model in molecular population genetics. In particular, the infinite sites model provides a powerful in-road for evaluating patterns of DNA sequence changes. This model considers each nucleotide site separately, and assumes that the mutation rate per site is sufficiently low that each site can mutate once at most. Said another way, each new mutation will change the nucleotide at a site that has never been altered previously and will not be altered again. These mutations are irreversible. Therefore, any given nucleotide site will have either one or two alleles in a population (Figure 2.2), and we will be able to observe all mutational changes because they will always occur at unique locations in the DNA sequence. In practice, we generally observe it to be true that polymorphic sites are **bi-allelic** for the vast majority of nucleotide sites in most species. This property of permitting only two alleles at a site is similar to the recurrent mutation model. When the sites that can be hit by mutation are perfectly linked to one another, with no recombination, then many properties of the infinite alleles model also apply to the infinite sites model. As with all models, however, the infinite sites model is a simplification of reality. Consequently, its assumption that the locations of new mutations will always be unique will, in fact, break down when we consider long timescales of divergence, high mutation rates, or extremely large populations.

2.3.4 *Finite sites models*

Species with very high mutation rates and/or very large effective population sizes may be able to retain extreme amounts of polymorphism; this includes some types of bacteria, viruses, fungi, nematodes, and small marine organisms (see Figure 3.2). A consequence of this hyperdiversity is that individual nucleotide sites might have three or even all four possible bases represented as variants, not just the more usual two variants, which violates the assumptions of the infinite sites model. Similarly, the accumulation of sequence changes between species can involve multiple substitutions occurring in turn at the same nucleotide site one after another. Finite sites models allow nucleotide sites to be "hit" by mutation *more* than once, which is closer to reality, given that no nucleotide site is ever shielded from the possibility of future mutation. However, finite sites mutation models require more parameters to be specified than does the infinite sites model. In other words, finite sites models better describe the details of reality, but at the cost that they are more complicated and less tractable with basic mathematics. For our purposes, and for simplicity, we will not focus on this kind of mutational model, although current molecular population genetics research is placing renewed emphasis on them. However, we will revisit the idea of repeated substitutions of mutations in the molecular divergence between species in section 5.3.2.

2.3.5 *Stepwise mutation model*

Changes to some loci proceed not by changes in nucleotide identity, but in the length in nucleotides of the locus itself. The stepwise mutation model was first introduced to help describe the distribution of allozyme electrophoretic sizes, based on the assumption that mutations would change the charge of a protein by one unit. Allozymes are the similar proteins encoded by different alleles of a gene, which can often be differentiated by their electric charge with gel electrophoresis because amino acid differences make the protein more negative or less negatively charged (see section 3.2). The stepwise mutation model also is very useful for understanding the production of microsatellite alleles. Microsatellite alleles differ in the number of repeats of a given sequence motif. New mutations change the number of repeats in the sequence by -1 or $+1$ repeat, for example, from AAAAAAA to AAAAAA (A_7 to A_6). Derivatives of this standard stepwise mutation model have been developed in order to account for complicating factors in microsatellite evolution. For example, by including mutations that change the allele length by more than one repeat unit, by instituting higher mutation rates for longer repeat alleles, or by defining maximum allele sizes.

Loci that evolve by the stepwise mutation model can produce alleles that are **identical by state**, such that they have the same repeat length, but *not* **identical by descent**, meaning that copies of the same length microsatellite got that way from different ancestral copies (Figure 2.8). This is one form of **homoplasy**. That is to say, an allele of a given length could arise convergently via different evolutionary trajectories so that it may not be possible to infer a unique genealogical history to it. For example, both a TAC_9 and a TAC_{11} allele could mutate to create a TAC_{10} allele. If we observe copies of all three alleles in a population, we wouldn't know whether the TAC_{10} allele originated from a TAC_9, TAC_{11}, or some copies from each of TAC_9 and TAC_{11}, or even perhaps TAC_{10} is ancestral to TAC_9 and TAC_{11}. This problem of homoplasy means that microsatellite loci generally perform poorly for analyzing divergence between species. The fast mutation rates and high polymorphism, however, can make microsatellites very powerful **molecular markers** for understanding demography on contemporary timescales and so they remain popular molecular markers in forensics and conservation genetics.

2.3.6 *Birth-death model of gene duplication*

It is less intuitive how to think about the mutational origin of duplicated DNA sequences than it is for point mutations or short tandem repeat mutations. One way is to treat the mutational origin of a duplicate gene copy as a "birth" of new sequence information. Secondary mutations then accumulate by genetic drift in both copies over time by other

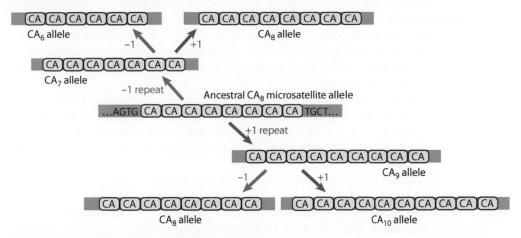

Figure 2.8 Homoplasy in microsatellite mutation.
With a stepwise mutation model of microsatellite evolution, the number of repeat motifs can increase or decrease by one with each mutational event. An ancestral allele length can evolve through distinct evolutionary paths and reach identical allelic states, as for the three examples of the CA_8 allele shown. This identity in state that occurs without identity by descent is a form of molecular homoplasy: not all copies of the CA_8 allele in the population have the same evolutionary history.

mutational mechanisms until one of the copies becomes a non-functional **pseudogene** (for example, due to a nonsense mutation). That is, in the absence of a force to maintain both duplicates, we expect that one copy will degrade into a non-functional pseudogene by subsequent mutation and genetic drift. This "pseudogenization" can be thought of as a gene "death," so that only one copy or the other of the original gene remains. This degeneration and eventual loss of one copy or the other is the most likely outcome: the **duplication-degeneration-loss model** of gene duplicate evolution.

For duplicate genes to be retained, one possibility is that selection could favor extremely high expression of their products (Figure 2.3), as for the multiple copies of genes encoding members of the ribosomal complex. But such perfect gene redundancy will not be a stable state in most cases. Instead, divergence in the regulatory architecture between the copies could lead to specialized expression for each copy, so that their complementary activity enables the persistence of both copies, an idea typically referred to as **sub-functionalization**. Sub-functionalization of gene copies would lead to the evolution of proteins that specialize on different aspects of the job done by the original protein, for example, in the timing or location of expression, or in biochemical activity. Finally, one of the duplicate copies could acquire a new and selectively favored function through mutation, while the other experiences purifying selection for the original protein function. **Neo-functionalization** refers to a novel function evolving in one of the duplicates, and also can lead to evolutionary maintenance of both copies.

2.3.7 *Biological complexities in mutation*

I have said that mutations arise randomly, but that some kinds of mutations are more likely than others. How can this be? The notion that "random" does not mean "equally probable" might seem confusing at first. But, to be random, alternative possible events do not have to be equally likely. For example, if transition mutations are twice as likely as transversions at a given site, because of the chemical details of how DNA is constructed and looked after by the cell, then it is like rolling one through four on a six-sided die giving you a transition, whereas a roll of five or six gives you a transversion. There is still randomness. Likewise, some parts of the genome are more mutable, which gives rise to differences along the length of a chromosome in the kind and number of mutations it experiences. These differences are often referred to as spatial heterogeneity or **mutation rate heterogeneity** among loci.

There is one key feature of randomness that we care about most: mutations are random with respect to their potential effect on fitness. This means that regardless of whether a possible change might have a beneficial or deleterious or no effect on survival and reproduction, the likelihood that mutation arises at any given site is not biased by that potential fitness effect.

More subtle features of the mutation process are not captured in any of the models of mutation that we have discussed. For example, it may be the case that heterozygous indels foster higher rates of single nucleotide mutation in the DNA that immediately flanks the indel polymorphism. Although such **indel-associated mutation** of heterozygous regions represents a fascinating piece of the molecular evolutionary process, throughout the remainder of this book we will focus on the mainstays of mutational models in molecular population genetic analysis.

Molecular population genetics deals with heritable changes in DNA. But there are other forms of inheritance that can also influence fitness. In particular, changes to the **chromatin** state of DNA through biochemical modification of histone proteins or even DNA itself can be passed on from parent to offspring, for example when methyltransferase proteins add methyl groups to these components of chromatin. These chromatin modifications can alter gene expression. Because these alterations do not directly change the DNA sequence itself, but nevertheless can sometimes be inherited across generations, they are called **epigenetic** changes or **epimutations**. These epigenetic changes are very labile, meaning that they can be reset across generations much more rapidly than genetic mutations. The potential role of epimutations in evolution represents an ongoing and contentious area of research and, unlike genetic mutations to DNA, does not yet have a well-developed molecular evolutionary framework for analysis.

Because it is heritable mutations that we care about for evolution, we can mostly ignore **somatic mutations** to non-germline tissue because they won't get passed on to offspring. Despite the potentially devastating effects of somatic mutation to an individual organism's viability and fertility, with human cancers as a heart-wrenching example, those mutated somatic cells will not themselves propagate to the next generation. The evolutionary relevance of somatic mutations does arise in two kinds of cases. First, in some organisms that reproduce asexually, there is no distinct separation of germline tissue and so in some sense we can think of inheritance of mutations that arise in somatic tissues. Second, some heritable genotypes might be predisposed to producing high rates of somatic mutation. For example, some alleles of the human *BRCA1* gene predispose women (and men) to breast cancer because the DNA repair protein that it encodes is disrupted and can lead to somatic mutations that induce cancerous tumors. If such **mutator** alleles affect an organism's reproductive capacity, for example by causing reduced viability prior to reproductive age, then selection could still act on it.

Further reading

Baer, C. F., Miyamoto, M. M. and Denver, D. R. (2007). Mutation rate variation in multicellular eukaryotes: causes and consequences. *Nature Reviews Genetics* 8, 619–31.

Bataillon, T., Duan, J. J., Hvilsom, C. et al. (2015). Inference of purifying and positive selection in three subspecies of chimpanzees (*Pan troglodytes*) from exome sequencing. *Genome Biology and Evolution* 7, 1122–32.

Fondon, J. W. and Garner, H. R. (2004). Molecular origins of rapid and continuous morphological evolution. *Proceedings of the National Academy of Sciences USA* 101, 18058–63.

Jauch, A., Wienberg, J., Stanyon, R. et al. (1992). Reconstruction of genomic rearrangements in great apes and gibbons by chromosome painting. *Proceedings of the National Academy of Sciences USA* 89, 8611–15.

Keightley, P. D. and Lynch, M. (2003). Toward a realistic model of mutations affecting fitness. *Evolution* 57, 683–5.

Kibota, T. T. and Lynch, M. (1996). Estimate of the genomic mutation rate deleterious to overall fitness in *E. coli. Nature* 381, 694–6.

Kimura, M. and Crow, J. F. (1964). The number of alleles that can be maintained in a finite population. *Genetics* 49, 725–38.

Lai, C. Y., Baumann, L. and Baumann, P. (1994). Amplification of *trpEG*: adaptation of *Buchnera aphidicola* to an endosymbiotic association with aphids. *Proceedings of the National Academy of Sciences USA* 91, 3819–23.

Lynch, M., Ackerman, M. S., Gout, J. F. et al. (2016). Genetic drift, selection and the evolution of the mutation rate. *Nature Reviews Genetics* 17, 704–14.

Messer, P. W. and Petrov, D. A. (2013). Frequent adaptation and the McDonald-Kreitman test. *Proceedings of the National Academy of Sciences USA* 110, 8615–20.

Mosher, D. S., Quignon, P., Bustamante, C. D. et al. (2007). A mutation in the myostatin gene increases muscle mass and enhances racing performance in heterozygote dogs. *PLoS Genetics* 3, 779–86.

Pritchard, J. K., Stephens, M. and Donnelly, P. (2000). Inference of population structure using multilocus genotype data. *Genetics* 155, 945–59.

Redon, R., Ishikawa, S., Fitch, K. R. et al. (2006). Global variation in copy number in the human genome. *Nature* 444, 444–54.

Rival, P., Press, M. O., Bale, J. et al. (2014). The conserved PFT1 tandem repeat is crucial for proper flowering in *Arabidopsis thaliana. Genetics* 198, 747-U391.

Romiguier, J., Lourenco, J., Gayral, P. et al. (2014). Population genomics of eusocial insects: the costs of a vertebrate-like effective population size. *Journal of Evolutionary Biology* 27, 593–603.

Rosenblum, E. B., Rompler, H., Schoneberg, T. and Hoekstra, H. E. (2010). Molecular and functional basis of phenotypic convergence in white lizards at White Sands. *Proceedings of the National Academy of Sciences USA* 107, 2113–17.

Sanjuan, R., Moya, A. and Elena, S. F. (2004). The distribution of fitness effects caused by single-nucleotide substitutions in an RNA virus. *Proceedings of the National Academy of Sciences USA* 101, 8396–401.

This, P., Lacombe, T., Cadle-Davidson, M. and Owens, C. L. (2007). Wine grape (*Vitis vinifera* L.) color associates with allelic variation in the domestication gene *VvmybA1. Theoretical and Applied Genetics* 114, 723–30.

van't Hof, A. E., Campagne, P., Rigden, D. J. et al. (2016). The industrial melanism mutation in British peppered moths is a transposable element. *Nature* 534, 102–5.

CHAPTER 3

Quantifying genetic variation at the molecular level

Knowing how new mutations arise and what kinds of mutations can occur tells us about the types of alleles that we can expect to see in a population. However, it does *not* tell us much about how often we should expect to see them. How do we measure those mutations **segregating** as alternative alleles in the population? What we need is a framework for quantifying genetic variation. Populations are where evolution starts. Mutations arise within some individual in a population and so we need to think about those mutations as the spark of population genetic variation. Classic population genetics models provide the logic behind allele frequency quantification, with an emphasis on scenarios involving one locus with two alleles. We will now think about how this kind of logic applies to problems in molecular population genetics.

Quantifying genetic variation is important because it represents one of the basic ingredients to making biological sense about the evolutionary process in natural and experimental populations. As Richard Lewontin wrote in 1974, "The whole of population genetic theory remains an abstract exercise unless the frequencies of alternative alleles at various loci can be determined in different populations and at different times in the history of a given population." In this chapter, we will work through how to paint specific measurements onto the abstractions of population genetics, to help us translate Picasso into da Vinci.

Of course, it is old news that populations have variation. I like Alfred Wallace's comment on this point from way back in 1889: "wherever variations are looked for among a considerable number of individuals of the more common species they are sure to be found; that they are everywhere of considerable amount." Since then, thousands of genetic studies of quantitative traits, a.k.a. "quantitative genetics," have demonstrated that significant *heritable* variation among individuals is present within natural populations. But what fraction of loci in a typical individual are heterozygous? What proportion of loci in the genome of a species are polymorphic? We can't answer these questions from observations of phenotypic variation. It is very difficult to attribute genetic variation in traits to specific genes or to infer how many genes contribute to variation in a trait. Even in those cases of dimorphisms for which standard genetic crosses can demonstrate a simple genetic cause, for example the one or two loci that control shell color in scallops (see Figure 1.3), we do not know how representative they will be for the genome overall.

Much of the DNA sequence variation in genomes might not even affect trait variation at all. In contrast to statistical measures of the genetic variation that underlies phenotypes, modern molecular techniques provide an easier and more direct route to characterizing

A Primer of Molecular Population Genetics. Asher D. Cutter, Oxford University Press (2019).
© Asher D. Cutter 2019. DOI: 10.1093/oso/9780198838944.001.0001

heritable genetic variability in populations. Therefore, we first start by examining DNA variability itself, waiting until later to concern ourselves with its potential consequences for phenotypes. Surveying variation at the molecular level lets us get a direct and quantitative answer to how much of the genome is polymorphic. As we will see, genetic variation depends intimately on the rate of mutational input and on the genetically effective population size (Box 3.1). Consequently, molecular genetic variation helps us to learn about these fundamental properties of populations and how they can tell us about evolution.

Box 3.1 EFFECTIVE POPULATION SIZE (N_E)

The size of a population seems a straightforward thing: can't you just count up the number of individuals? Unfortunately, this **census population size** (N) does not necessarily reflect how many genetically distinct sources of gametes will contribute to each generation. What matters for evolution is how alleles behave, how they change in abundance, how random chance can cause predictable changes in allele frequencies. This genetic size, the **effective population size** (N_e), is usually much smaller than the census size because of factors like unequal sex ratios, inbreeding, variability among individuals in how many offspring they have, population subdivision, population size changes, and natural selection. The effective size N_e is the size that an idealized population would have to be to mimic the random fluctuations in allele frequencies that we see in our real population of census size N. In terms of how allele frequencies change from generation to generation, a large census population with non-random mating may behave like a smaller random-mating population. So we can use this smaller effective population size as a common frame of reference. Worded with some jargon: genetic drift in our real population of size N operates at the same rate as for an ideal Wright-Fisher population of size N_e. In short, it lets us convert our messy real-world population to the common reference population used in mathematical models of population genetics to understand evolutionary change.

The **Wright-Fisher population** model is based on a simplified, generic lifecycle to use as a mathematically convenient point of comparison. It presumes a constant size, random mating, selective neutrality, and non-overlapping generations. The **Moran model** is similar, but better approximates populations with overlapping rather than discrete generations. The key distinction between them is that genetic drift proceeds twice as fast in the Moran model.

How does one go about determining the value of N_e from data? It turns out that there are several ways to think about calculating N_e. For a simple ideal population, they all will give the same value, but not so for some more complicated population histories. First, the "inbreeding effective size" is the classic view of how to depict the idea of N_e as first enunciated by Sewall Wright in 1931. It summarizes the size of the corresponding ideal population with an equivalent probability of identity by descent of different alleles. The rate of decline in heterozygosity due to genetic drift can be written mathematically in terms of the inbreeding coefficient (F) and F tells us the probability of identity by descent: $H_S = 2pq (1 - F)$ and $F = 1/ (2N_e)$. The inbreeding effective size is essentially identical to the "coalescent effective size," derived from coalescent theory (see section 5.2). As a result, the inbreeding and coalescent effective size mostly reflects the long-term N_e of a population and will tend to reflect the smallest size over that timespan (i.e. the **harmonic mean** population size). The coalescent effective size is the notion of N_e that is most often used in molecular population genetics.

(Continued)

Box 3.1 CONTINUED

Second, the "variance effective size" determines N_e from the variability of changes in allele frequencies over time: $N_{e(v)} = pq/ (2 \cdot \text{Var}[\Delta p])$. The variance effective size best reflects the strength of genetic drift over recent generations, so is valuable to think about the current population state when population census sizes have grown recently or fluctuate through time. Third, the "eigenvalue effective size" is computed from the rate of change in heterozygosity over time. Finally, linkage disequilibrium between either closely spaced or distant loci can be used to calculate N_e in a way that reflects population size on different timescales into the past (see Chapter 6).

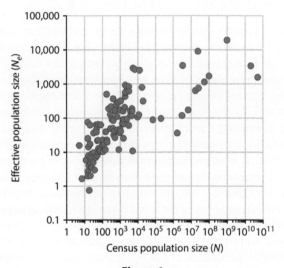

Figure 1

Figure 1 Heterozygosity decreases faster and homozygosity increases faster in small populations. These genotypic shifts will be more likely to influence alleles that affect fitness in small populations, as well. Given that most fitness-affecting mutations are deleterious, this genetic load can create problems for species of conservation concern. On the flip side, methods of estimating N_e that underestimate how big a population is in the present also would tend to underestimate the input of beneficial mutations into it and the ability of selection to mete out adaptive change. Analysis of 115 species shows how species with larger census sizes tend to have larger effective sizes, though the range of census sizes spans many more orders of magnitude (redrawn from Palstra and Fraser 2012).

3.1 Heterozygosity

Heterozygosity is a straightforward and common way to think about molecular genetic variability: the fraction of individuals that have different alleles at a locus is its single-locus heterozygosity. Conveniently, the notion of heterozygosity can apply to any kind of locus, including different nucleotide variants if our locus of interest is just a single nucleotide site. First, let's think about heterozygosity generally and then we will apply it to particular kinds of molecular variants.

If we look at a single locus, x, that has two possible alleles in a population of diploid organisms (allele "1" and allele "2"), then its heterozygosity (H_x) is simply the number of individuals that we observe to have different alleles at the locus (k_{12}) among all of the k individuals that we care to look at for our population sample. That is:

$$H_x = k_{12}/k \tag{3.1}$$

and, more generally, for a single locus that is multi-allelic, we can calculate its heterozygosity as:

$$H_x = \sum_{i \neq j} \frac{k_{ij}}{k}, \tag{3.2}$$

where k_{ij} is the number of diploid individuals with different alleles i and j among the k total number of individuals sampled. For example, for locus A with alleles A_1 and A_2, k_{12} is just the number of individuals with the A_1A_2 genotype. Or, for the B locus with alleles B_1, B_2, and B_3, k_{ij} would include individuals with genotypes B_1B_2, B_1B_3, and B_2B_3.

Box 3.2 HARDY-WEINBERG PRINCIPLE

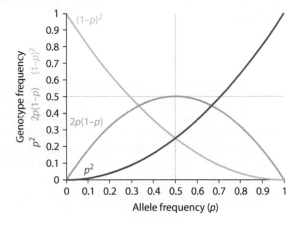

Figure 2

Figure 2 When alleles segregate in meiosis in a Mendelian fashion, we can make a robust prediction about the genotype frequencies of homozygotes and heterozygotes, provided that individuals mate at random with one another so that there is random union of gametes in the population. Specifically, we expect that an allele with frequency p in a population will be found homozygous in a fraction p^2 of individuals and heterozygous in a fraction $2p(1-p)$ of individuals (H_{exp}). When there are two alleles, it means that we should see the three possible genotypes in a ratio of $p^2 : 2p(1-p) : (1-p)^2$. It takes just a single generation of random mating to take even maximally skewed genotype frequencies to achieve this ratio. This population genetic prediction is known as the **Hardy-Weinberg principle**, after Godfrey Hardy and Wilhelm Weinberg who illustrated the idea back in 1908. To interpret this result as a long-term equilibrium depends on many simplifying assumptions, including no mutation, selection, or genetic drift, but in the short term from one generation to the next, the Hardy-Weinberg genotype frequencies will be quite accurate for reasonably large

(Continued)

Box 3.2 CONTINUED

populations. Even though drift will change allele frequencies over time, random mating so rapidly yields the predicted genotype frequencies that we can confidently expect them to be what we expect from the current allele frequencies in the population. In this vein, we should think of Hardy-Weinberg genotype frequencies as a constantly changing dynamic "steady state" that rapidly tracks the slower process of changes in allele frequencies in the population.

Keep in mind that when we collect a set of individuals to then look at heterozygosity, we aim to have a **random sample** of individuals from the population. Random in this sense means that every individual in the population had an equal chance of having been collected for analysis. This random sample is a subset of the population that is small enough that the collection procedure does not in itself alter the population in any substantive way. By being a random sampling of genetic information from the entire population, it allows us to use statistical principles to draw inferences about the entire population from the measured properties of the small subset of individuals.

The **observed heterozygosity** for the population *overall* is the average frequency of heterozygotes ($\bar{H}_{obs}$) when we look across m different loci. In other words, $\bar{H}_{obs}$ is the arithmetic average of H_x across loci:

$$\bar{H}_{obs} = \frac{\sum\limits_{x=1}^{m} H_x}{m}. \tag{3.3}$$

It is also useful to calculate the **expected heterozygosity** ($H_{\exp}$), also sometimes called gene diversity. $H_{\exp}$ represents the probability that two alleles (or variants) selected randomly from the population will differ from each other. That is, the expected heterozygosity lets us answer the following question: if individuals mate at random, how frequently should we expect to see organisms that contain different alleles, given that we know the overall abundance of each allele individually? Expected heterozygosity for two alleles is just what we would predict for the frequency of heterozygous individuals under Hardy-Weinberg equilibrium for a single locus with two alleles with frequencies p and $1 - p$ (Box 3.2):

$$H_{\exp} = 2p(1-p) = 1 - p^2 - (1-p)^2 = 1 - \left[p^2 + (1-p)^2\right] \tag{3.4}$$

and, for a locus with any number of alleles:

$$H_{\exp} = 1 - \sum_j p_j^2. \tag{3.5}$$

This is the same as saying that expected heterozygosity at a locus is just one minus the frequencies of all of the homozygotes. This description of heterozygosity in terms of homozygotes simply comes from the fact that the expected frequency of each homozygote genotype A_jA_j is just $p_j \cdot p_j = p_j^2$, given random mating (Box 3.2). The term **homozygosity** (G) just refers to one minus heterozygosity: $G_{obs} = 1 - H_{obs}$, $G_{\exp} = 1 - H_{\exp}$.

It is easier to calculate *expected* heterozygosity than *observed* heterozygosity, in practical terms, because we can do it from allele frequencies alone. By contrast, observed heterozygosity requires that we know whether every individual is heterozygous or homozygous, which is harder information to get hold of. Consequently, expected heterozygosity as a measure of genetic variation also can be applied to haploid organisms like bacteria that only have one allele at a given locus in a given individual: organisms that have an undefined "observed heterozygosity." However, interpreting H_{exp} necessarily would be more abstract for haploid organisms. Just as we would obtain an overall estimate of observed heterozygosity for a population by averaging H_{obs} across loci, we can similarly calculate an overall expected heterozygosity ($\bar{H}_{exp}$) by averaging H_{exp} across loci:

$$\bar{H}_{exp} = \frac{\sum\limits_{y=1}^{m} H_y}{m}. \tag{3.6}$$

When we want to infer the expected heterozygosity of a population from a **sample**, we can use Nei's unbiased estimator for a single locus:

$$\hat{H}_{exp} = \frac{n}{n-1}\left(1 - \sum_j p_j^2\right), \tag{3.7}$$

where n is the number of copies of the locus that we sampled and p_j is the frequency of allele j calculated from that sample of individuals. This factor of $n/(n-1)$ is just a standard statistical correction for the fact that, in practice, we would have measured allele frequencies with some degree of error because we used just a subsample of size n and not every single individual in the population. For samples of two alleles from k diploid individuals, n would equal $2k$ because we would be including two copies of each allele from each individual.

Box 3.3 WRIGHT'S F: FIXATION INDEX, INBREEDING COEFFICIENT, PROBABILITY OF IDENTITY BY DESCENT

You can think about what F means in several ways. It tells us how much more or less heterozygosity we actually find in a population relative to what we would expect from random mating, with $F = 0$ meaning that we see what we would expect from Hardy-Weinberg conditions and values of $+1$ or -1 being extreme departures. It tells us how much excess homozygosity the population has. It tells us how predictably you could know the identity of one allele given that you had already sampled another allele at the locus; the correlation should be zero with random mating and Hardy-Weinberg conditions. It tells us the probability that two alleles sampled from the population will be identical because they came from the same parents, relative to when mating and sampling are random. It tells us the rate of decline in heterozygosity due to non-random mating like inbreeding or self-fertilization.

You'll note that we use Hardy-Weinberg genotype frequencies as our reference point for F. This means that anything that skews genotype frequencies will lead to $F \neq 0$. Most commonly we think about non-random mating, like inbreeding. However, population subdivision and migration in a species also represent common real-world differences from the assumptions of the Hardy-Weinberg model. It is also possible that negative-assortative mating could lead to $F < 0$, though we should expect to see this influence only at the loci controlling traits used as the basis for mating preferences and at loci closely linked to them. Most typically, we observe values of $F \geq 0$.

(Continued)

Box 3.3 CONTINUED

With random mating, the probability that two alleles would be sampled from the same copy from a parent in the preceding generation is $F = 1 / (2N_e)$. With 100% self-fertilization reproduction, that probability of identity by descent from a heterozygous parent is $F = \frac{1}{2}$. This tells us that heterozygosity will decrease in a population just by the process of random sampling, also known as **genetic drift**, at a rate of $1 / (2N_e)$ each generation; heterozygosity will decrease by half each generation if all individuals reproduce by self-fertilization. When heterozygosity decreases, it means that homozygosity increases which will lead to fixation of one of the alleles. Hence, the fixation index F tells us about the approach of neutral alleles to fixation by genetic drift. We will develop this idea of identity by descent further back in time with the process of coalescence in section 5.2.

Box 3.4 ALLOZYME POLYMORPHISM EXAMPLE

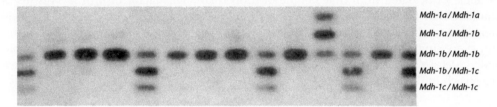

Mdh-1a / Mdh-1a
Mdh-1a / Mdh-1b
Mdh-1b / Mdh-1b
Mdh-1b / Mdh-1c
Mdh-1c / Mdh-1c

Figure 3

The gel stain in Figure 3 shows the protein electrophoretic pattern for NAD-dependent malate dehydrogenase (*Mdh-1*) from 14 horseshoe crabs (*Limulus polyphemus*; Figure 4) from a 1970 study by Robert Selander and his colleagues. The 14 vertical columns represent different individuals and the rows represent different mobilities of 3 alternative isoforms of the protein dimer (isoforms *a*, *b*, *c*; heterozygotes dimerize distinctly from homozygotes). Heterozygotes have three bands, homozygotes have one band. Based on these data, what is the H_{exp} for *Mdh-1*? What is F? Gel image published with permission of the publisher John Wiley and Sons: Selander et al. (1970). Genetic variation in the Horseshoe Crab (*Limulus polyphemus*), a phylogenetic "relic," *Evolution*, 24 (2): 402–14. Horseshoe crabs © Sean Crane/Minden Pictures/Getty Images.

Genotype	Number of individuals	Genotype frequency	H-W expected genotype frequency	Allele	Allele frequency
Mdh-1a/Mdh-1a	0	0.000	0.001	Mdh-1a	0.036
Mdh-1a/Mdh-1b	1	0.071	0.056		
Mdh-1a/Mdh-1c	0	0.000	0.013		
Mdh-1b/Mdh-1b	8	0.571	0.617	Mdh-1b	0.786
Mdh-1b/Mdh-1c	5	0.357	0.281		
Mdh-1c/Mdh-1c	0	0.000	0.032	Mdh-1c	0.179

Figure 4

As a quick aside about nomenclature in equations, the "bar" above a variable like $\bar{H}_{\exp}$ means "average" and a "hat" above a variable like $\hat{H}_{\exp}$ means "expectation." The word "expectation" means it is the value that you calculate from data in an attempt to estimate the true value that the population has, that we could know for certain only with perfect information. I will often use these bits of math jargon when I am first setting up and defining any variables with equations, but will generally not be so explicit when talking about them in the text, unless it is especially key to some point.

You may recall from classic population genetics theory that the **probability of identity by descent** is Wright's statistic F, also known as the **inbreeding coefficient** or the **fixation index** (Box 3.3). As you might guess, the inbreeding coefficient relates to heterozygosity in a predictable way:

$$F = \frac{H_{\exp} - H_{obs}}{H_{\exp}}. \tag{3.8}$$

It tells us how much the observed heterozygosity differs from what we would expect from Hardy-Weinberg equilibrium. If we want to draw inferences about the population as a whole, then we would use F averaged across loci because H would be averaged across loci. As a result, this depiction of F would *not* describe the probability of identity by descent for any given locus. Instead, it is an indicator of how well the population conforms to an idealized randomly mating population, with respect to the incidence of heterozygous genotypes across the genome. At equilibrium, $H_{obs} = H_{\exp}$, so $F = 0$. If no individuals are heterozygous at any loci despite the presence of allelic variability ($H_{obs} = 0$), then $F = 1$, meaning that the genetic variation in the genomes of individuals in the population differs very strongly from what we should expect to see in the equilibrium situation.

In other words, the population would have non-Hardy-Weinberg genotype frequencies. One possible cause of $F > 0$ could be inbreeding in the population, a form of non-random mating (Box 3.3).

In talking about genetic variation in terms of heterozygosity, I have slipped in the implicit assumption that we are working with a diploid organism, such as most animals and plants and other eukaryotes. However, organisms with haploid genomes, like bacteria and many yeasts, also have genetically variable populations. In these cases, we can still use expected heterozygosity as a metric of genetic diversity. The caveat is that "heterozygosity" is more of a figurative term for haploids, referring just to the probability that any two alleles at a locus will be different when we pick them randomly from individuals in the population. As we will see in section 3.3, this is not much of a limitation for many analyses that work directly with DNA sequence polymorphism.

3.2 Allozyme variation

Now that we have the conception of heterozygosity in mind in a somewhat abstract way, how can we apply it specifically to genetic variation using **molecular markers**? A simple empirical way to quantify molecular variation is to use gel electrophoresis to look for differences in the mobility of protein isoforms. This approach was the pioneering method to assess molecular variability in the 1960s and so it may seem old-fashioned. It is. But as Brian and Deborah Charlesworth wrote in 2017, "It is hard today to grasp the revolutionary nature of the discovery of molecular variation." Such protein **allozymes** serve as a convenient, and historical, starting example to illustrate the revolutionary principles of quantifying molecular genetic variation.

Proteins are composed of amino acids, each of which has a distinct mass and carries a positive or negative or no charge, and so each amino acid contributes to the combined size and charge of the protein. Any non-synonymous DNA sequence variant that alters the net protein charge will then cause the different protein variants to move at different rates through an electrical field on a gel. Proteins with such size and charge differences, which can be visualized with **protein electrophoresis**, are called allozymes. We can then calculate the heterozygosity for such allozyme molecular markers by counting up the number of copies of each allele in a sample from a population (Box 3.4). Proteins can be chosen by an experimenter at random and analyzed in this way to infer the proportion of loci that are polymorphic.

Using the protein electrophoretic approach to quantifying molecular diversity, Richard Lewontin and John Hubby (1966) found that about one-third of the allozyme loci that they looked at were polymorphic in the fruit fly *Drosophila pseudoobscura*, and that about 12% of the loci were heterozygous in any given fly. One drawback to indirectly quantifying polymorphisms in DNA using allozymes, however, is that usually it depends on having enzymatic interaction with a visible dye, limiting the number and kinds of proteins that may be assayed in practice. Another more serious drawback to allozymes is that protein electrophoresis only reveals non-synonymous or indel differences that also result in a net charge or size change to the protein. As a result, there may be substantial DNA sequence variation among individuals that is not captured by allozyme differences. In addition, electrophoresis can tell us that alleles of the protein are different, but gives us no way of measuring how different they are.

3.3 DNA sequence variation

The advent of DNA sequencing technology as a widespread tool starting in the 1980s made it possible to identify directly the heritable nucleotide variants at a locus. It realized the vision laid out in 1920 by Edgar Altenburg and Hermann Muller that:

> It would accordingly be desirable, in the case of man, to make an extensive and thorough-going search for as many factors as possible that could be used in this way, as identifiers. They should, preferably, involve character differences that are (1) of common occurrence, that are (2) identifiable with certainty, and that are (3) heritable in a simple Mendelian fashion. It seems reasonable to suppose that in a species so heterozygous there must really be innumerable such factors present.

How prescient they were! We can now quantify the number of nucleotides that differ among individuals by examining **homologous** DNA sequences in a sample of individuals. Individuals that share identical sequences for a given locus have the same **haplotype**. Direct sequencing allows the detection of many kinds of genetic variants, including SNPs, indels, and microsatellites, and reveals much more genetic variation than can be observed with allozymes.

For our purposes here, it is not necessary to understand the interesting details of the biochemistry and molecular biology that allow inference of a nucleotide sequence. Moreover, DNA sequencing represents a subject of ongoing and rapid technological development, now permitting individual research laboratories to sequence entire genomes rather than just individual genes. Each technology has its own set of technical constraints or biases that are important to manage when implementing them to accurately determine DNA sequences. What we care about most in the context of molecular population genetics is simply that one may confidently determine the DNA sequence for any given locus of interest in any given individual and, consequently, identify nucleotide differences between different copies of that locus from a set of individuals sampled from a population.

As an illustrative example, let's consider the *Adh* locus in the fruit fly *Drosophila melanogaster*, which was the focus of the very first molecular population genetic analysis of DNA sequence variation. In 1983, Martin Kreitman sequenced 2,721 nucleotides of 11 copies of the alcohol dehydrogenase (*Adh*) gene in *D. melanogaster*. This sequence included the four exons of the gene plus three introns and some upstream and downstream untranscribed sequence. Considering only SNPs, he found nine distinct haplotypes (Figure 3.1); after also incorporating indel polymorphisms, the sequences described 11 haplotypes that differed in allelic state for 43 sites. In other words, the sequence of every single copy was different! Limiting the data to the 768 nucleotides (or **base pairs**, abbreviated **bp**) of coding sequence, there were 8 unique haplotypes and 14 sites were polymorphic.

What is the heterozygosity for the complete *Adh* sequence? In this case, every haplotype copy in the sample is different, so the expected heterozygosity is 100%. By contrast, only two allozyme alleles had been identified in previous studies with samples of hundreds of individuals. If finding this amount of DNA variation from just 11 copies of a gene is common (Figure 3.2)—and it turns out that it is indeed very common—then heterozygosity measured for DNA sequences in this way will not be a very useful metric for comparing between loci or populations. To learn something about evolutionary pressures

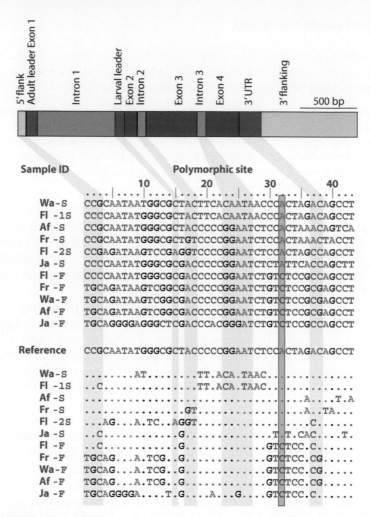

Figure 3.1 Table of polymorphism.

This data matrix summarizes the single nucleotide polymorphisms from Kreitman's (1983) classic study on *Adh*. Sites that were monomorphic or that contained indels are not shown. In this **table of polymorphism**, each row represents a haplotype from one of the 11 individuals, each column is one of the 43 polymorphic sites. The bottom table shows the same information with the "reference" variant indicated with a "." to help visualize the distinct variant states across sites. The shading of polymorphic sites indicates which portion of the *Adh* gene they derive from, using the gene structure indicated across the top (the darkest gene regions are coding exons). The adaptively significant non-synonymous polymorphism from Exon 4 is highlighted in the table of polymorphism.

operating on different loci, we need to be able to discriminate among them quantitatively, which is not possible if every locus has $H_{exp} = 1$! Instead, to circumvent this problem, it is useful to treat individual nucleotide sites as the focal point for assessing genetic variability.

3.3.1 *Nucleotide polymorphism and diversity: θ_W and θ_π*

When we quantify diversity in DNA, it makes sense to use individual nucleotides as the point of reference for polymorphism. The number of nucleotide sites that are polymorphic

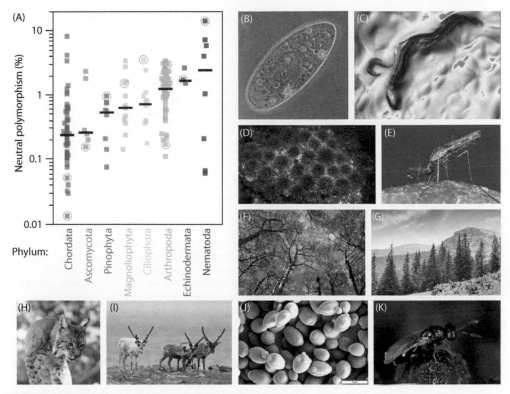

Figure 3.2 Polymorphism across the tree of life.
Species vary by orders of magnitude in the amount of nucleotide diversity that their populations contain. Species from some phyla tend to be more genetically variable than others, with smaller organisms tending to have larger population sizes and more diversity. Example organisms indicated by dotted circles are shown in the photos (*Paramecium aurelia, Caenorhabditis brenneri, Strongylocentrotus purpuratus, Anopheles gambiae, Picea glauca, Populus tremula, Lynx lynx, Rangifer tarandus, Saccharomyces cerevisiae, Nasonia vitripennis*). (*A*) Data updated and redrawn from Leffler et al. (2012); (*B*) *P. aurelia* photo by Barfooz, reproduced under the CC BY-SA 3.0 license; (*C*) Nematode photo by Lucia Kwan; (*D*) Purple sea urchins by Ed Bierman, reproduced under the CC BY 2.0 license; (*E*) Mosquito © Everett Historical/Shutterstock.com; (*F*) *P. tremula* photo by Kari, reproduced under the CC-BY 2.0 license; (*G*) Spruce forest © AlinaMD/Shutterstock.com; (*H*) *L. lynx* Eurasian Lynx photo by Dogrando, reproduced under the CC BY-SA 2.0 license; (*I*) Reindeer © Wolfgang Kruck/Shutterstock.com; (*J*) *S. cerevisiae* SEM image by Mogana Das Murtey and Patchamuthu Ramasamy, reproduced under the CC BY-SA 3.0 license; (*K*) *N. vitripennis* wasp photo by Gernot Kunz for Bioplanet.

across all copies of a locus that we examine, *S*, is commonly referred to as the **number of segregating sites** (Figure 3.3). The term "segregating" simply means that the site (or locus) under consideration has two or more variants present in the population; it is used by population geneticists in reference to the process of independent segregation of alternative alleles during meiosis. Given *S*, we may then calculate the *proportion* of sites that differ between copies: **nucleotide polymorphism**. For a sequence of total length *L* with *S* nucleotide sites being polymorphic, the proportion of polymorphic sites is simply $P_S = \frac{S}{L}$. However, P_S will tend to be greater for samples that include sequences from more individuals because more sites will be found to differ as more sequences get added. Therefore, it is useful to have a metric that accounts for sample size so that we can compare across loci or populations that have different sample sizes.

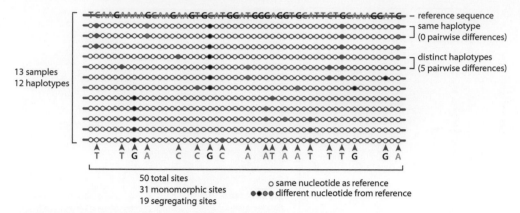

Figure 3.3 Segregating sites.
We can visualize the monomorphic and polymorphic sites in a random sample of allelic DNA sequences from a population. This example shows 12 unique haplotypes among the 13 copies of the 50-bp-long locus, as a result of the distinct patterns of polymorphisms among them. The proportion of polymorphic sites is $P_S = 19 / 50 = 0.38$. Using a value of our sample size correction factor $a = 3.10$, we would compute $\theta_W = 0.38 / 3.10 = 0.126$. If it were fair to assume that the population is at equilibrium and that mutations to any of the 50 sites in this locus would be selectively neutral, then this number would provide an estimate of the scaled mutation rate $(4N_e\mu)$.

A version of P_S that properly accounts for sample size is commonly referred to as **Watterson's θ_W** (note that the Greek letter θ is pronounced "thay-ta" or "thee-ta"). This measure of nucleotide variation standardizes P_S appropriately by dividing it by the harmonic number $a = \sum_{i=1}^{n-1} \frac{1}{i}$ for a sample of n chromosome copies:

$$\hat{\theta}_W = P_S/a = S/(La).$$
(3.9)

Later on, in Chapter 5 (Equation 5.4), we will see how θ_W can also be extracted from coalescent theory to see how a relates to genealogies. Note that a value for θ_W can be reported per locus ($\hat{\theta}_W = S/a$) or per site ($\hat{\theta}_W = P_S/a$), depending on the relevant context for reporting nucleotide variation. Measures of variation *per site* allow the easiest and most direct comparisons across loci, across populations, and across species; we will assume that θ_W is measured per site from now on.

Another common measure of molecular variation is Masatoshi Nei and Wen-Hsiung Li's metric for **nucleotide diversity**, θ_π: the average proportion of nucleotides that differ between a randomly chosen pair of gene copies. These differences are the nucleotide mismatches between pairs of sequences. In essence, θ_π is the DNA sequence analog of expected heterozygosity applied on a per-nucleotide basis. θ_π is often referred to as just "π" (pronounced "pie"), but here I use the variable θ_π to try to make its connection to other aspects of molecular variation more obvious.

To compute θ_π, first let's count up the differences between haplotypes i and j so that the proportion of sites that differ between just that pair of haplotypes is denoted by π_{ij}. We will then repeat this procedure for all pairs of haplotypes and combine it with the relative frequencies of those haplotypes in our sample of n sequences because some haplotypes may be represented more than once in our collection of individuals (i.e. the frequency of haplotypes i and j are p_i and p_j, respectively). For k different haplotypes, the nucleotide diversity of the population is:

$$\theta_\pi = \sum_{i=1}^{k}\sum_{j=i}^{k} p_i p_j \pi_{ij} = 2 \cdot \sum_{i=2}^{k}\sum_{j=i}^{i-1} p_i p_j \pi_{ij}. \tag{3.10}$$

I have led you through the logic of Equation 3.10 to emphasize that θ_π incorporates some information about the frequencies of distinct haplotypes. But it is mathematically equivalent to, instead, count up the pairwise mismatches between every possible combination of sequence copies in our sample, some of which will represent identical haplotypes if k is less than n:

$$\theta_\pi = \sum_{i=1}^{n}\sum_{j=i}^{n} \pi_{ij} = 2 \cdot \sum_{i=2}^{n}\sum_{j=1}^{i-1} \pi_{ij}. \tag{3.11}$$

Equations 3.10 and 3.11 are just different ways of calculating the same thing: one counts up to k for differences among distinct haplotypes, the other counts up to n for differences among individual sequence copies but some pairs of those sequence copies might have the same haplotype and so have no SNP differences. For an unbiased estimate for n samples, which is the basis of calculations from observed data, we need to modify this calculation to make our standard statistical adjustment for sample size as we did with $\hat{H}_{exp}$ (see section 3.1):

$$\hat{\theta}_\pi = \frac{n}{n-1}\theta_\pi. \tag{3.12}$$

I have led you through the usual explanation of how to calculate θ_π by comparing pairs of haplotypes or sequence copies (Equations 3.10 and 3.11). To make its connection to heterozygosity more obvious, we can also calculate θ_π by looking across sites. In other words, we can calculate the expected heterozygosity (H_{exp}) for each site along the sequence and glue the values together to arrive at an equivalent value for θ_π. Remember that from Equation 3.5, $H_{exp} = 1 - \sum_j p_j^2$ for a locus. For our DNA sequence, we calculate H_{exp} separately for all S polymorphic sites, which leads to an alternative to using Equation 3.12:

$$\hat{\theta}_\pi = \frac{n}{n-1}\sum_{i=1}^{S} H_{exp(i)} = \frac{n}{n-1}\sum_{i=1}^{S}\left(1 - \sum_{j=1}^{n} p_{j(i)}^2\right). \tag{3.13}$$

This equation says that for each polymorphic site i, we calculate the expected heterozygosity for that site ($H_{exp(i)}$), then sum up those heterozygosities across all S polymorphic sites. Note that n is not necessarily the number of different individuals, but the number of sequence copies. For example, if we sequenced both copies of a locus from a diploid organism in a sample of 10 individuals, we would have 20 sequences; therefore, $n = 20$. As a consequence, there could be anywhere from $k = 1$ to 20 unique haplotypes among these 20 sequences. As another example, if we sampled all copies of an X-chromosome locus from each of 7 female lions (diploid for the X-chromosome) and 9 male lions (haploid for the X-chromosome, also termed "hemizygous"), then $n = 2*7 + 1*9 = 23$ copies of the locus. Just like for θ_W, we can calculate θ_π either on a per-locus or on a per-site basis. To arrive at the per-site values, we must divide the numbers that we get from Equations 3.12 and 3.13 by the length of the locus, being sure to pay attention to the appropriate class of sites (e.g. synonymous sites only) and to include both polymorphic and monomorphic sites in the length of the locus.

For Kreitman's *Adh* data, we find that $\theta_\pi = 0.0066$ and $\theta_W = 0.0055$ ($P_S = 0.016$, $a = 2.929$). In general, θ_π ranges between 0.004 and 0.02 in *D. melanogaster*. So, given that a typical gene is about 1,000 bp long, then a random pair of homologous sequences for a gene are expected to differ at 4–20 sites (1,000 bp*0.004 to 1,000 bp*0.02). It turns out that humans tend to have much less genetic variation than flies, with θ_π on average being about only 0.00081 (or, roughly 1 nucleotide difference every 1,200 bp between an allelic pair of sequence copies). At the other end of the diversity spectrum are organisms like the nematode *Caenorhabditis brenneri*, named after Nobel prize-winning geneticist Sydney Brenner, that have $\theta_\pi = 0.14$, representing roughly 1 nucleotide difference every 7 bp between a homologous pair of selectively unconstrained DNA sequences (Figure 3.2).

It probably has not escaped your notice that we have just gone through two similar but distinct ways of quantifying DNA sequence variation: θ_W and θ_π. This is neither redundant nor a mistake, as these metrics differ in a subtle but important way. θ_W differs from θ_π in how it is calculated because θ_W depends just on the *number* of segregating sites, whereas θ_π depends on the *frequencies* of variant sites in addition to the number of polymorphic nucleotides. This seemingly slight difference in how polymorphism is measured has important implications that we can use to our advantage in understanding the evolution of particular loci. In an equilibrium population, the two measures will be equivalent ($\theta_W = \theta_\pi$), a relationship that forms the basis of some tests of neutrality that we will discuss in Chapter 8. Predictions about θ_W and θ_π being equal under neutrality depend on the infinite sites model of mutation, so violation of this mutational model or of the selective neutrality of the sites under consideration in the real world also can lead to these metrics having different values.

We can calculate these measures of sequence variability for any class of sites. However, when we focus specifically on silent sites (non-coding DNA and synonymous sites), then θ_W and θ_π are particularly useful. This is because the expected heterozygosity with the infinite sites model under neutrality is simply $4N_e\mu$ (where N_e is the effective population size and μ is the point mutation rate per generation in a diploid). This combined parameter, $4N_e\mu$, is also sometimes called the **population mutation rate** or the **scaled mutation rate**, θ, because it is the rate of mutational input "scaled" by the mutational target size in the population. The distinction between θ and θ_W and θ_π is that θ_W and θ_π (or, more formally, $\hat{\theta}_W$ and $\hat{\theta}_\pi$) are empirical **estimators** of the "true" **population parameter** θ.

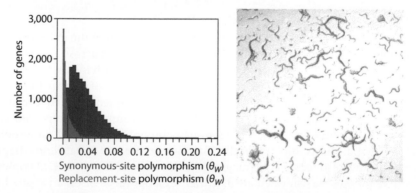

Figure 3.4 Polymorphism at synonymous versus replacement sites.
Synonymous sites contain a higher density of segregating sites than do replacement sites. Among 18,132 autosomal protein-coding genes in the nematode *Caenorhabditis remanei*, the θ_W measure of polymorphism averages 3.6% for synonymous sites but only 0.74% for replacement sites in a sample of 49 sequence copies for each gene. (photo courtesy J. Bermudez)

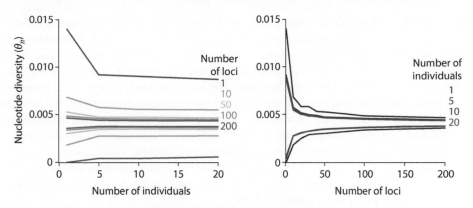

Figure 3.5 Sensitivity of estimating polymorphism to sample sizes.

Estimates will have more uncertainty in calculations of the average nucleotide diversity for a population when we use few individuals or few loci. The accuracy depends more strongly on the number of loci than on the number of individuals included. However, you need larger sample sizes of individuals to accurately summarize the site frequency spectrum than to estimate nucleotide diversity. (data for 95% confidence intervals redrawn from simulations by Dutoit et al. 2017.)

These estimators are also sometimes referred to as **summary statistics**, metrics that provide a single-number representation of a biological feature of the population. If we had an independent measure of the per-nucleotide mutation rate, for example from mutation accumulation studies, then we would be able to infer the effective population size from observed DNA sequence diversity by assuming that $4N_e\mu = \theta_W$ or that $4N_e\mu = \theta_\pi$ (see Chapters 4 and 7).

Because mutations in sites that affect the amino acid sequence are more likely to alter the function of the encoded protein, they tend to experience stronger purifying selection, which acts to eliminate mutations from the population. Consequently, despite the fact that roughly ¾ of coding sites are non-synonymous, as a byproduct of the structure of the genetic code (see Box 2.2), most of the observed variant sites between coding sequences usually will *not* occur at replacement positions in the sequence. That is, θ_W for replacement sites alone generally will be lower than θ_W calculated for synonymous sites alone (Figure 3.4).

We measure diversity from data, so how much data is enough to do so accurately? There are three main pieces to "how much": the number of individuals, the number of loci, and the length of those loci. For many questions, it is most valuable to have as much independent information as possible, which generally means maximizing the number of loci from across the genome at the expense of fewer individuals and shorter stretches of sequence (Figure 3.5). For simply estimating nucleotide diversity for a species, the rule of thumb for a long time has been to use about ten copies of each locus and to examine about ten loci. The optimal length of sequence needed to assess diversity at a locus depends on how polymorphic the species is: the more variation, the more polymorphic sites per nucleotide and the shorter the sequence needs to be in order to determine adequately the density of SNPs. In practice, technical limitations of Sanger sequencing of DNA generally defined the sequence length to be about 700 bp, unless you were examining the full length of a given gene because of some special interest. To define accurately the site frequency spectrum will require a larger sample size for each locus (section 3.3.2), preferably having at least 20 copies; to confidently detect a statistically significant departure from neutrality, typically you would want about 50 copies.

Since high-throughput DNA sequencing has become mainstream (section 3.5), these rules of thumb about "how much" have become a bit old-fashioned. Researchers can routinely sequence entire genomes, or at least all of the protein-coding genes in the genomes, for dozens or even thousands of individuals (e.g. Figure 3.4). Such a plethora of data is great in many ways. However, this genomic scope makes it all the more important to think about the structural complexity of chromosomes when we analyze and interpret all that data, including the role of recombination and its interaction with selection (see Chapters 6 and 7). Genome-scale analysis of polymorphism also presents researchers with the problem that, inevitably, some sites will have missing data for some individuals. While we won't dwell on this issue, it is important to keep in mind that this absence of DNA sequence information from some individuals must be taken into account to properly calculate molecular population genetic metrics.

3.3.2 *The site frequency spectrum*

In our description of θ_π as a measure of molecular variability in a population, I pointed out that it incorporates the frequencies of different variants into how it is calculated. We can take this idea of variant frequencies one step further, to summarize nucleotide polymorphism entirely in terms of their frequencies. From this perspective, we will focus on the rarer of the two variants at any given polymorphic site and first count up the number of variants that occur in just a single instance in our sample of individuals. These variants are **singletons**, with frequency $1/n$ where n is the number of copies of the locus that we have for analysis, and we count how many sites in our DNA sequence have these singleton variants. For example, one of the polymorphic sites might have one copy with a G variant whereas all the other copies in our sample have a T variant at that position in the DNA sequence. We then repeat this procedure to count polymorphic sites with "doubletons," for which the rarer variant has a frequency of $2/n$; tripletons with frequency $3/n$; and so on (Figure 3.6). We keep going until we reach the frequency class $n/2$, because we are looking just at the rarer copy and so the highest variant frequency possible is 50%. The collection of these variant frequencies in a histogram provides a way to visualize the

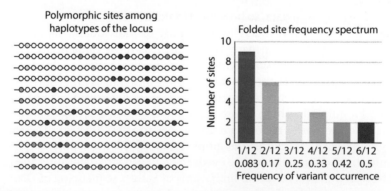

Figure 3.6 Molecular variation encapsulated in the site frequency spectrum.
The folded frequency spectrum can be inferred by counting separately the number of sites that are polymorphic for each class of variant frequencies, focusing on the rarer of the two variants found at any given position. Singleton variants are highlighted in red in the diagram of haplotypes, and represent 36% of the 25 polymorphic sites, where each circle represents a nucleotide position along a horizontally arranged DNA sequence. Cooler colors represent minor allele variants that have more intermediate frequencies in the sample of 12 sequences.

frequencies of all the polymorphic sites while ignoring their spatial configuration along the segment of DNA (Figure 3.6).

The variant allele that is rarer in the population is also often referred to as the **minor variant**, because it comprises a minority of the alleles at that site, in contrast to the more abundant **major variant**. We can combine all these counts of "minor" variant frequencies together into a histogram of **minor allele frequencies** (MAF) using either the absolute counts of sites or the relative abundance of sites out of the total number of polymorphic sites. Specifically, we visualize this summary of molecular variation as the **site frequency spectrum** (Figure 3.6), abbreviated as "SFS" and sometimes also called the **allele frequency spectrum** or **variant frequency spectrum**. "Spectrum" is just a colorful word for "distribution." So, the site frequency spectrum just represents the distribution that shows the abundances of variants that occur at distinct allele frequencies in the population. We generally expect singletons to be the most common class of variants in the distribution, with fewer and fewer sites that have alleles of more intermediate frequency (Figure 3.6). In Chapter 4 I will explain why this prediction is generally true for equilibrium populations.

By focusing on the frequencies of the rarer of the alternative variants at a polymorphic site, the possible range of the minor variant frequencies is necessarily constrained: they can range from a minimum near zero to a maximum of 50%. This way of constructing the SFS gives us what is termed the **folded site frequency spectrum**, where we "fold" or "reflect" the distribution back on itself at the 50% point when we use minor frequency variants.

An alternative way to construct the SFS is, instead, to distinguish **derived variants** versus **ancestral variants** rather than minor versus major variants. The derived allelic state corresponds to new mutations in the population, relative to the past ancestral population. Distinguishing derived from ancestral requires some additional information; specifically, we would need DNA sequences from related species to help determine the ancestral state (see Chapter 5). The SFS that results from derived variant frequencies can thus range from near zero up to 100%, technically excluding the extreme values of 0 and 1 because those represent monomorphic sites in the population. This broader distribution is called the **unfolded site frequency spectrum**, as it spans the full range of possible variant frequencies.

The distinction of "folded" versus "unfolded" SFS yields a difference in range (maximum variant frequency of 50% vs. ~100%) and a difference in conferring an evolutionary basis to the variants (minor/major vs. derived/ancestral). When we do not distinguish ancestral and derived states, we must "fold" the upper half of the SFS back on itself. So, a derived variant with a frequency of 90% would be represented as a "high frequency variant" in the unfolded SFS, whereas it would be represented by the complementary SNP (at frequency 10%) in the folded SFS that focuses on whichever variant is rarer. In some ways the unfolded distribution might seem superior because it implies more information by virtue of distinguishing ancestral versus derived allelic states. However, the folded SFS is much simpler to calculate and also is not prone to errors that can inevitably arise from mistakes in the inference of which variant is derived versus ancestral.

Summarizing molecular variation with the SFS is obviously more complex than with a single number like we can do for θ_W and θ_π. More to the point, the SFS is a vector, a list of numbers. If you have two populations, you can make the SFS even more complex by constructing a matrix for the "**joint site frequency spectrum**": the values in each cell of the matrix indicate how common each frequency class is within each of the populations (see Figure 8.3). Conveniently, however, the frequency spectrum has all the information we need to calculate θ_W and θ_π, implying that θ_W and θ_π represent a special-case simplification of the full SFS.

3.4 Simple sequence repeat variation

While SNPs are an abundant and informative kind of molecular variation, they are not the only kind of variability introduced into genomes by mutation. The simple sequence repeats of microsatellite loci also have been very useful molecular markers. Microsatellites are a common feature of most genomes, individual loci tend to be highly variable in a population because of high mutation rates, heterozygotes and homozygotes can readily be differentiated, their mutational dynamics can be modeled, and historically they have been relatively inexpensive to assay and analyze with well-developed software tools. Microsatellite alleles can be **genotyped** in a number of ways, with gel or capillary electrophoretic methods capable of detecting length differences. In some circumstances, direct DNA sequencing can also be used to identify alternative microsatellite alleles. It is generally presumed that microsatellite loci are neutral with respect to natural selection because they are most commonly found in non-coding regions of the genome. As a result, microsatellites are often used to study **phylogeography** and characteristics associated with demographic change, like migration, population structure, or population expansions and bottlenecks. Microsatellites are especially useful for understanding these demographic dynamics on very recent timescales, because of their high polymorphism and high mutation rate. These same beneficial features, however, become a problem in studying molecular population genetic questions that depend on longer timescales because of the resulting homoplasy of alleles that produce identical sequences from independent mutations.

Heterozygosity at a microsatellite is quantified in the same way as for allozymes (Box 3.4). In addition, the stepwise mutation model predicts that the variance in allele lengths (V) will equal $2N_e\mu$ for a population at neutral equilibrium, where μ is the microsatellite mutation rate per generation. V is calculated from empirical data in the same way you calculate the variance among any list of values, so N_e can be estimated from V if we know what is the mutation rate at the locus.

3.5 Other measures of molecular diversity

The molecular population genetics literature is rife with diverse means that researchers have devised to quantify genetic variability. Some historical methods that exploited interesting molecular biology were used uncommonly, such as single-strand conformational polymorphisms of DNA, and are largely obviated by modern technologies. Other historical methods were applied extensively, if also superseded in the present day. One common means of quantifying molecular diversity in the past was via **fragment length polymorphisms**. In this method, restriction enzymes, or **endonucleases**, are used to cut DNA at highly specific nucleotide recognition sites. Consequently, if individuals differ at a base associated with that restriction enzyme recognition sequence motif, the enzyme will cut one variant but not the other. This results in one long sequence and two short sequences for the alternative alleles at that locus, which can be visualized by gel electrophoresis. These kinds of polymorphisms are called **restriction fragment length**

polymorphisms or **RFLPs** because they result in different-sized DNA fragments due to the presence or absence of restriction enzyme cut sites.

AFLPs, or **amplified fragment length polymorphisms**, can be detected in an analogous way to RFLPs, except that a bit more molecular biology is required. With this method, genomic DNA is cut simultaneously with two restriction enzymes (for example, *EcoRI* and *BamHI*) so that some fragments will have an *EcoRI* site at one end and a *BamHI* site at the other. Adapter DNA molecules of known sequence are then ligated to the fragments, allowing the fragments to be amplified by the **polymerase chain reaction (PCR)** using primer oligonucleotides. AFLP markers are scored as the presence or absence of a band on a gel, but are difficult to use to calculate heterozygosity because it is difficult to associate bands with specific loci. The modern incarnation of AFLP uses massively parallel DNA sequencing in a procedure called **RADtag** (restriction-site associated digest tag markers), for which endonucleases are used to allow the investigator to sequence a subset of the genome even in the absence of any reference genome assembly in a sequence database.

Other high-throughput platforms employ clever molecular biological protocols for using **genotyping chips** that have oligonucleotide probes bound to glass slides or microbeads to assess sequence differences between a sample and the probe that is present on the "chip." These genotyping methods can use microarrays or oligonucleotide tiling arrays, among other technological approaches, often referred to as "SNP chips" although some can also quantify copy number variation caused by deletions and duplications. A limitation of some probe-based genotyping methods is the phenomenon of **ascertainment bias**, such that alleles can only be detected if they are already known to be present as polymorphisms. When new alleles cannot be identified, it presents a systematic bias in the measurement of genetic variation.

For many purposes in molecular population genetics, it is critical to accurately determine the nucleotide changes between individuals and so indirect molecular methods are at best a compromise. Incredible advances in sequencing technology over the past decade have led to **massively parallel DNA sequencing** that is sufficiently cost-effective to acquire genome-scale DNA sequence information by even modest research laboratories. Some platforms produce "short read data" comprised of billions of short sequence fragments 100–250bp long from a genome (e.g. Illumina sequencing-by-synthesis). This approach to determining nucleotide sequences can be applied to genomic DNA as well as transcribed RNA in order to determine the coding gene complement of a genome, the **transcriptome**. When the transcribed portions of genomes are used in population genetic analysis, they are sometimes referred to as "**exome**" re-sequencing studies, where the "re-" prefix to sequencing indicates that many individuals had their DNA sequenced in order to identify polymorphisms. The term "exome" is a portmanteau of "exon" and "genome" and, more generally, the addition of the "-ome" suffix to a biological word implies a reference to the entire genomic complement of whatever biological prefix is used. Other technologies generate "long reads" that are many kilobases in length, albeit with lower per-nucleotide fidelity. These different technologies require careful attention to the details of how to determine reliably the nucleotide at any given position, with rapid advances in both the molecular biology and the data analysis. All of these high-throughput sequencing approaches push the burden of researchers from data acquisition to data analysis, requiring sophisticated data-processing algorithms. Consequently, the mainstreaming of genome-scale approaches puts a premium on training in bioinformatics and computing for all biologists.

3.6 Genetic variation among multiple populations

Up to now, we have considered genetic variation almost entirely in the context of species that are made up of a single well-mixed group of individuals. Real-world distributions of individuals, however, often do not neatly match the notion that they make up just one population in which any given individual is equally likely to mate with any other. Oftentimes, instead, populations are structured. **Population structure** occurs when genes from one or more **subpopulations**, also commonly called **demes** or **local populations**, do not get exchanged freely with other subpopulations (Figure 3.7). Individuals within each subpopulation can be thought of as randomly mating, but the total collection of individuals in the species as a whole cannot. How much this matters, in terms of how much the total population differs from what we would expect from species-wide random mating, depends on **migration**. The magnitude of genetic effects of structure depends on how much **gene flow** through migration connects the subpopulations, in addition to the influence of various forms of natural selection and demographic size change. The presence of structure is also often referred to as deviation from **panmixia**. Given that the genetic variation of species will be affected by population structure, how can we think about it quantitatively? This is the question that we will deal with for the remainder of this chapter.

3.6.1 *Quantifying population genetic structure with* F_{ST}

Population structure is a simple idea, but how can we quantify it? What we need is a metric that would allow us to describe how much of the genetic differences that occur within a species can be found within any single subpopulation and how much distinguishes different subpopulations from each other. We already have a simple way of summarizing genetic differences using observed and expected heterozygosities, so a metric based on this idea makes for a good starting point. Specifically, we can make use of heterozygosity as it gets incorporated into Wright's inbreeding coefficient F (Box 3.3) to quantify **genetic differentiation** between subpopulations of a species.

We previously defined Wright's F in terms of observed and expected heterozygosities for a single population (Box 3.3). It turns out that this idea can be generalized to relate to **subdivided populations** with **Wright's hierarchical F statistics** (Box 3.3), or **fixation indices**, usually denoted F_{IS}, F_{ST}, and F_{IT}. The subscripts represent: I = individuals, S = subpopulations, T = total population (Figure 3.7). So, F_{ST} is the fixation index for the subpopulations relative to the total population. As we will develop in this section, F_{ST} is especially useful in characterizing population subdivision because this metric has another convenient interpretation: the fraction of all genetic differences in the species that correspond to differences among populations.

Conceptually, the F statistics describe the reduction in heterozygosity relative to Hardy-Weinberg expectations that is due to differences between individuals within a subpopulation (F_{IS}), differences between subpopulations (F_{ST}), and differences between individuals pooled across the total species (F_{IT}). A slightly different formulation of these ideas can be used with gene diversity measures of heterozygosity so that DNA sequence information can be interpreted in this context (G_{ST} instead of F_{ST}); for our purposes the differences are minor, so we will simply treat them as conceptually equivalent.

First, let's start with F_{IS}. F_{IS} is equivalent to our previous definition of the inbreeding coefficient F (Box 3.3), extended to the case of multiple subpopulations. So, F_{IS} essentially

Figure 3.7 Subdivided populations with partial genetic connectivity.

Islands like those in Lake Ontario provide subdivided populations for their inhabitants, as do separated lakes for aquatic organisms in northern Ontario. When the individuals of a species are structured into two or more subpopulations, with less exchange of genetic material than would be expected for random mating across the entire species range (i.e. restricted gene flow), then genetic drift and selection can operate partially independently in each subpopulation. Wright's hierarchical F-statistics permit quantification of the fraction of the total genetic diversity in the species (T) that distinguishes subpopulations (S) from one another (F_{ST}). With information about diploid genotypes, then one can quantify the extent of non-random mating within subpopulations (F_{IS}; individuals I) or the overall departure of the species from random mating (F_{IT}). The number of migrants moving between the populations each generation is Nm, with higher rates of migration making the subpopulations genetically more similar (thicker arrows indicate more migration). (A) Photo courtesy of Ian Coristine/1000IslandsPhotoArt.com. (B) Satellite image obtained from http://www.canmaps.com/topo/nts50/orthoimage/052f13.htm and used under the terms of Natural Resources Canada: https://open.canada.ca/en/open-government-licence-canada

summarizes the amount of non-random mating that takes place within subpopulations. It is the average deviation in heterozygosity within subpopulations, defined as:

$$F_{IS} = \frac{H_S - H_I}{H_S},$$ (3.14)

where H_I is the observed heterozygosity averaged across subpopulations and H_S is the expected heterozygosity within each subpopulation (a "**local sample**") averaged across all of the subpopulations. Specifically, for a study system in which there are D demes such that the relative size of the jth deme is c_j and the observed heterozygosity of that j^{th} deme is H_{oj}, then:

$$H_I = \sum_{j=1}^{D} c_j H_{oj}.$$ (3.15)

Likewise, given the expected heterozygosity of each deme (H_{ej}), we calculate the overall expected heterozygosity of demes by averaging across them:

$$H_S = \sum_{j=1}^{D} c_j H_{ej}.$$ (3.16)

So, for a species with a single population (i.e. $D = 1$, so $c_1 = 1$), you can work out the algebra to see that F_{IS} reduces down to our previous definition of Wright's inbreeding coefficient F (Box 3.3). Usually it is assumed for simplicity that all demes are of equivalent size, so each $c_j = 1/D$. The F_{IS} statistic is useful for understanding the peculiarities of what is going on *within* subpopulations, in terms of how much deviation there is from the assumptions of Hardy-Weinberg equilibrium. If each subpopulation is effectively a randomly mating population, then we would expect $F_{IS} = 0$. Inbreeding within subpopulations, for example, would cause $F_{IS} > 0$. F_{IS} can potentially vary between -1 and $+1$, and its calculation requires that we know the diploid genotype information to identify the observed heterozygous versus homozygous variants. Consequently, while F_{IS} commonly was calculated in studies using allozymes or microsatellites, studies based on DNA sequence polymorphisms do not typically compute F_{IS}.

In contrast to F_{IS}, F_{ST} describes how population structure itself leads to deviations in heterozygosity. This is often the statistic that biologists are most interested in, because F_{ST} indicates how drastic the effects of subdivided populations are likely to be on patterns of genetic variation and how genetically distinctive the different populations are from one another. F_{ST} also is the simplest F-statistic to calculate, because F_{ST} only requires knowing allele frequencies; F_{IS} and F_{IT} both need information about genotype frequencies. This reliance solely on allele frequencies also is useful because F_{ST} can be used to describe differentiation among subpopulations of haploid organisms that lack diploid genotypes. We can compute F_{ST} as:

$$F_{ST} = \frac{H_T - H_S}{H_T},$$ (3.17)

where H_T is simply the expected heterozygosity for the entire metapopulation, by pooling all individuals together and ignoring the identity of demes (a "**pooled sample**"). Again, because both H_T and H_S are *expected* heterozygosities, only the allele frequencies are needed to obtain numerical values. F_{ST} can potentially vary between 0 and $+1$. Low migration rates among demes can lead F_{ST} to be high, as we will see in more detail in section 3.6.2.

An important alternative interpretation of F_{ST} is that it represents the fraction of the total genetic variation that is due to differences among subpopulations. We can see this connection explicitly by expressing F_{ST} in terms of how one calculates it from nucleotide polymorphism data using the θ_π measure of diversity (see section 3.3.1):

$$F_{ST} = \frac{\theta_{\pi\_pool} - \theta_{\pi\_local}}{\theta_{\pi\_pool}} = 1 - \frac{\theta_{\pi\_local}}{\theta_{\pi\_pool}}, \qquad (3.18)$$

where θ_{π_pool} indicates the average number of pairwise differences among *all* haplotypes sampled from all demes pooled together, whereas θ_{π_local} is computed as the diversity of local samples *within* demes that is then averaged across the demes. Because $\theta_{\pi\_local}/\theta_{\pi\_pool}$ represents the fraction of all variation that occurs within demes, one minus that value represents the between-deme portion.

In practice, slightly more complicated formulae must be used to calculate F_{ST} from real-world data to properly account for unequal sample sizes among demes, though the conceptual application is identical. We can also imagine a version of F_{ST} that follows the stepwise mutation model (see section 2.3.5), usually called R_{ST}, that can be applied to microsatellite loci. Montgomery Slatkin defined R_{ST} in terms of the variance in allele lengths as $R_{ST} = (V_T - V_S)/V_T$ (section 3.5), which you can see mirrors the form of F_{ST} as in Equation 3.17.

Finally, F_{IT} is defined as:

$$F_{IT} = \frac{H_T - H_I}{H_T}. \qquad (3.19)$$

This metric quantifies the F-statistic that summarizes what we will see if we ignore the presence of population subdivision altogether. It represents the overall deviation from Hardy-Weinberg expectations for the species from having pooled individuals from all the subpopulations. We will examine this situation in more detail with a numerical thought experiment demonstrating the Wahlund effect in section 7.3. As a prelude, the Wahlund effect can arise from mixing data from different populations together by accident to give a false impression of deviation from Hardy-Weinberg expectations. F_{IT} can potentially vary between -1 and $+1$. Note also that these three F-statistics do not sum to 1. They are related to each other as $(1 - F_{IT}) = (1 - F_{IS}) \cdot (1 - F_{ST})$, or rearranged, as:

$$F_{ST} = 1 - \frac{1 - F_{IT}}{1 - F_{IS}} = \frac{F_{IT} - F_{IS}}{1 - F_{IS}}. \qquad (3.20)$$

I said earlier that biologists are often most interested in quantifying F_{ST} among the F-statistics. So, what kind of values do we see for F_{ST} in nature, and how high is "high"? Most organisms that have been studied show average F_{ST} values less than 0.1 and fewer than 10% of species have values higher than 0.4. For example, Galapagos lava lizards have $F_{ST} = 0.44$ across islands and roe deer of France have an average $F_{ST} = 0.01$ among populations (Meirmans and Hedrick 2011).

When thinking about genetic differentiation between populations or between closely related species, it also is important to distinguish relative and absolute measures of genetic change. Genetic "differentiation" generally reflects unequal allele frequencies between populations, corresponding to *relative* measures, whereas genetic "divergence" is the **genetic distance** between populations in *absolute* terms. Relative measures are sensitive to the amount of polymorphism within populations. In particular, F_{ST} represents a relative measure of differentiation, whereas metrics like K_S aim to capture absolute divergence in terms of cumulative changes between populations or species (see section 5.3.1 and

Box 8.6). A limitation of F_{ST} is its sensitivity to small values of polymorphism. When H or θ_π is very small there will be few polymorphic sites to use in calculating metrics of genetic variation, making the values subject to stochastic variation that can affect the magnitude of F_{ST}. Consequently, F_{ST} will be biased toward high values for species with very low genetic variation or in regions of the genome with unusually low polymorphism, as can be caused by linked selection (a.k.a. genetic hitchhiking, background selection; see section 7.1).

To quantify "absolute" genetic distances between populations, we can use the d_{xy} measure that was introduced by Masatoshi Nei and Wen-Hsiung Li. The d_{xy} metric captures the average sequence distance between haplotypes found in one subpopulation compared to the haplotypes sampled in another subpopulation. The calculation of d_{xy} is equivalent to θ_π for a pooled sample, except that none of the within-deme sequence comparisons are included in the calculation. As a result, d_{xy} does not depend on polymorphism within demes in the way that F_{ST} does. You can calculate d_{xy} separately for synonymous sites, which, if we had only a single sequence from each subpopulation, would make the value of d_{xy} equivalent to K_S (see section 5.3.1).

Researchers have devised a whole slew of other metrics besides F_{ST} to quantify population differentiation (see Table 3.1), although F_{ST} (or equivalently, G_{ST}) provides the most widely applied measure. Other measures of genetic structure include things like the nearest-neighbor metric S_{nn} or Jost's D that captures allelic differentiation. The d_a metric, which is derived from d_{xy}, also is a relative measure of differentiation. It is the value of d_{xy} after subtracting off the amount of polymorphism expected to have been present in the common ancestor of the two populations being compared, which is intended to capture the split time of the populations (see section 5.3.1 for the K_S analog and more explanation of the logic). Mathematically, we would write this as $d_a = d_{xy} - \theta_{\pi(anc)}$, where $\theta_{\pi(anc)}$ is that ancestral polymorphism which is usually presumed to be the average of the observed θ_π values for each subpopulation. The number of fixed differences, d_f, also is a relative measure of differentiation because it depends on the time since populations split and coalescence times. In any case, a potential limitation of any of these metrics of differentiation, including the F-statistics, is that they require the biologist to specify the members of the subpopulations. In some cases, it can be difficult to assign certain individuals discretely to one population and other individuals to another.

Table 3.1 Common metrics of population differentiation and divergence.

Metric	Description	Requires >1 sample per population?	Relative vs. absolute
F_{ST}, G_{ST}, R_{ST}	Fraction of overall genetic variation due to differences between populations	Yes	Relative
K_S	Total silent-site divergence between a representative sequence copy from each of two populations or species	No	Absolute
d_{xy}	Average pairwise difference between haplotypes from one population to another	No	Absolute
d_a	Net sequence difference between populations after accounting for ancestral polymorphism	Yes	Relative
d_f	Fixed differences between populations	Yes	Relative

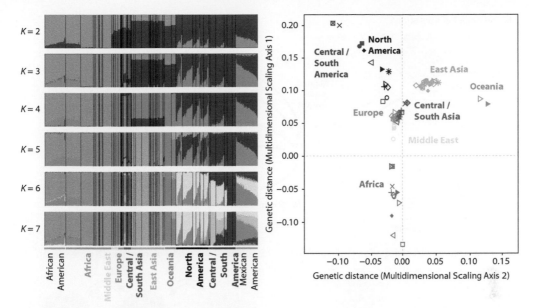

Figure 3.8 Visualizations of population structure.

Genomic haplotypes of individuals can be clustered using linkage disequilibrium information into similar genetic groups, given an assumption that the species has K genetic groups (subpopulations). This approach, used by the STRUCTURE software, can be useful to visualize genetic differentiation when subpopulations are not discretely defined (Pritchard et al. 2000), as for this global sample of humans. Statistical methods like Principal Components Analysis (PCA) or Multidimensional Scaling (MDS) that reduce the dimensionality of genetic variation can ordinate genetic differences, often recapitulating geographic trends (plots for human variation redrawn from Verdu et al. 2014).

The difficulty of assigning population membership can be especially acute in species that are distributed continuously over their range, but where mating does not happen totally randomly across space. Some approaches in landscape genetics aim to capture the non-discreteness of a species range in a **continuum model** of migration and subdivision. An alternative approach to characterizing differentiation is to use linkage disequilibrium (see Chapter 6) among many molecular markers to estimate the number of subpopulations represented in the data as well as the genetic composition of each individual that derives from each of the subpopulations (Figure 3.8; see section 8.2.3). Another visualization approach is to use the statistical procedure known as principal component analysis (PCA) to extract all the genotypic correlations into a much simpler subset of data dimensions to look for genetic clustering. In essence, these more sophisticated approaches try to let the data *tell us* how many genetically distinct groups there are, but for our purposes we will not delve into the statistical details of how they try to do it.

3.6.2 *Migration models and effects on population subdivision*

As the simplest case of population structure, we can imagine that there are exactly two isolated subpopulations, a two-island model of migration (Figure 3.7). A related scenario would be a mainland-island model, also termed a continent-island model, in which there is migration only in one direction: from the larger mainland source to the smaller island destination. Both of these situations are special cases of considering a finite number of subpopulations that can exchange migrants.

Island model of migration

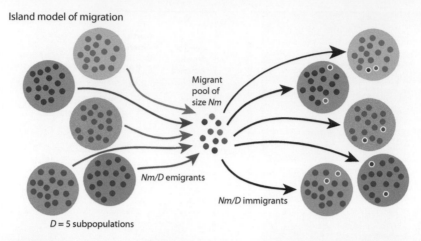

Stepping-stone model of migration (1-dimensional)   Stepping-stone model of migration (2-dimensional)

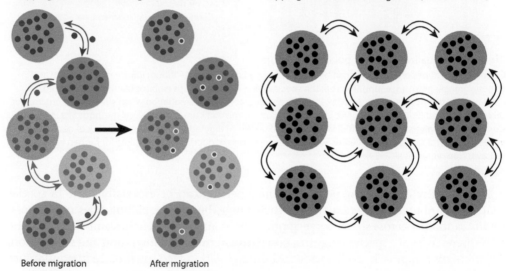

Before migration   After migration

Figure 3.9 Island and stepping-stone models of migration.
In an island model of migration (top diagram), $N_e m/D$ migrant haplotypes enter a "migrant pool" from each of D subpopulations (each subpopulation having size N_e). The haplotypes then randomly re-assort among the subpopulations as immigrants, potentially immigrating back into their original source population. In contrast, migrant haplotypes only move between nearest-neighbor demes in a stepping-stone model of migration (bottom diagrams). In the one-dimensional example, migrant individuals in their new population are indicated with a white outline, colored based on island of origin. Only the migration paths are shown for the two-dimensional stepping-stone diagram (black).

Given that a species is composed of more than two subpopulations, it becomes important to think about *how* those populations connect to one another through migration. One useful model of how migration works is the **stepping-stone model**, in which demes share migrants only with neighboring demes (Figure 3.9). This type of connectivity will cause adjacent demes to be genetically more similar (lower F_{ST}). The simple linear stepping stone model can be modified to allow a two-dimensional matrix of subpopulations with only the nearest neighbors connected by migration (Figure 3.9). In general, stepping stone

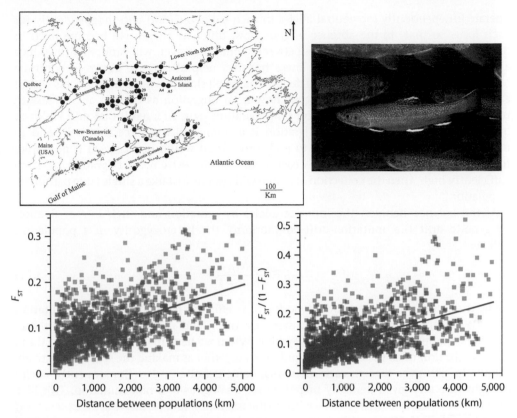

Figure 3.10 Isolation by distance among subpopulations.
A stepping-stone migration process leads to more distant subpopulations being genetically more dissimilar. These brook trout (*Salvelinus fontinalis*) sampled from distinct coastal locations of eastern Canada show increasing genetic differentiation (higher F_{ST}) with increasing physical geographic distance between locations (Castric and Bernatchez 2003). Mathematical theory predicts a linear relationship between $F_{ST} / (1 - F_{ST})$ and distance, provided that migration rate is a linear function of distance between locations (data redrawn from Rousset 1997). Map republished with permission of the Genetics Society of America from Castric and Bernatchez (2003). The rise and fall of isolation by distance in the anadromous brook charr (*Salvelinus fontinalis* Mitchill). *Genetics* 163, 983–96. Permission conveyed through Copyright Clearance Center, Inc. Brook Trout © M Rose/Shutterstock.com.

migration results in **isolation by distance** (Figure 3.10), with demes that are physically farther apart in geographic space being progressively more differentiated (higher F_{ST}).

There are many other models that one might conceive of, with varying degrees of abstraction and complexity, all aiming to describe how gene flow connects subpopulations. However, we will focus on the **island model** (Figure 3.9). In the island model of migration, any of an infinite number of "island" subpopulations can share migrants with any other island, and there is no "mainland" source population. The island model of migration has a long history in population genetics, owing to its mathematic simplicity, and it consequently forms a cornerstone for understanding key features of the influence of structure and migration on genetic diversity. We will now go into the island model in more detail.

Genetic drift will tend to cause a random allele at a polymorphic locus to become fixed in any given subpopulation, eventually (Box 3.5). This process of genetic drift will operate independently for neutral alleles in each subpopulation and independently for each locus so that, in the absence of gene flow, different alleles will become fixed in different subpopulations. Gene flow between demes, however, will randomly introduce alleles from other subpopulations. These transfers help alternative alleles to continue being polymorphic within any given deme, which will slow down the drift process within each deme. You can think of this gene flow from migration as causing an increase in the effective population size of a local subpopulation (Box 3.1); recall that drift is slower in larger populations (Box 3.5). Migration is a homogenizing force that will tend to make subpopulations more similar to each other than would be expected if drift were to operate totally independently on each deme. In the extreme, if migration rates are sufficiently high, then the collection of demes will operate just like a single large panmictic population.

At the balance between the input of alleles by mutation and their loss or fixation by genetic drift (i.e. mutation-drift equilibrium), the heterozygosity in a population will be:

$$H_{exp} \cong \frac{4N_e\mu}{4N_e\mu + 1} \tag{3.21}$$

under the infinite alleles mutational model (N_e = effective population size, μ = mutation rate). What about migration? Remember that the island model envisions an infinite number of subpopulations, which is key here. When we have many demes, we can think of the collection of all the other subpopulations together as making the species effectively infinite in size, even though each individual deme is finite in size. Consequently, alleles brought into any given deme by migration are analogous to a new mutation. This idea means that we can simply replace μ with m (the migration rate) to arrive at an expectation for **migration-drift equilibrium**. For a subpopulation at migration-drift equilibrium under the island model, the expected heterozygosity will therefore be:

$$H_S \cong \frac{4N_e m}{4N_e m + 1}. \tag{3.22}$$

This mathematical sleight-of-hand assumes that the migration rate is substantially greater than the mutation rate, otherwise we would substitute μ with $m + \mu$ when the values of m and μ are of similar magnitude. For this calculation, we presume all subpopulations to be of the same size (N_e). The compound parameter $N_e m$ is key: $N_e m$ represents the *number* of individual migrants that are genetically effective in moving to another subpopulation each generation. Equivalently, $1/(N_e m)$ represents the average number of generations between each migration event of one individual. What you can see from this relationship between heterozygosity and migration is that it takes very little migration to produce very high levels of heterozygosity (Figure 3.11).

When we assume the species is comprised of many demes, we can make another convenient approximation by continuing to use the infinite alleles mutation model. Specifically, we can approximate the value of $H_T \cong 1$, because essentially any pair of alleles will be different when we sample a small subset of them from the effectively infinite collection of individuals constituting the species as a whole. For local populations, however, H_S will be *less* than 1, it will be the value calculated from Equation 3.22. Putting these values of H_S and H_T together with our knowledge that $F_{ST} = (H_T - H_S) / H_T$, we can infer that:

$$F_{ST} \cong 1 - \frac{4N_e m}{4N_e m + 1} = \frac{1}{4N_e m + 1}. \tag{3.23}$$

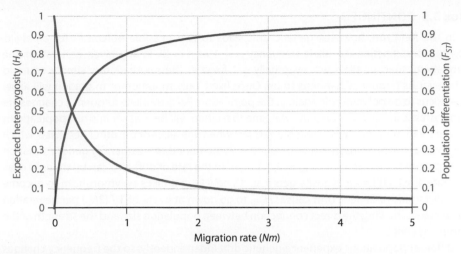

Figure 3.11 The effect of migration on heterozygosity and F_{ST}.
Just a small number of breeding migrants each generation will exert strong effects on the equilibrium heterozygosity and differentiation of subpopulations in a species with population structure. For example, with only one migrant per generation on average, expected heterozygosity would be ~80% of that expected for a panmictic population of equivalent total size, and only ~20% of the species' total variation would be partitioned as differences between populations (i.e. F_{ST} ~0.2).

This lets us connect genetic differences as measured from data to the biological process of migration. By having specified a particular mutation model and a particular migration model, it thus allows us to relate the amount of genetic differentiation in a species (F_{ST}) directly to population size and migration rate. After some algebraic rearrangement, we can solve for $N_e m$:

$$N_e m \cong \frac{1}{4} \left(\frac{1}{F_{ST}} - 1 \right). \tag{3.24}$$

This is a very neat result! It means that if we can estimate F_{ST}, then we can infer the migration rate between subpopulations. Recall that quantifying F_{ST} from molecular data is relatively easy.

Box 3.5 GENETIC DRIFT

The zygotes of one generation get formed from a subsampling of gametes from the parental population. But that subsample inevitably is an imperfect representation of the alleles found in the parental population. This **sampling error** results in chance fluctuations in allele frequencies across generations, known as genetic drift. The random chance that a given allele goes up in frequency is just the same as the chance that it goes down: 50-50. Repeating this discrete sampling generation after generation describes a **Markov chain**. The amount that the allele frequency changes each generation depends on the size of the population (N_e): smaller populations have more drastic changes. Quantitatively, we can think of the drawing of two alternative alleles in gametes as Bernoulli binomial sampling, so the expected change in allele frequency p corresponds to the variance of the **binomial distribution**: $2p (1 - p) / (2N_e)$. In other words, smaller populations will have a larger binomial sampling variance, making genetic drift a stronger force in smaller populations.

(Continued)

Box 3.5 CONTINUED

In the long run, however, genetic drift leads to two possible outcomes for an allele: fixation or extinction (i.e. allele frequency $p = 1$ or $p = 0$). In any given generation, the probability that an allele will eventually get fixed in the population is simply equal to whatever its frequency happens to be. Once fixed, only mutation or migration will allow that locus to be polymorphic again in the population. Because allele frequency changes are more drastic in small populations, the time to fixation will be quicker in small populations. Genetic variation at a locus inevitably gets lost when one allele or the other gets fixed, which means that drift eliminates genetic variation. Remember that we quantify genetic variation as expected heterozygosity, $H_{exp} = 2p(1-p)$, so the rate of drift also can be thought of as the rate at which heterozygosity gets lost. It's how long it takes polymorphism to become monomorphism. Genetic drift causes H_{exp} to go down at a rate of $1 / (2N_e)$ per generation on average, showing the direct connection between population size and the strength of the sampling effect.

Different populations experience genetic drift independently, so the frequency changes for neutral alleles will be uncorrelated. As one allele or another fixes in each subpopulation, drift makes the subpopulations diverge from one another: ancestral genetic variation that was present within populations gets "converted" into between-population variation. This genetic differentiation will arise faster for small populations.

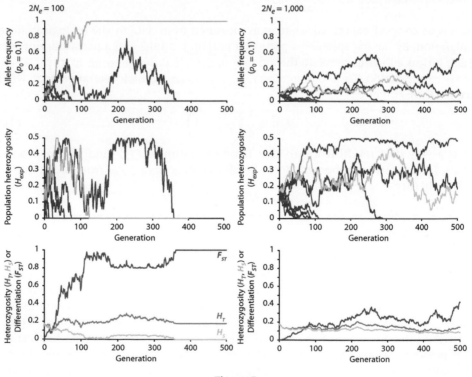

Figure 5

Figure 5 Markov chain simulations of genetic drift show the stochastic changes in allele frequency. For 10 subpopulations without gene flow, each starting with the rarer allele at $p_0 = 10\%$ frequency, we can see the more drastic fluctuations in smaller populations ($2N_e = 100$ vs. 1,000; an example subpopulation is highlighted in a distinct color for each in the top two rows of graphs). When one allele gets fixed, heterozygosity in that population is lost and over time average subpopulation heterozygosity (H_S) declines. The expected heterozygosity for the pooled sample of the total metapopulation (H_T), however, remains close to its initial value. As allele frequencies diverge among subpopulations, the genetic differentiation among them increases (increasing F_{ST}).

An important caveat to inferring N_em from F_{ST} in this way is that, for the inferred values of N_em to have any merit, the assumption of many demes connected by the island model of symmetric migration must be met well by the real populations. Also, it is not possible to know from F_{ST} values alone whether the populations continuously exchange migrants, whether migration occurs in bursts over time, or whether the populations have separated with no further gene flow. This last case is, effectively, a scenario of divergence, illustrating the blurred line between differentiation with gene flow and divergence without gene flow. There are additional technical issues in accurately estimating N_em, for example, in having correctly identified population membership and the timescale of population isolation and migration. The key points that I want to illustrate here are the conceptual link between the amount of population genetic differentiation among demes (F_{ST}) and the degree of genetic exchange (N_em) and that it is possible to infer migration rates from molecular population genetic data. Using the kind of logic that we have just worked through, one can take even more sophisticated approaches, using coalescent theory or other means, to more accurately understand migration and its effects on evolutionary change at the molecular level.

Further reading

Altenburg, E. and Muller, H. J. (1920). The genetic basis of truncate wing: an inconstant and modifiable character in *Drosophila*. *Genetics* 5, 1–59.

Avise, J. C. (2004). *Molecular Markers, Natural History, and Evolution*, 2nd ed. Sinauer: Sunderland, MA.

Castric, V. and Bernatchez, L. (2003). The rise and fall of isolation by distance in the anadromous brook charr (*Salvelinus fontinalis* Mitchill). *Genetics* 163, 983–96.

Charlesworth, B. and Charlesworth, D. (2017). Population genetics from 1966 to 2016. *Heredity* 118, 2–9.

Cruickshank, T. E. and Hahn, M. W. (2014). Reanalysis suggests that genomic islands of speciation are due to reduced diversity, not reduced gene flow. *Molecular Ecology* 23, 3133–57.

Cutter, A. D., Jovelin, R. and Dey, A. (2013). Molecular hyperdiversity and evolution in very large populations. *Molecular Ecology* 22, 2074–95.

Dutoit, L., Burri, R., Nater, A. et al. (2017). Genomic distribution and estimation of nucleotide diversity in natural populations: perspectives from the collared flycatcher (*Ficedula albicollis*) genome. *Molecular Ecology Resources* 17, 586–97.

Jordan, M. A. and Snell, H. L. (2008). Historical fragmentation of islands and genetic drift in populations of Galapagos lava lizards (*Microlophus albemarlensis* complex). *Molecular Ecology* 17, 1224–37.

Kreitman, M. (1983). Nucleotide polymorphism at the alcohol-dehydrogenase locus of *Drosophila melanogaster*. *Nature* 304, 412–17.

Leffler, E. M., Bullaughey, K., Matute, D. R. et al. (2012). Revisiting an old riddle: what determines genetic diversity levels within species? *PLoS Biology* 10, e1001388.

Lewontin, R. C. (1974). *The Genetic Basis of Evolutionary Change*. Columbia University Press: New York, pp. 95–6.

Lewontin, R. C. and Hubby, J. L. (1966). A molecular approach to the study of genic heterozygosity in natural populations. II. Amount of variation and degree of heterozygosity in natural populations of *Drosophila pseudoobscura*. *Genetics* 54, 595–609.

Meirmans, P. G. and Hedrick, P. W. (2011). Assessing population structure: F_{ST} and related measures. *Molecular Ecology Resources* 11, 5–18.

Palstra, F. P. and Fraser, D. J. (2012). Effective/census population size ratio estimation: a compendium and appraisal. *Ecology and Evolution* 2, 2357–65.

Pritchard, J. K., Stephens, M. and Donnelly, P. (2000). Inference of population structure using multilocus genotype data. *Genetics* 155, 945–59.

Rousset, F. (1997). Genetic differentiation and estimation of gene flow from *F*-statistics under isolation by distance. *Genetics* 145, 1219–28.

Selander, R. K., Yang, S. Y., Lewontin, R. C. and Johnson, W. E. (1970). Genetic variation in the horseshoe crab (*Limulus polyphemus*), a phylogenetic "relic." *Evolution* 24, 402–14.

Simonsen, K. L., Churchill, G. A. and Aquadro, C. F. (1995). Properties of statistical tests of neutrality for DNA polymorphism data. *Genetics* 141, 413–29.

Verdu, P., Pemberton, T. J., Laurent, R. et al. (2014). Patterns of admixture and population structure in native populations of northwest North America. *PLoS Genetics* 10, e1004530.

Wallace, A. (1889). *Darwinism: An Exposition of the Theory of Natural Selection with Some of its Applications*. Macmillan and Co: London, pp. 71–4.

CHAPTER 4

Neutral theories of molecular evolution

The last two chapters gave us the nuts and bolts of molecular population genetics: how mutations arise and how to quantify them in DNA. But how should we think about what happens to those DNA variants over time: the conversion of new mutations into population polymorphisms and then into fixed differences between species? How do genomes evolve, one generation to the next to the next to the next, to produce the divergent DNA sequences of different species? Is there a baseline framework that could help us see what to expect, something that uses simple ideas, as simple evolutionary processes as possible? We want such a foundational conceptual framework to guide our expectations for how evolution proceeds. We'd also like to use it as a reference point to help tell us when the simplest evolutionary processes just aren't enough to explain data, and which kinds of more complicated evolutionary forces we need to invoke in a given situation. The mathematics of classic population genetics of the early 1900s provides one important pillar of evolutionary understanding, but here we want to integrate the fact of DNA as heritable material as closely as possible into our view of evolution. For this goal, we must look to the later part of the 1900s, to DNA-inspired evolutionary theory.

4.1 The standard neutral model

Prevailing thought in evolutionary biology, until the 1960s, viewed natural selection as the key force that controlled how much genetic variation was present within populations (the **balance theory**). Then it was discovered, by looking directly at the molecular basis of genetic variation, that there was too much genetic variation in most populations to be explained easily by such an extreme **adaptationist** view of the problem. DNA sequences differed in just too many positions throughout the genome for it to be plausible that selection could selectively maintain all of them simultaneously (see Chapter 3). As James Crow put it, "it became increasingly clear that nucleotide and amino acid substitutions are marching to a different drummer than that of traditional morphological evolution." So, how else might we explain such a cornucopia of genomic diversity?

Along came Motoo Kimura who, in combination with Jack King and Thomas Jukes, proposed a radical new notion in 1968 and 1969, most of which we now take for granted. They proposed what came to be known as **the Neutral Theory of Molecular Evolution** or, simply, "**Neutral Theory**," as a way to account for the high observed levels of polymorphism. And they based their ideas on some of the simplest of evolutionary principles: mutation and genetic drift. In other words, the influence of selection on

A Primer of Molecular Population Genetics. Asher D. Cutter, Oxford University Press (2019).
© Asher D. Cutter 2019. DOI: 10.1093/oso/9780198838944.001.0001

evolution at the *molecular* level might not have the same pre-eminence as selection at the *phenotypic* level with the converse also being true: genetic drift may often be the evolutionary boss in DNA sequence change, despite what is surely a limited role for genetic drift in phenotypic evolution. The Neutral Theory provides another hypothesis alongside the balance theory and mutation-selection balance to explain the **paradox of variation**: why is there so much genetic variation within populations? The Neutral Theory proposes that there can be so much variation because the molecular differences are selectively neutral relative to one another.

As with all theoretical models, the Neutral Theory has at its core a simplification of reality that we can use as a convenient reference point. The **standard neutral model (SNM)** provides a powerful and essential baseline for understanding molecular evolutionary change, and so that is where we will start. Whenever I refer to the "Neutral Theory" without qualification, it generally means that we are talking about this reference scenario of the SNM. But as we progress in thinking about evolution at the molecular level, we shall see that the SNM is just a starting point and that we can build more sophisticated and realistic neutral models. Without getting too far ahead of ourselves, it is important to keep in mind that the SNM provides an example **null model**. We use null models to test how well they can explain real-world patterns, starting with the most simple of assumptions. We can then make refined null models that build in more complexity to further test specific hypotheses about how the real world might not match a simpler null model.

So, what is the SNM? What does it assume and what does it predict? The SNM takes as its starting point that there is a population comprised of a finite number of randomly mating hermaphrodites with non-overlapping generations, and also that it is at demographic equilibrium; that is, the species is neither growing nor shrinking in size nor made up of subpopulations. This kind of idealized population should be a relatively familiar idea to you at this point. Such an ideal population also is often referred to as a Wright-Fisher population, because this is the same set of assumptions devised by those classic geneticists to define a reference population in which the census population size and the genetically effective population size are equivalent (see Box 3.1).

An assumption more specific to the SNM holds that new mutations have either very strong fitness effects or none at all, relative to their alternate alleles. The idea is that strongly deleterious mutations will quickly get eliminated from the population and so will be very rare or absent in a typical sampling of individuals. As a result, such deleterious mutations would make at most a negligible contribution to measurements of polymorphism. Such purged deleterious mutations also would therefore never be able to contribute to divergence between species. Strongly selected beneficial mutations, on the other hand, would quickly get fixed, although the SNM presumes that they arise only very rarely and so for practical purposes can be ignored entirely. In sum, the SNM presumes that mutations conferring fitness effects would not contribute to genetic variability or to differences between species because of the directional force of selection. This view directly contrasts with the balance theory view that proposes, instead, that most of the variation is actively maintained by selection.

What are the consequences of mutations with fitness effects that do *not* contribute to genetic variability? The logical outcome of the SNM is that the only kind of mutations left to make up molecular variation in a population are those alleles that are selectively neutral! As Jack King and Thomas Jukes (1969) described the situation, "Those alleles which do become fixed through drift are not a random selection of all substitutional mutations, but alleles which have been 'selected' for innocuousness." Therefore, the SNM proposes that most alleles that segregate as polymorphisms within a population are selectively neutral. As a consequence, genetic drift will determine the evolutionary fate of most alleles that actually make up the DNA sequence variation that we see in populations (see Box 3.5; see also Box 3.1).

There are just two possible fates for any given allele that is subjected only to genetic drift: it will ultimately be lost from the population or be fixed, to make up all copies of the locus in the population. Of course, only those polymorphisms that become fixed will actually contribute to DNA **sequence divergence**, the fixed nucleotide differences between isolated populations and species. So, what does the Neutral Theory have to say about this divergence? The SNM presumes that strongly deleterious mutations will be purged and never rise to fixation, and that beneficial mutations are so exceedingly rare that we can ignore them. Consequently, the Neutral Theory predicts that most sequence differences that accumulate between species will derive, again, from the remaining kind of mutational input: neutral mutations. Therefore, the SNM proposes not only that most polymorphisms will be selectively neutral, but also that most molecular differences between species will be selectively neutral.

It is important to keep in mind that the Neutral Theory does not claim that natural selection is unimportant in evolution, nor does it claim that most *new* mutations are neutral. In fact, selection is a key player in the Neutral Theory: a central proposition to the Neutral Theory is that most new mutations are subjected to purifying selection because they are deleterious (see section 2.2), with the rest being selectively neutral. It does make the assumption that a sufficiently small fraction of new mutations are beneficial that we can ignore them, to a first approximation, in understanding polymorphism and divergence at the molecular level. The essence of this view of the distribution of mutational effects is roughly correct, albeit in a simplified form, as we saw empirically in section 2.2. If this assumption bothers you, though, just wait until section 4.2 where we will relax it.

While it may seem counterintuitive that selection plays a crucial role in something called "Neutral Theory," that's how it works. As Kimura (1983) later wrote in the introduction of his famous book, "One possibility would be to rename the theory the 'mutation-random drift theory', but the term 'neutral theory' is already widely used and I think it better not to change horses in midstream." Regardless of its name, the logic is that, despite being so numerous, most deleterious mutations are rapidly eliminated from populations by selection, and therefore are unable to contribute either to polymorphism (differences within a population) or to divergence (differences between populations or species). As a result, the main predictions of the Neutral Theory are that (1) all or most of the molecular genetic variation that is observed in a population is selectively neutral and (2) all or most of the fixed differences in DNA between species are selectively neutral. It also creates a framework to explain the abundance of molecular variation in natural populations. One of the most important attributes of Neutral Theory, however, is that it provides a way of creating specific null models that we can use as a reference point for testing alternatives.

4.1.1 *Neutral divergence and the molecular clock*

Now that we have an idea of the underpinnings of the Neutral Theory and the SNM, how can we use this view of molecular evolution to learn something new? The first big idea that we will explore has to do with divergence, to answer the question: do substitutions accumulate between species in a predictable way? Having posed this leading question, it should come as no surprise that the answer is "yes." So, let's work through the logic for why DNA changes will accrue predictably between species, with this predictable divergence commonly referred to as a **molecular clock** (Figure 4.1; see section 5.3).

The speed at which new mutations get fixed, separately, in each of two populations is what defines the rate of evolutionary change between them. When a new mutation gets fixed or completely established in a population, we say that there has been a **substitution** of the ancestral allele with the new derived allele. These substitutions create fixed differences in **orthologous** sequences to make up what we call **divergence** between species. The rate at which this divergence accumulates is the rate of evolution. How can we predict this **substitution rate**?

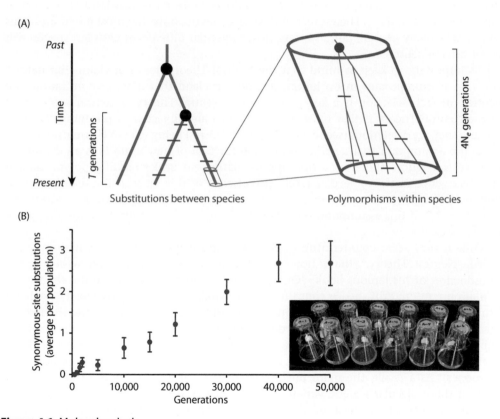

Figure 4.1 Molecular clock.

The idea of using molecular changes to date the passage of time actually belongs to Emile Zuckerkandl and Linus Pauling, who proposed in 1962 that changes to protein sequences seem to accumulate linearly with evolutionary distance between organisms. But this original notion of a molecular clock was phenomenological, without a sound theoretical basis being widely known, until Motoo Kimura developed Neutral Theory that provided a logical justification for why molecules should accumulate changes in a predictable way (A). Amazingly, however, Sewall Wright had already theoretically derived the idea that the rate of divergence equals the rate of mutational input under neutrality way back in 1938. But, of course, DNA was not known to be the genetic material at that time and there were no data that could be applied to it, and so the importance of the result was neglected until Kimura's research. (B) The linear accumulation of neutral substitutions from new mutation during the experimental evolution of isolated and diverging populations of E. coli over the course of 50,000 bacterial generations demonstrates the molecular clock idea in practice (data redrawn from Tenaillon et al. 2016, photo of bacterial cultures used in experimental evolution courtesy B. Baer).

New neutral mutations are special in that their dynamics of change over time depend only on genetic drift. This special property means that we can easily think about the probability that such a new neutral mutation will have as its ultimate fate *fixation* by drift, rather than *loss* by drift. The eventual **probability of fixation** for any neutral allele is simply its frequency, so for a brand new allele that probability is $1/(2N)$ (see Box 1.3). In the SNM, the population size is stable so that $N = N_e$. We also know how many new neutral mutations arise each generation in the population: it is just the mutation rate (μ) times the number of mutational "targets" in the diploid population ($2N$), which therefore equals $2N\mu$. Don't forget that the mutation rate here corresponds to neutral mutations. For simplicity, you can think of sites in the DNA at which any change would be selectively neutral, but as we shall see in section 4.2 it also applies to sites at which just a fraction of changes are selectively neutral.

Now for the clever part that Motoo Kimura figured out: the rate of evolution, what we call the fixation rate or substitution rate (ΔK), is just equal to the number of new mutations arising in a population each generation ($2N\mu$) times the probability of fixation for a new mutation ($1/(2N)$; note our units of time in generations). Again, remember that in this SNM, the effective and census population sizes are the same ($N_e = N$). So, mathematically, this looks like $2N\mu * 1/(2N)$, which, after algebraic simplification just equals μ:

$$\Delta K = 2N\mu \cdot \frac{1}{2N} = \mu. \tag{4.1}$$

In words, this equation says that substitutions will accumulate at a rate equal to the mutation rate. Despite seeming so simple, this result is so very important that I will say it again: the rate of *divergence* at selectively unconstrained sites is equal to the rate of neutral *mutation*. It does not depend on population size or use some complicated formula. This exceptionally elegant conclusion sets the foundation for much of how we think about molecular evolution and also provides the rationale for a molecular clock: by measuring sequence divergence we can measure the amount of time that has elapsed between two lineages (Figure 4.1; see section 5.3).

An important feature of the molecular clock that we learn from the Neutral Theory is that the accumulation of neutral substitutions does not depend on the population size. You can see this independence of population size mathematically when the "$2N$" terms cancel out in Equation 4.1. This independence of population size might seem counterintuitive, especially for a branch of biology with "population" in its name! The idea is that while large populations have more copies of each locus and so provide more targets for mutational input, genetic drift also works more slowly in large populations so it takes longer for an allele to become fixed or lost. By contrast, genetic drift in small populations rapidly converts new mutations into fixed differences, but there are few copies of each locus and so the total input of new mutations is low. These countervailing forces exactly balance each other out so that neutral mutations accumulate as substitutions independently of the population size. Remember that this result holds only for the accumulation of *neutral* mutations, so it assumes that we are dealing with changes to sites that are selectively unconstrained, no matter how big or small the population size might be (see section 4.4).

Another point is that the substitution rate represents the rate of accumulation of fixed differences between two reproductively isolated populations or species. Using these substitutions as a measure of the time that has elapsed since the separation of the species presumes that the speciation time was sufficiently long ago that we may ignore the fact that the population of the common ancestral species itself would be genetically variable.

In species that separated recently, however, this simplification will not hold. We will discuss this caveat and others in more detail when we consider genealogies (see section 5.3).

By measuring time with molecular change, we usually must compare orthologous DNA sequences from different species that are alive in the present day. We usually don't have a time series of DNA to watch directly the dynamics of evolutionary change. Only under special circumstances can we measure the changes over time in a single population, like with **experimental evolution** studies or in historical time series of clinical samples that document rapid pathogen evolution (Figure 4.1). So, when we compare nucleotide differences between two species, the fixed differences will have accumulated independently in the ancestry of each of those two species since their common ancestor existed as a single population (Figure 4.1). We can write *this* idea down mathematically, as well. The number of fixed differences between two species (K) that separated from one another T generations ago will be: $K = 2\mu T$. The "2" comes from the two species and the independent accumulation of neutral mutations as substitutions at rate μ in each of them. Now we have a way to take the *idea* of the molecular clock to *data* to let us measure time with changes to DNA (see section 5.3).

For example, the neutral divergence between orthologous genes of the Chinese swan goose and domestic chickens averages $K = 0.3$. This value means that three out of every ten nucleotide sites in a selectively unconstrained locus, on average, will have mutated and become fixed in the population history of one species or the other since their common ancestor, which we can now observe as sequence differences between the orthologous copies. If the point mutation rate in these birds is three mutations out of every billion nucleotides each generation ($\mu = 3 \times 10^{-9}$), then we would estimate the timing of the divergence between them to be $T = 50$ million generations ago. Presuming that geese and chickens typically pass through one generation per year, then this would indicate a shared common ancestor 50 million years ago (official estimates by Lu et al. (2015) place the date at 64.6 million years).

What else do we need to keep in mind for using molecular divergence as a measure of time? First of all, the units of "time" with a molecular clock are mutational changes, when measured from the comparison of homologous sequences of different species. To convert molecular changes into other units, like generations or years, we need to know the neutral mutation rate in generations or years and then make the appropriate conversion. Because the molecular clock predicts substitutions to accumulate at rate μ, this means that the average time between successive fixations of neutral alleles is $1/\mu$, which also can be converted to units of generations or years with relevant information about the neutral mutation rate (see Box 2.5). In any case, these calculations presume that the mutation rate per generation has remained constant and that the number of generations per year also has remained constant.

Kimura showed that the average time it takes for a new neutral mutation to eventually become fixed by genetic drift is $4N_e$ generations on average. But drift will make most new mutations go extinct, and the average time to loss is approximately $2 \cdot \ln(2N_e)$ generations. This time to loss is much quicker. For example, for a population with N_e of 1,000, the time to loss would be only about 15 generations compared to 4,000 generations for the expected time to fixation. Even more extreme, new mutations in a population with N_e of 10 billion would take fewer than 50 generations to get lost by drift.

Remember that the Neutral Theory, in the narrow sense, presumes either that mutations exert no effect on fitness, that is, they are selectively equivalent to pre-existing alleles, or that they are strongly deleterious. We can extend this logic to divergence in the coding sequence of genes, which will let us quantify the fraction of replacement mutations that

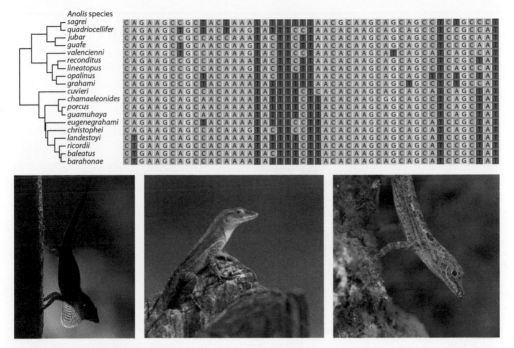

Figure 4.2 Sequence alignment.

The first step to calculating divergence is to make a sequence alignment. These data show the phylogeny and corresponding multiple-sequence alignment for orthologous portions of the mitochondrial NADH dehydrogenase 2 gene (*ND2*) from 19 species of anoles (data redrawn from Mahler et al. 2016). *Anolis sagrei, A. lineatopus,* and *A. christophei* (left to right) are found on islands in the Caribbean, where the genus has undergone an adaptive radiation (photos courtesy L. Mahler).

are neutral versus selectively constrained. The first step is to get sequences for orthologous loci from two or more species, and to align them in a **sequence alignment** (Figure 4.2). We can then calculate how many replacement sites there are (see Box 5.3) and how many of those sites differ among species.

Under the logic of the SNM, those sites that differ in an alignment of orthologous genes from a pair of species will represent selectively neutral mutations that became fixed. Reciprocally, the sites in the alignment that are identical represent sites at which either no mutation occurred or where selection eliminated a deleterious mutation. We can distinguish the last two possibilities by looking at synonymous sites, which we can usually presume to be totally unconstrained by selection. The fraction of synonymous sites that have substitutions tells us at what fraction of sites we should expect for differences between species to arise from mutation and drift alone (see section 7.1.3). In general, of course, purifying selection acts on replacement changes to preserve the amino acid sequences that make up proteins, which results in far fewer substitutions per replacement site than per synonymous site. For mammals, roughly 20% of replacement-site mutations can be inferred to be effectively neutral by this method (or conversely, selective constraint suggests mutations to 80% of replacement sites are deleterious; Box 4.1). Estimates of selective constraint in non-coding sequence suggest that roughly half of non-coding sites experience purifying selection in *D. melanogaster* (Andolfatto 2005). As we will see in section 4.2, however, these proportions will differ across species depending on their population size because of nearly neutral mutations at replacement sites.

Box 4.1 SELECTIVE CONSTRAINT DIFFERENCES ACROSS SITE TYPES

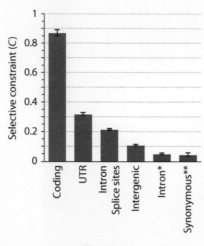

Figure 1

Figure 1 Changes to some features of the genome will be more likely to affect fitness than others. By analyzing DNA sequence change separately for different classes of sites, we can quantify how sensitive they tend to be to mutational perturbation and, correspondingly, how strongly selection constrains their molecular evolution. Selective constraint (C) is a measure of DNA sequence conservation, the converse of divergence (see section 7.1.3). In murid rodent genome comparisons, as a nice example, 4-fold degenerate synonymous sites** and intron sequences* (excluding splice sites) within coding genes show very low constraint, implying that nearly all changes to such sites are effectively neutral in these species (data redrawn from Gaffney and Keightley 2006). Changes to sites that would alter the amino acid encoded within a gene, however, are more constrained, with selection retaining the identity of 80% or more of coding replacement sites.

4.1.2 *Neutral predictions for polymorphism*

The amount of neutral genetic variation in a population represents the balance between the input of new mutations and the loss of this polymorphism by genetic drift (see Boxes 1.2 and 3.5). We can predict from the Neutral Theory how much variation we should see at mutation-drift equilibrium for DNA sequences. We can use the infinite alleles mutation model to get at how much polymorphism there will be quantitatively at this balance between the input of variation by mutation and its loss by drift. Heterozygosity will be our metric of polymorphism (H), which is one minus homozygosity (see section 3.1). And we will turn the problem on its head by using the idea that the probability of two alleles being homozygous ($1 - H$) also is the probability that they are identical by descent (F, see Box 3.3), which means that $F = 1 - H$ and $H = 1 - F$. So, homozygosity in generation t is F_t. F_t is a function of genetic drift causing homozygosity in previous generations (the term in brackets in Equation 4.2), weighted by the likelihood that mutations will not create heterozygosity by inputting new unique alleles (μ; so the lack of mutation input into a haplotype is $1 - \mu$ in Equation 4.2). We can write this idea down mathematically as:

$$F_t = \left[\left(\frac{1}{2N_e}\right) + \left(1 - \frac{1}{2N_e}\right)F_{t-1}\right] \cdot (1 - \mu)^2. \tag{4.2}$$

A way to think about what this equation means is that there are two ways for a pair of alleles to be identical by descent: they can derive directly from a single common ancestor (the first term, $1/(2N_e)$), or, they can derive from two ancestors (with probability $1 - 1/(2N_e)$) that were themselves identical by descent (F_{t-1}; together yielding the second term). For the two alleles to be homozygous, then neither can have mutated (the third term, $(1 - \mu)^2$). The equilibrium value derives from Equation 4.2 by solving for F when $F = F_t = F_{t-1}$, which is:

$$\hat{F} = \frac{1}{1 + 4N_e\mu}. \tag{4.3}$$

But we are actually interested in H, which is $1 - F$. After some algebraic rearrangement, this turns out to be:

$$\hat{H} = \frac{4N_e\mu}{1 + 4N_e\mu}. \tag{4.4}$$

We can often expect that N_e will be very large and μ will be very small, which lets H in Equation 4.4 be simplified further to be well-approximated as $H = 4N_e\mu$.

This quantity $4N_e\mu$ has a special place in population genetics. It is so central to how mathematical models of evolution work and how we measure molecular variability in nature that it has its own Greek letter: θ (pronounced "thay-tah" or "thee-tah," see section 3.3.1). This θ is the thing that θ_π and θ_W aim to estimate from DNA sequence data (see section 3.3.1). From having made this theoretical prediction for H and θ, we can see that in situations with higher mutation rates or with larger population sizes, those populations will have more genetic variation at the locus. Motoo Kimura also used diffusion theory, borrowed from physics, to make this prediction about θ. This prediction from Neutral Theory tells us how much genetic diversity we should expect to see for loci with mutational dynamics that match the infinite alleles model. But what about thinking about mutation and genetic drift at DNA nucleotide sites, which will be better served by the infinite sites model of mutation?

The expected heterozygosity is $4N_e\mu$, where we have the diploid effective population size N_e and neutral mutation rate μ. Recall that for DNA sequences, "expected heterozygosity" can be thought of as the proportion of sites that are polymorphic between a randomly chosen pair of sequences from the population (estimated as θ_π from real data; see section 3.3.1). Because this prediction about heterozygosity depends on the SNM, it is a prediction about the **equilibrium** amount of polymorphism in a stable population for loci that are unaffected by selection.

Why does the equilibrium neutral heterozygosity equal $4N_e\mu$? The mathematics of diffusion theory are beyond our scope for this book, but in section 5.2 we shall walk through a way to derive this same result from coalescent theory. In the meantime, here is a heuristic way to think about why the SNM gives us this expectation for how much sequence polymorphism a locus should have. We can think about heterozygosity by comparing a pair of gene copies. At equilibrium, the input of mutations and their loss or fixation by drift exactly balance out to leave us with those neutral mutations that genetic drift has not yet made extinct or fixed. We know that *each* sequence copy will accumulate mutations at rate μ, so the pair of haplotypes accumulates 2μ mutations per generation since they shared a common ancestor. Given that the probability that a random pair of haplotypes from the population will be identical by descent is $1/(2N_e)$, then the average time since the pair of haplotypes shared a common ancestral sequence will be the reciprocal value: $2N_e$ generations. So, the average amount of difference between a pair of haplotypes will be 2μ times $2N_e$: $4N_e\mu$. These neutral variants that currently are sitting in evolutionary limbo, between loss and fixation, represent the **standing variation** in the population. This heuristic description is not a formal proof, but I hope that it helps to illustrate the point that the amount of standing neutral variation in a population represents the balance between new mutational input and the inevitable loss or fixation of those new variants by genetic drift.

We have framed these predictions of the Neutral Theory in terms of looking at a single population. We can also make predictions about what we should see when looking across species. Specifically, our expectation that polymorphism will depend on population size means that species with larger populations should have more neutral genetic variation. When we go out in the world and measure them, we find that larger populations do indeed tend to have higher polymorphism (Figure 4.3; see also Figure 3.2). But, as we shall see in Chapter 7, a slew of other factors including population size changes make the pattern among species a bit more messy than a simple linear relationship between population size and the amount of molecular variation that is contained within a species. Richard Lewontin pointed out in 1974 that the range of census population sizes that different species span appears much larger than the range of effective population sizes implied by genetic diversity (Box 3.1). This paradox of variation runs counter to the simple predictions of Neutral Theory. Understanding the causes of why species have as much or as little genetic diversity as they do provides one of the grand motivations for molecular population genetics to this day.

In addition to making predictions about nucleotide polymorphism, the Neutral Theory also predicts a specific neutral distribution for the site frequency spectrum (SFS; see section 3.3.2). When we look at DNA in a population, many sites will be polymorphic and we can calculate the frequency of each variant at each polymorphic site. How many mutations are there that exist as singletons, being the only copy of that particular allele variant at that site in our sampling of the locus among individuals? That is, if we collect n copies of DNA sequence for a selectively unconstrained locus from a population, we can ask how many polymorphic sites should have their rarer, minor variant present as a

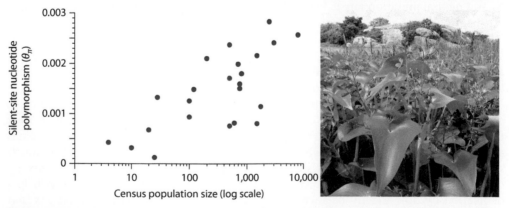

Figure 4.3 Neutral polymorphism depends on population size.
Larger populations of *Eichhornia paniculata* contain more nucleotide diversity (data redrawn from Ness et al. 2010). (photo courtesy R. Ness and S. C. H. Barrett)

singleton (frequency $f_1 = 1/n$), or as a doubleton (two copies, $f_2 = 2/n$), etcetera up to the maximum possible value of 50% ($f_{n/2} = n/2$; Figure 4.4). Qualitatively, the **neutral SFS** is characterized by many low-frequency variant sites and an increasingly small number of sites that have variants at higher frequency. This inevitably arises because new mutations are rare and rare neutral alleles will rarely rise to high frequency in the population. Specifically, if there are S polymorphic sites among our n copies of the locus, then we expect each variant frequency class i to have a relative abundance of:

$$x_i = \frac{1}{i \cdot \sum_{j=1}^{n/2} \frac{1}{j}}. \qquad (4.5)$$

As a result, we would predict there to be $\xi_i = S^* x_i$ sites for which the rarer variant is present in i of our n copies of the locus. The Greek letter ξ is pronounced "zy" to rhyme with "sky." The expected value of ξ_i also relates to the population mutation rate θ, such that $\xi_i = \theta/i$, and $\theta = 4N_e\mu$.

This depiction of the neutral SFS may sound rather abstract, so let's make it concrete with an example. If we found $S = 20$ polymorphic sites after having sampled $n = 10$ copies of our locus, then the Neutral Theory would predict that singleton variants

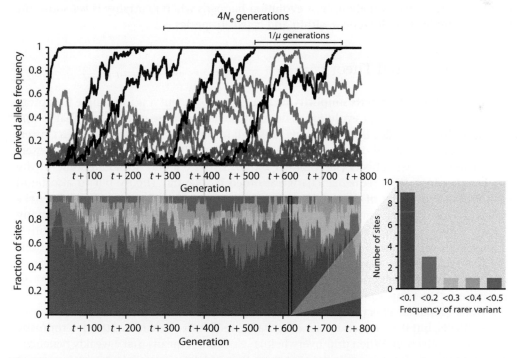

Figure 4.4 Minor allele frequency distribution from mutational input.

As new neutral mutations arise and change in frequency by genetic drift, the Neutral Theory predicts that they will fix at a rate of $1/\mu$ after having spent $4N_e$ generations as polymorphisms, on average. Most variants will be rare and eventually lost, leading to a predictable shape of the "folded" site frequency spectrum of polymorphic variants (see section 3.3.2). The top graph shows simulation results across 800 generations of mutation and genetic drift for a region of DNA with free recombination; fixed substitutions are in black and polymorphic variants in purple. The bottom graph colors the fraction of sites by the minor variant frequency for each timestep, with an example folded SFS as a histogram on the right for the timestep at $t + 621$ generations.

($i = 1$, so $f_1 = 1/10 = 0.1$) would be represented by $x_1 = 43.8\%$ of the polymorphic sites, so that on average we ought to observe 8.76 sites for which there is 1 copy of the rarer singleton allele and 9 copies of the other, more common major allele (i.e. $S*x_1 = 20*0.438 = 8.76$ sites). Similarly, we would predict that $x_5 = 8.75\%$ of the sites in the sequence would be polymorphisms in the highest frequency class (i.e. $S*x_5 = 20*0.0875 = 1.75$ sites), where the two variant alleles at a given site are equally abundant among the 10 copies of the locus, each variant having 5 copies apiece, and so each variant would be at an allele frequency of $f_{n/2 = 5} = 0.5$. These quantitative predictions presume that the infinite sites mutation model applies, so that there are exactly two alternative variants at any given polymorphic site. Because this is a prediction about the neutral SFS, we must reiterate that the prediction applies only to polymorphic sites that are selectively indistinguishable, so that changes in the relative abundance of the different variant nucleotides are affected only by genetic drift. Similarly, the SNM assumes that we are dealing with a single population that is at demographic equilibrium.

These simple and quantitative predictions from the Neutral Theory might seem humble, but they turn out to be exceptionally powerful. The SNM gives us a concrete reference point that we can use to check against real data to ask whether the most basic of evolutionary processes, mutation and genetic drift, might be sufficient to explain patterns in real data. In Chapters 7 and 8, we will go into detail about how to compare real data to these predictions to learn about how evolution happens when neutrality is *not* sufficient to explain molecular differences within and between species.

4.2 Nearly Neutral Theory

The SNM is convenient in its simplicity for many purposes, but it is too simple to capture some aspects of molecular evolution. The most obvious feature of the SNM that is ripe for improving is the idea that mutations must fall into one of two discrete fitness classes, either selectively neutral or strongly deleterious. The reality is that mutations can exert a spectrum of fitness effects (see Chapter 2). So, is there a way to retain some of the key insights into DNA sequence change afforded by Neutral Theory but to also build in additional subtleties of nature? Tomoko Ohta did just that. She introduced "**Nearly Neutral Theory**" that considers a continuous range of mutational effects, allowing many slightly deleterious mutations. Within the conceptual framework of the Nearly Neutral Theory, we can still use the SNM as a special case: the SNM applies to those mutations that have their evolutionary dynamics dominated by the influence of genetic drift. Such mutations are called effectively neutral. As we shall now explore, the fraction of mutations that are effectively neutral depends on the effective population size (see Box 3.1).

Mutations that are effectively neutral have slightly deleterious or slightly advantageous fitness effects, but these effects are so weak that genetic drift overwhelms the deterministic influence of selection. When drift overwhelms selection, it means that a weakly beneficial allele might actually *decrease* in frequency or that a deleterious allele might *increase* in frequency (Figure 4.5). Genetic drift causes the stochastic changes in allele frequency that can lead to these non-adaptive outcomes. Such mutations have nearly neutral fitness effects, so that we refer to their changes in allele frequency as "effectively neutral," being controlled almost entirely by genetic drift.

Researchers sometimes use the phrase "**neutral site**" or "**neutral locus**" as a shorthand to refer to nucleotide positions that are selectively unconstrained. This shorthand implies that such sites have effectively neutral polymorphisms or sites at which any

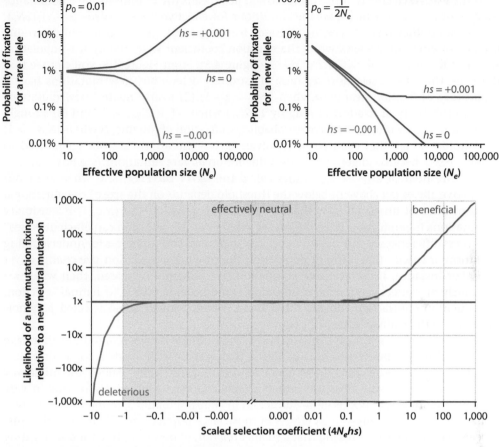

Figure 4.5 Mutation fixation probabilities depend on population size.
When an allele is rare (frequency $p_0 = 1\%$) or new (frequency $p_0 = 1/(2N_e)$), then the likelihood that it will ultimately become fixed or lost depends mostly on the sign of selection (s) and the strength of selection relative to population size. Motoo Kimura showed that the probability of fixation (P_{fix}) of an allele is: $P_{fix} = \frac{1-e^{-4N_ehsp_0}}{1-e^{-4N_ehs}}$, where h indicates the dominance coefficient. The bottom graph shows the ratio of $P_{fix}/(1/(2N_e))$, which summarizes how likely an allele is to become fixed relative to how likely fixation will be from genetic drift alone (i.e. drift alone will lead to a probability of fixation equal to $1/(2N_e)$). When the relative effects of a mutation ($|4N_ehs|$) are less than 1, then it is effectively neutral.

mutation would be effectively neutral. This sloppy terminology is technically nonsense, but nevertheless a commonplace jargon: it is only *relative to the alternative allele* that mutations and variants can really be thought of as neutral or effectively neutral. As a telling case in point, neutral polymorphisms can occur at sites that experience selective constraint: an A/C nucleotide polymorphism might be selectively neutral but an A/G polymorphism selectively distinguishable at the same nucleotide position.

We can actually be more quantitative about what makes alleles effectively neutral in their evolutionary dynamics. Specifically, we can write down a mathematical approximation of when drift overwhelms selection, which occurs when the selection coefficient $|s|$ is less than $1/(4N_eh)$. The absolute value of s indicates that this result is true regardless of whether the mutation is beneficial ($s > 0$) or deleterious ($s < 0$). For simplicity, we can presume **additive** effects of the alleles which just means the mutations have dominance coefficient $h = \frac{1}{2}$, in which case the conditions for effective neutrality are $|s| < 1/(2N_e)$. Presented another way, after some algebraic rearrangement, genetic drift will determine the fate of alternative alleles when the selection coefficient weighted by the population size is sufficiently small, when $2N_e|s| < 1$ (Figure 4.5). From Figure 4.5, you can see that there are three basic selection regimes: (1) strongly deleterious new mutations have a probability of fixation near zero when $2N_e s << -1$; (2) nearly neutral new mutations have a probability of fixation of roughly $1/(2N_e)$ when $-1 < 2N_e s < 1$; and (3) strongly advantageous new mutations have a probability of fixation of roughly $2s$ when $2N_e s >> 1$. You should notice that even strongly advantageous mutations are not guaranteed to become fixed, because of genetic drift's influence on *all* rare mutations (see Box 3.5).

This threshold of $2N_e s$ is sometimes called the **drift barrier**, and whether or not alternative alleles fall above or below the threshold depends on the size of the mutational effect in units of fitness (s) as well as the effective population size (N_e). This sensitivity of mutations to population size, in terms of whether or not selection can efficiently "see" the fitness differences of alternative alleles, has important consequences for understanding evolution of species that differ in population size or when population size changes over time. For instance, if the effective population size suddenly increased, then what was previously an "effectively" neutral mutation could become visible to natural selection. A population contraction, on the other hand, would convert some selected polymorphisms into the realm of effective neutrality.

A consequence of selection being unable to discriminate between mutations that have mutational effects below the drift barrier is that species with smaller population sizes ought to have a greater fraction of effectively neutral mutations. Stated the other way around, mutations with even tiny fitness effects will *not* be effectively neutral in very large populations. Selection is more efficacious in large populations, so selection can "see" mutations with even tiny fitness effects in those large populations. So, if we look across species, we should expect to find that a smaller fraction of new mutations and segregating

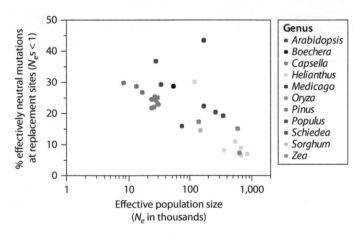

Figure 4.6 Near-neutrality of mutations.

Genetic drift will more drastically change allele frequencies in smaller populations, even for alleles that affect fitness. Because the efficacy of selection in distinguishing fitness differences between alleles depends on the combined effects of s and N_e, species with smaller N_e will have a larger fraction of changes being effectively neutral. These data from plant species of differing population size illustrate how effectively neutral changes are more common in species with smaller populations (data redrawn from Hough et al. 2013).

polymorphisms will be effectively neutral in species with large populations (Figure 4.6). This potential for species to differ in the accumulation of changes by drift has important implications for how to apply the molecular clock. Specifically, it is crucial to use sites at which changes are the "most neutral" by being the most selectively unconstrained, otherwise purifying selection will cause them to evolve more slowly than we would otherwise anticipate from genetic drift alone.

As an example, in most mammals, changes to synonymous sites can safely be thought of as effectively neutral because big-bodied mammals typically have rather small populations (see Figure 3.2). Sometimes only the **four-fold degenerate** synonymous sites are used, as a subset of synonymous sites (see Box 2.2). But in species with very large population sizes, like many insects, nematodes, and bacteria, selection on highly expressed genes can actually lead to tiny but evolutionarily distinguishable fitness differences between alternate synonymous codons. Such selection for **codon usage bias** must then be taken into account when performing analyses that would otherwise presume selective neutrality of synonymous sites (see Box 8.4), including for molecular clock calculations.

Molecular population geneticists generally have in mind the Nearly Neutral Theory when they talk about molecular evolution, even if loosely using the phrase "neutral theory." This casual talk simply reflects the fact that the Nearly Neutral Theory is a much better depiction of evolution at the molecular level, yet we can still often treat effectively neutral mutations ($2N_e|s| < 1$) as if they were truly neutral in the sense of the SNM ($s = 0$). In other words, when a population geneticist refers to "neutral polymorphism" or "neutral divergence," they really are using shorthand for polymorphisms or fixed differences that are "effectively neutral." However, it is important to keep clear in your mind the consequences of the stricter assumptions of the SNM regarding fitness effects of mutations as well as the implications of population size for nearly neutral mutations.

4.3 More complex models of selective neutrality

With the nice predictions of the SNM, and the nearly neutral amendment to it, it would be very convenient if we could stop there. We could use these neutral null models to test against real data and, if they don't match, then conclude that non-neutral evolution must be happening. That is to say, we'd like to be able to detect perturbations in the genome caused by natural selection.

But it turns out that other neutral evolutionary forces besides genetic drift will tweak molecular evolution away from the null model that the SNM provides. The key culprit is non-equilibrium (a.k.a. non-standard) demography. Non-standard demography includes population expansions and contractions and bottlenecks, non-random mating, subdivided populations, as well as gene flow and admixture of populations. These non-standard demographic effects matter for two key reasons. First, we can use the population genetic patterns to learn about demography. This idea is the view of **phylogeography** where deciphering organism life history constitutes the goal of molecular population genetics in itself. Second, ignoring the influence of demography can lead us to draw incorrect conclusions about natural selection in genomes. We shall go into detail in Chapter 7 about the effects that demography has on population genetic patterns. What I would like to emphasize here is that we can get even more sophisticated with our selectively neutral models by incorporating information about demography into them. Usually this requires using computer simulations of the evolutionary process (see Box 8.2). The key point is that the SNM is just a starting reference scenario. By integrating the SNM with additional selectively neutral evolutionary features, like population size change, we can construct the most appropriate null model that we can think of.

The flexibility of computer simulations that mimic neutral evolution under specific demographic scenarios of a population's history is very useful and conceptually appealing (see Box 8.2). But it is not all roses. It may be that our imperfect knowledge will lead us to mis-specify that history, or we may be constrained by data limitations, or there may be just too much stochastic evolutionary noise to detect the evolutionary signal that we may be most interested in. These causes and more can lead an analysis to **fail to reject neutrality**, meaning that our null model appears to be sufficient to explain the patterns in the data. Whatever the reason, it is important to accept the fact that a failure to reject a null model does not mean that the null model is the correct evolutionary scenario. Even if the null model is sufficient to explain the data, other models might also be sufficient. A failure to reject a given null model simply means that we are unable to find better support for an alternative scenario with the data and analysis that we have in hand. We are usually inclined to follow Occam's razor, known as **parsimony** in scientific circles: the idea that the simplest explanation is most likely to be correct. Depending on how much faith you have in the plausibility of your null model, the simpler neutral model may or may not provide the best explanation for patterns in data.

4.4 Misconceptions about Neutral Theory

When it was first proposed, the Neutral Theory was highly controversial. That initial controversy has since subsided as Neutral Theory melded with Nearly Neutral Theory and was validated with truckloads of DNA sequence information. Some renewed controversy is brewing in scientific circles as many studies now suggest that recurrent adaptive molecular evolution appears more prevalent than was initially supposed by Kimura and Ohta (see section 8.3.3). I have avoided dwelling on the controversies, instead aiming to present the SNM as one of many useful neutral models that we commonly use in molecular population genetics. Nevertheless, it is worth staving off some common points of confusion that can lead to misconceptions about applying neutral models to DNA sequence analysis.

First, the Neutral and Nearly Neutral Theories of molecular evolution do *not* say that natural selection is unimportant. In fact, selection is intrinsic to their logic. Remember that the Neutral Theory presumes that there are many strongly deleterious mutations that get weeded out by purifying selection, and that this is a critical piece of how we conceive of the SNM. You could actually interpret this as an exceptionally adaptationist view of nature: to presume that all fitness-affecting mutations are deleterious is like saying that the organism is exquisitely well-adapted to its environment, so that any new mutations that alter gene function will cause it to move away from its adaptive optimum. Similarly, the Nearly Neutral Theory envisions a wide distribution of fitness effects of mutations, with the portion that selection can effectively sift between being sensitive to population size. Despite having the word "neutral" in their names, these evolutionary theories inherently incorporate natural selection into how evolution works at the molecular level.

A related issue is that just because a polymorphism is selectively neutral does not mean that it will be unaffected by selection. Selection at linked sites can influence neutral polymorphisms that are not themselves the direct targets of selection, as we will talk about in more detail in Chapter 7. This distinction between direct and indirect selection can be especially important for drawing conclusions about demographic history based on loci that occur in genomic regions with low recombination. Moreover, changes to population size can shift the boundary for whether alleles are effectively neutral or highly visible to natural selection.

Second, the Neutral and Nearly Neutral Theories of molecular evolution do *not* say that most new mutations are selectively neutral. In fact, deleterious mutations could very well make up the majority of mutations according to the assumptions of the theories. And with Nearly Neutral Theory, it is conceivable that a large enough population size could make it so that virtually no place in the genome is free from the influence of selection, by shifting $N_e s$ to the point that essentially no mutations fall into the zone of effective neutrality. These theories make claims about polymorphisms and species differences. Namely, they argue that most of the nucleotide variants that we see segregating within populations are selectively neutral with respect to one another, and also that genetic drift was responsible for most fixed differences between species. This follows from the logic that selection is efficient in eliminating deleterious mutations and in fixing beneficial mutations, so they never really contribute much to extant variation. The theories do, however, presume that beneficial mutations are very rare, being vastly outnumbered by both deleterious and effectively neutral mutations, so that we can mostly ignore new beneficial mutants in thinking about molecular evolution, at least to a first approximation. We will return to this idea about the incidence of beneficial mutations and substitutions in Chapter 8 when we go into detail about tests of neutrality.

Third, the Neutral Theory does not say that selectively unconstrained sites are the only kind of site that can receive neutral mutations. The proportion of mutations that are neutral may be lower at sites subject to selective constraint, but that proportion need not be zero.

Lastly, the Neutral and Nearly Neutral Theories of molecular evolution do *not* say that most phenotypes evolve via genetic drift. These theories are about DNA sequence change, and make no claims on the connection between genotype and phenotype. They accept that it is possible that those rare beneficial mutations are what drive all phenotypic change between species, but generally are agnostic about phenotypic evolution because it depends on the specific mapping of genotype on phenotype in the face of selection.

Further reading

Andolfatto, P. (2005). Adaptive evolution of non-coding DNA in *Drosophila*. *Nature* 437, 1149–52.

Crow, J. F. (1987). Anecdotal, historical and critical commentaries on genetics: twenty-five years ago in *Genetics*: Motoo Kimura and molecular evolution. *Genetics* 116, 183–4.

Gaffney, D. J., Blekhman, R. and Majewski, J. (2008). Selective constraints in experimentally defined primate regulatory regions. *PLoS Genetics* 4, e1000157.

Gaffney, D. J. and Keightley, P. D. (2006). Genomic selective constraints in murid noncoding DNA. *PLoS Genetics* 2, 1912–23.

Hough, J., Williamson, R. J. and Wright, S. I. (2013). Patterns of selection in plant genomes. *Annual Review of Ecology, Evolution, and Systematics* 44, 31–49.

Kimura, M. (1968). Evolutionary rate at molecular level. *Nature* 217, 624–6.

Kimura, M. (1983). *The Neutral Theory of Molecular Evolution*. Cambridge University Press: New York.

King, J. L. and Jukes, T. H. (1969). Non-Darwinian evolution. *Science* 164, 788–98.

Kreitman, M. (1996). The neutral theory is dead. Long live the neutral theory. *Bioessays* 18, 678–83.

Lanfear, R., Welch, J. J. and Bromham, L. (2010). Watching the clock: studying variation in rates of molecular evolution between species. *Trends in Ecology and Evolution* 25, 495–503.

Lewontin, R. C. (1974). *The Genetic Basis to Evolutionary Change*. Columbia University Press: New York.

Lu, L., Chen, Y., Wang, Z. et al. (2015). The goose genome sequence leads to insights into the evolution of waterfowl and susceptibility to fatty liver. *Genome Biology* 16, 89.

Mahler, D. L., Lambert, S. M., Geneva, A. J. et al. (2016). Discovery of a giant chameleon-like lizard (*Anolis*) on Hispaniola and its significance to understanding replicated adaptive radiations. *American Naturalist* 188, 357–64.

Ness, R. W., Wright, S. I. and Barrett, S. C. H. (2010). Mating-system variation, demographic history and patterns of nucleotide diversity in the tristylous plant *Eichhornia paniculata*. *Genetics* 184, 381–U105.

Tenaillon, O., Barrick, J. E., Ribeck, N. et al. (2016). Tempo and mode of genome evolution in a 50,000-generation experiment. *Nature* 536, 165–70.

Zuckerkandl, E. and Pauling, L. (1962). Molecular disease, evolution, and genetic heterogeneity. In: Marsha, M. and Pullman, B. (eds) *Horizons in Biochemistry*. Academic Press: New York, pp. 189–225.

CHAPTER 5

Genealogy in evolution

Is a picture worth a thousand words? Then let's save our breath a little bit and think about how to paint a picture of evolution. We want to connect the dots between ancestors and descendants to see how history changes over time, how gene sequences split apart and reconnect. Trees will be the landscape. But we will use gene trees and species trees. That is, we can use "tree thinking" to understand molecular differences within and between species. We have already seen how it is simple to compute and interpret "summary statistics" of molecular diversity from a set of DNA sequences, like the θ_π and θ_W metrics that we introduced in Chapter 3. This simplicity makes them powerful tools for understanding evolutionary change. These metrics, however, inherently neglect the historical relationships of loci in terms of shared ancestry of similar haplotypes and orthologous genes. A complementary way of thinking about molecular diversity and divergence includes an explicit conception of these relationships through their genealogical history. We will now explore this genealogical tree-based approach for thinking about molecular population genetics.

5.1 Gene trees forward-in-time within species

Branching tree diagrams provide an intuitive way of visualizing the evolutionary relationships between alleles, haplotypes, individuals, and species. They show who is related to whom, how they are related, and how long ago. In particular, **gene trees** are branching diagrams that summarize the "descent with modification" of a genetic feature within a species or among multiple species (Figure 5.1). For now, let's start by considering gene trees that correspond to haplotypes within a population. The **tips** of a tree represent the present-day haplotypes that we have sampled from the population and that we can observe, also sometimes referred to as "leaves" or "external nodes" (Figure 5.2). By contrast, the **internal nodes** in the tree represent the haplotypes of ancestors that have subsequently accumulated mutations to create the distinct modern-day haplotypes. An internal node is the "common ancestor" shared among all the lineages that derived from it. **Branches**, the **edges** that connect nodes, represent time or genetic distance as measured by the number of mutations or substitutions that have occurred in a lineage.

Different branches drawn on a gene tree usually imply that evolutionary change distinguishes those branches since they split. In some representations of genealogies, however, splitting branches simply represent the passing of copies of an allele (or haplotype) to descendants (i.e. in a pedigree diagram without any mutation at the locus), but we will focus on the interpretation of trees showing evolutionary divergence. The splitting of branches at a node is caused by mutation creating a new and distinct haplotype. We usually envision nodes splitting in two, **bifurcating**, to create descendant lineages.

A Primer of Molecular Population Genetics. Asher D. Cutter, Oxford University Press (2019).
© Asher D. Cutter 2019. DOI: 10.1093/oso/9780198838944.001.0001

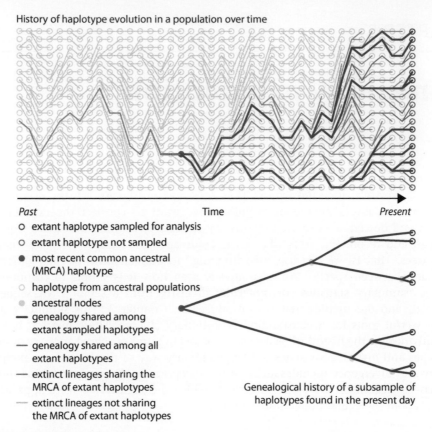

History of haplotype evolution in a population over time

Past   Time   *Present*

○ extant haplotype sampled for analysis

○ extant haplotype not sampled

● most recent common ancestral
   (MRCA) haplotype

○ haplotype from ancestral populations

● ancestral nodes

— genealogy shared among
   extant sampled haplotypes

— genealogy shared among all
   extant haplotypes

— extinct lineages sharing the
   MRCA of extant haplotypes

— extinct lineages not sharing
   the MRCA of extant haplotypes

Genealogical history of a subsample of
haplotypes found in the present day

Figure 5.1 Tracing genealogies.
Genealogies trace the ancestry of haplotypes that are found in the present day through their common
ancestors, all the way back to the sample's most recent common ancestor (MRCA). Because we only have the
haplotypes that survived to the present day and that we happen to have sampled, and we do not have actual
representatives of the ancestral haplotypes, we must infer the ancestral states. In this example, the MRCA of the
sample is also the MRCA of the population, but this will not be the case for all loci.

Sometimes multiple descendant lineages might form simultaneously in a **multifurca-
tion**, a possibility that some more complex models of evolution can account for; we will
mostly think about the simpler scenario where only bifurcations take place.

Haplotypes of common ancestors generally are unobserved by a researcher. Consequently, the exact combination of mutations that define ancestral haplotypes cannot be
known with total certainty, but can be inferred in many cases; that is, we can estimate the
ancestral state. The **most recent common ancestor (MRCA)** of a set of haplotypes
is represented by the single node that branches into the set of all present-day haplotypes
being considered (Figure 5.1).

Evolution proceeds forward in time. If you were to sample haplotypes in a time series,
looking at the DNA of 20 copies of a locus each generation for 40 generations, it might
reveal the genealogical relationships like in Figure 5.1. In this example, all the haplotypes
represented by green (and pink) nodes existed at some point in time, but we can't observe
them directly from a collection of haplotypes today (at the tip of the arrow). Yet the
genealogical relationships of the haplotypes sampled today can be traced back to a single
common ancestor (MRCA, pink). For our purposes, we will not go into the rich statistical

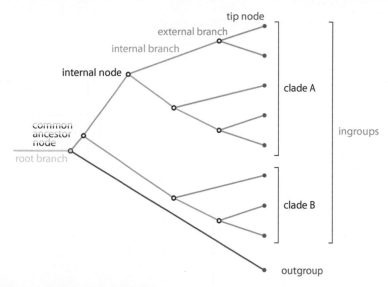

Figure 5.2 Trees and their nomenclature.
Tree diagrams relating DNA sequences to one another, whether alternative haplotypes within a species or orthologous genes between species, share much of the same terminology. Phylogenies of species would refer to clade A and B as being **reciprocally monophyletic**, but this term is not often used for within-species genealogies. A clade of lineages also are sometimes referred to simply as clusters or groups. The orientation of trees is arbitrary, with terminal nodes often positioned to the right (implying time flowing from past to present from left to right), at the bottom (implying past to present from top to bottom, often used with coalescent genealogies), or at the top (implying past to present from bottom to top, often used in phylogenies to mimic the time information recorded in geologic stratigraphy).

literature and computational details used to infer the genealogical relationships between haplotypes. Also keep in mind that the genealogical tree in the figure represents only a single locus. The dynamics of haplotype trajectories through time will be different for different loci because of stochasticity in the evolutionary process induced by differences among loci in mutational input and genetic drift, and because of the independence among loci conferred by recombination between them (Figure 5.3). Therefore, any one locus does not necessarily tell us everything that we might want to know about the population or species as a whole, it just tells us about the history of that locus. We can tell how representative a given locus is for the rest of the genome only by looking at many other loci.

By showing a genealogy, we can show a lot of information. We can represent the entire history of mutational differences among haplotypes on a genealogy. Consequently, it should come as no surprise that we can also calculate summary statistics of diversity from a genealogy. If we know the history of mutations, then we can count them up or compare them among tips of the tree to calculate diversity metrics (Figure 5.4; also see Figure 3.3). In a similar way, we can extract the site frequency spectrum (SFS) from information in a genealogical tree. It is pretty neat how there are a multitude of ways to visualize and summarize information about genetic diversity (Figure 5.4; also see Figure 4.4). For example, note that any mutation that occurs on an external branch of the genealogy represents a singleton variant in the SFS, whereas any mutation that occurs on an internal branch will be present in multiple haplotypes (Figure 5.4).

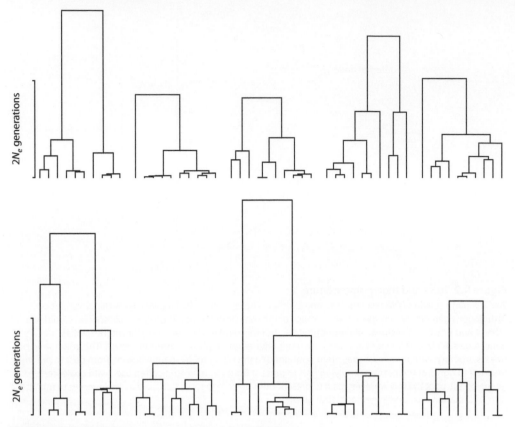

Figure 5.3 Stochasticity among coalescent genealogies.
The coalescent process models the probabilistic nature, the randomness, the stochasticity, that is inherent to how mutation and genetic drift operate in populations. As a result, repeated computer simulations of coalescence among a set of ten haplotypes will yield a wide diversity of coalescent branching times and times to the most recent common ancestor. This heterogeneity among gene trees also is found for real genes sampled from different parts of the genome.

5.2 Gene trees in reverse-time coalescence within species

Despite the fact that evolution proceeds only forward in time, with new haplotypes arising by mutation and splitting the gene tree, it is often useful to think of genealogies in reverse-time. It can take some getting used to and it can be confusing to think of time in reverse, but you can do it! Thinking in reverse time, we can envision the haplotypes we observe today reuniting with their closest relative in a common ancestor. In other words, as we trace the ancestry of those non-recombining haplotypes back in time, we encounter their ancestors. That is, each lineage moving back in time "loses" its unique mutations, so that only those shared mutations between similar haplotypes remain in the ancestral haplotype.

Note that we are not simply talking about the common ancestors of individuals (e.g. your mother and grandmother), but the common ancestors of sequence haplotypes that differ from each other due to the mutations that they have acquired. As this process of merging haplotypes is repeated again and again into the past, and the number of lineages

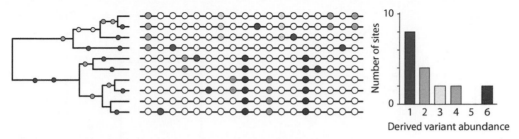

Figure 5.4 The many representations of mutational polymorphisms.
The information about mutations that occur on a genealogy (left) can equally be represented in a haplotype diagram (middle) or with a graph of the site frequency spectrum (right). Recall as well that we can also calculate summary statistics like the diversity metric θ_W from this same information.

decreases, eventually we are led farther back in time to the single most recent common ancestor of the entire set of haplotypes. This reverse-time view of genealogy is called the **coalescent process** (Figure 5.5). When we talk about evolution in a "coalescent framework" or a "coalescent context," it means that we are using this reverse-time viewpoint of how gene copies relate to one another.

In talking about **coalescence**, we will limit ourselves with several convenient and familiar simplifying assumptions. Namely, let's work with contiguous haplotypes for a selectively unconstrained locus that does not experience recombination in a population that is at demographic equilibrium (i.e. not growing/shrinking in size and not subdivided). Recall that we focus on neutral mutations to unconstrained loci so that we can exploit the concepts of the Neutral Theory, meaning that genetic drift and mutation are the only important evolutionary forces. In this idealized scenario, the census population size (N) and the effective population size (N_e) are equivalent (i.e. $N_e = N$). This scenario allows us to understand what happens in the absence of selection. Down the road, it will also allow us to test particular observed situations to see if they are consistent with the absence of selection, or if we must invoke natural selection as a cause for observed patterns—that is, it will permit us to test a null hypothesis based on the Neutral Theory. We will dive into these non-equilibrium issues and tests of neutrality in Chapters 7 and 8.

Another convenient aspect of the coalescent perspective is that it lets us ignore any haplotypes that might have existed in the past, but that have since gone extinct. In other words, we only have to worry about the haplotypes from the present day and their ancestors, we don't need to worry about the other zillions of haplotypes that existed at some point in the past; we don't worry about them because they didn't leave any descendants to survive into the present day. With coalescent theory, we start with the "survivors" and work back from there. This feature contrasts with forward-time thinking, for which we do not know the ultimate fate of any given haplotype: fixation versus loss. In reverse-time thinking, we *do* know the ultimate fate of any haplotype as we trace it back in time: coalescence. Every haplotype is fated to coalesce with some other haplotype by losing unique mutations as we proceed back in time. A final benefit of reverse-time thinking about genealogies in a coalescent context is that we can focus explicitly and solely on just a subset of copies of a locus from a population and yet still draw inferences about evolutionary properties of the entire population. This also is the approach we must take in practice with real data, which helps match the model to data.

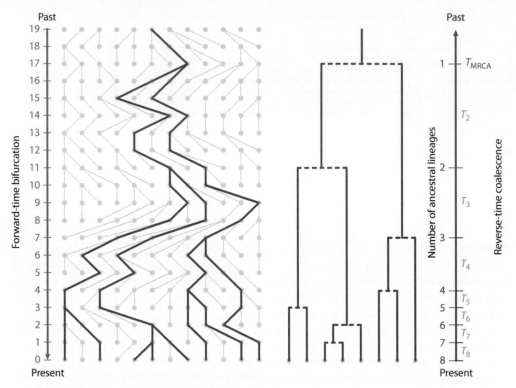

Figure 5.5 Contrasting forward-time and reverse-time genealogies.
In the coalescent process, as we trace evolutionary history back in time, each time a pair of haplotypes encounter their common ancestor the number of lineages decreases by one. In this example, we have sampled eight haplotypes in the present day, and we move backward in time by one unit at a time; here we show time as being discrete rather than continuous, where each time step corresponds to a different generation. This example shows a single coalescent event in each of the first four generations into the past, correspondingly reducing the number of remaining haplotypes to three over the period of time $T_5 + T_6 + T_7 + T_8$. Notice how coalescence takes longer when there are fewer haplotypes left to merge, and that these remaining haplotypes make up a small minority of haplotypes that existed at that point in time in the past.

Now that we have developed a basic sense of what reverse-time coalescence is, let's launch into a more technical depiction of this conceptual view of neutral evolution. The mathematical logic of coalescent theory was originally worked out by John Kingman in 1982, with important subsequent contributions by Richard Hudson and many others. If we sample k different haplotypes in the present day, then after $k - 1$ coalescent events, we will have found the single most recent common ancestor (MRCA) of the haplotypes (Figure 5.5). Why $k - 1$ coalescent events? Each coalescent event reduces the number of haplotypes by one, but coalescence stops only when there is a single haplotype, the MRCA. For example, if we had sampled $k = 4$ haplotypes, then going back in time we will have $4 \rightarrow 3 \rightarrow 2 \rightarrow 1$ haplotype to reach the MRCA, which involves three coalescent events (each arrow indicates coalescence). Three is 4 minus 1, and $k = 4$, hence it takes $k - 1$ coalescent events to reach the MRCA. Because the MRCA is just a single haplotype, it is sometimes called "*the* coalescent."

But how can we make some specific predictions about how to get from a set of k haplotypes down to just one haplotype over time? Firstly, we must answer the question: what is the probability that a coalescent event occurs in the generation immediately preceding our present-day sample (p_c)? Only haplotypes that are identical by descent can coalesce. Another way of asking this question is to ask: what is the likelihood that some pair of haplotypes in our sample are identical by descent and so would share the same common ancestor haplotype sequence in the preceding generation? It turns out that it is easiest to answer this question by turning it on its head. Specifically, we can more simply calculate the probability that individual pairs of haplotype copies are *not* identical by descent (p_{nc}). We then subtract p_{nc} from 1, which will give us what we actually want to know: the overall probability of being identical by descent and, therefore, of coalescence, $p_c = 1 - p_{nc}$.

To help solve this problem, we can apply the logic from statistical sampling theory for large populations such that we can assume that we have drawn haplotypes at random and **with replacement**. As a consequence of this random sampling procedure, the probability is $1/(2N_e)$ that two randomly selected sequences at a locus from a diploid population will be identical by descent (i.e. that they are the same haplotype). Naturally, the probability that they are *not* identical by descent is $1 - 1/(2N_e)$. Remember that because we are working under the standard neutral model, $N_e = N$. If we randomly picked a third sequence at the locus from the population, the probability that it would be identical by descent to one or the other of the previously sampled haplotypes is $2/(2N_e)$ and the probability that it is different from the first two sequences is $1 - 2/(2N_e)$; for a fourth sequence, $1 - 3/(2N_e)$; and so on.

Consequently, the probability that *no coalescence* occurs in the first generation—looking backward in time—for all k sequences that we have at our disposal for analysis will be the combined probability of all these possibilities, which we multiply together: $1 - p_c = \left(1 - \frac{1}{2N_e}\right) \times \left(1 - \frac{2}{2N_e}\right) \times \cdots \times \left(1 - \frac{k-1}{2N_e}\right)$. We can rewrite this series as $1 - p_c = 1 - \left(\frac{1}{2N_e} + \frac{2}{2N_e} + \cdots + \frac{k-1}{2N_e}\right)$ by ignoring terms that have higher powers of N_e in the denominator (which will be very small). This expression further simplifies algebraically to $1 - p_c = 1 - \left(\frac{1+2+\cdots+(k-1)}{2N_e}\right)$. We can then distill down this expression even more to give $1 - p_c = 1 - \left(\frac{k(k-1)}{4N_e}\right)$, which tells us the probability that none of the k haplotypes will coalesce with any other haplotype in the preceding generation. This, in turn, means that the overall probability that a coalescent event *does* occur between some pair of the k haplotypes in the preceding generation is approximately:

$$p_c = \frac{k(k-1)}{4N_e}, \tag{5.1}$$

which is accurate to first order $1/N_e$ (i.e. this considers terms with $1/N_e^2$ to be negligible). Another way of stating what this equation means is that it is the probability of going from k to $k - 1$ haplotypes over the course of one generation back in time.

Resurrecting our example of having sampled $k = 4$ haplotypes, Equation 5.1 tells us the probability that some pair of haplotypes will coalesce together so that there will be $k - 1 = 3$ haplotypes when we look one generation back in time. If N_e is 1,000, then $p_c = p_{4\rightarrow3} = 4*(4-1)/(4*1000) = 0.00033$. If k is large, that is, if we have sampled very many haplotypes, then there is a higher probability of coalescence in the immediate past than the more distant past, which we can see from the fact that the numerator $k(k-1)$ will be large. Coalescent events accrue more quickly when there are many lineages because there are many possible pairs of lineages that could coalesce. For example, if $k = 20$ instead of 4, then $p_{20\rightarrow19} = 20*(20-1)/(4*1,000) = 0.095$, which is a 285 times larger number than 0.00033! Similarly, a smaller N_e, which sits in the denominator of Equation 5.1, will also increase the probability of coalescence in the preceding generation.

These calculations are all fine for one generation, but what about looking farther back in time? Let's call the time at the present day t_0 and the time one generation ago t_1. The mathematical expression in Equation 5.1 gives the probability of coalescence taking place in the time interval between t_0 and t_1 (Figure 5.5). What if no coalescent event occurs between t_0 and t_1, what then would the probability of coalescence be if we look back *another* generation, between t_1 and t_2? It would be exactly the same, because there would have been no reduction in the number of haplotypes from t_0 to t_1, so k would still have the same value. If a coalescent event *did* occur between t_0 and t_1, what would the probability of coalescence be between t_1 and t_2? For this calculation, we would simply replace k with $k - 1$ in Equation 5.1. This iterating of an equation with new values is called a **recursion**, meaning that we can keep plugging in the new revised value of k every time we decrement its value by one from having had a coalescent event.

Again going back to our simple example of $k = 4$ haplotypes that we sampled in the present day, we can iterate through the probabilities for each cycle of coalescence: $p_{4 \to 3}$, $p_{3 \to 2}$, $p_{2 \to 1}$. Again, you should see that the bigger k is (i.e. the more unique haplotypes that we have in hand), then the greater the likelihood that there will be a coalescent event in a given time interval. Consequently, as we carry out this coalescent process recursively through time, the value of k keeps getting smaller, as does the corresponding probability of coalescence. Note also that if there were only 2 haplotypes (so $k = 2$), then the probability of coalescence reduces to $1/(2N_e)$—the same probability of identity by descent for a pair of haplotypes that we started with!

We can also derive a mathematical expression for the probability that a coalescent event occurs *exactly* t generations ago. This is the combined probability that there will be *no coalescence* in any of the first $t - 1$ generations looking back in time until there is finally a coalescence occurring only in generation t. This means that we have to multiply the probabilities of no coalescence together for the first set of generations, and then finally multiply this value by the probability of a single coalescence. This combination of events works out to be $(1 - p_c)^{t-1}$ times $1^* p_c$. We can multiply the probabilities in this way because the coalescence process from generation to generation is "memoryless" and this **Markov property** lets us combine probabilities independently across generations. So, the probability of coalescence from k to $k - 1$ haplotypes occurring some specific time that was t_x generations ago is $(1 - p_c)^{t_x - 1} p_c$, where $p_c = \frac{k(k-1)}{4N_e}$. It turns out that, in statistical terms, this mathematical expression corresponds to a geometric distribution. And a geometric distribution has an average expected value that is the inverse of the "probability of success," which is the coalescence probability p_c. As a result, the expected **time to first coalescence** (T_k) from k to $k - 1$ haplotypes is $1/p_c$, which we can write as:

$$\hat{T}_k = \frac{4N_e}{k(k-1)}. \tag{5.2}$$

Using coalescent theory, we have worked out a way that we can relate changes of genealogies to the passage of time.

Now that we have a way of estimating how much time it will take between each coalescent event, how much time will it take for all of our haplotypes to coalesce into their singular common ancestor? For a given locus, the number of haplotypes in the population will often be very large. Consequently, it is reasonable to approximate the expected **time to the most recent common ancestor (TMRCA)**, or coalescence time, as the cumulative amount of time elapsed to take us from a very large number of m haplotypes back to only 1. When the total number of haplotypes is large, then the sum total of individual coalescence times to go from the many m haplotypes down to just $k - 1 = 1$ haplotype approaches $4N_e$ generations:

$$\hat{T}_{\text{MRCA}} = \frac{4N_e}{m(m-1)} + \frac{4N_e}{(m-1)\cdot(m-2)} + \cdots + \frac{4N_e}{2(2-1)} = \sum_{k=2}^{m} \frac{4N_e}{k(k-1)} = 4N_e \text{ generations,}$$

(5.3)

where this equation "counts down" the number of coalescent events all the way from m haplotypes down to the last pair of haplotypes. This means that we expect the TMRCA to be approximately $4N_e$ generations ago (Figure 5.6), if there are a large number of haplotypes in the population in the present day. For any given *pair* of haplotypes, however, the average coalescence time is $2N_e$ generations (Figure 5.6).

Now remember that it is mutations that get recorded in a genealogy and that show the history of the origins of polymorphisms that we observe among the tips of the gene tree. We can also use coalescent theory to extract expectations for this genetic diversity. For example, the number of segregating sites is proportional to the total **tree length** of the coalescent genealogy, which is just the cumulative number of mutations on all the branches. This means that we can calculate the θ_W metric of sequence polymorphism in a different way than we did in section 3.3.

What we need to calculate θ_W is the total tree length L, which is simply the cumulative length, in units of mutations, of all b branches: you just have to add up all the mutations that arose on each of the branches in the genealogy. A bifurcating tree with k tips will have $2*(k-1) = b$ branches. Stating this mathematically, $L = \sum_{k=b}^{2} k \cdot T_k \cdot \mu = \sum_{k=b}^{2} \frac{k \cdot 4N_e \cdot \mu}{k(k-1)}$, which we can rearrange to $L = 4N_e\mu \sum_{k=2}^{b} \frac{1}{k-1} = 4N_e\mu \sum_{k=1}^{b-1} \frac{1}{k}$. Written this way, the tree

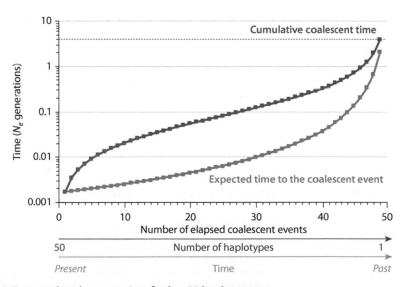

Figure 5.6 Expected coalescence time for $k = 50$ haplotypes.
The time between coalescence events is much shorter when there are many haplotypes than when there are few haplotypes (blue line). The total expected coalescence time summing across all coalescent events in reaching the most recent common ancestor converges on $4N_e$ generations ago (red line). The expected coalescence time for the final coalescent event is just as long as the sum of all the preceding coalescence events put together ($2N_e$ generations). Note the log scale for time on the y-axis for a diploid organism.

length L is just the count of the number of nucleotide differences among haplotypes in the sample, which is equivalent to the number of segregating sites in the sample of haplotypes (i.e. $L = S$ from section 3.3.1), with the term in the summation being equivalent to the sample size correction factor a from section 3.3.1. That's pretty nice! Also remember that the population mutation parameter θ equals $4N_e\mu$ when the population is at equilibrium, and, that we can estimate θ from sequence polymorphism data. If we are trying to estimate $4N_e\mu$, then we arrive at an equivalent equation for Watterson's estimator of polymorphism that we discussed in section 3.3:

$$\theta_W = L/\sum_{k=1}^{b-1}\frac{1}{k} = S/\sum_{k=1}^{b-1}\frac{1}{k} \tag{5.4}$$

This equation shows us that we can use genealogies to calculate population genetic summary statistics. And, reciprocally, we can use these summary statistics calculated from sequence data to tell us something about the underlying genealogy.

While it is extremely useful to think of genealogies from the reverse-time perspective of coalescence, this approach is not without limitations. It is challenging to incorporate selection into the coalescent process without burdensome assumptions. While a virtue of coalescence is the ease of computer simulations for individual loci, such evolutionary simulation becomes cumbersome when trying to analyze the molecular evolutionary history of entire chromosomes or genomes. As a result, researchers have shifted to developing computationally efficient forward-time evolutionary simulations to address some genome-scale molecular population genetic questions. Coalescent theory also presumes that the sample size used for analysis is much smaller than the effective population size of the species being studied. High-throughput DNA sequencing technologies have allowed researchers to obtain extremely large sample sizes that then violate this assumption: there will be too many rare variants observed in a large sample relative to the number expected when samples are small relative to population size. This result means that coalescent theory will not connect well to data for such large sample size datasets, though John Wakeley and Tsuyoshi Takahashi (2003) showed how it can be exploited to estimate both the mutation rate and effective population size rather than just their product, the population mutation rate.

Based on the results of coalescent theory that we have just worked through, we now have a way to connect theory, data, and genealogies. That is, we can use DNA polymorphism information to estimate the effective population size or to infer the expected time to the MRCA for a given locus, provided that we know the per-generation mutation rate—and that the assumptions of the standard neutral model have been met reasonably well.

5.3 Gene trees, species trees, and phylogeny among species

So far in this chapter, we have talked about using gene trees to describe haplotypes of polymorphic sequences within a species. However, gene trees also can help to illustrate the accumulation of DNA sequence changes between species. You may, in fact, already be familiar with this idea of gene trees to relate different species. Even in the absence of an understanding of genetics and DNA, Charles Darwin appreciated this idea in writing in his 1842 "sketch" for his most famous book: "There is much grandeur in looking at the existing animals either as the lineal descendants of the forms buried under thousand feet of matter, or as the coheirs of some still more ancient ancestor." Depending on the question at hand, we might be most interested in the phylogeny relating distinct species—

and therefore knowing with confidence the topology and branch lengths of the species tree—or, instead, we might be more interested in the genetic distances captured in any one given gene tree irrespective of speciation history. This distinction might seem subtle; after all, shouldn't gene trees and species trees have the same shape? Not always. But before we discuss this idea, first let's work through some more basic issues in species divergence.

Box 5.1 SPECIATION AND THE BIOLOGICAL SPECIES CONCEPT

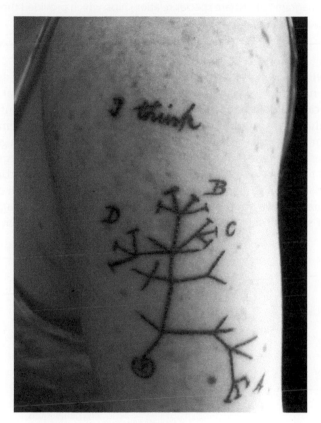

Figure 1

There are too many ideas of how to "define" a species to enumerate here, so I will just emphasize one: the "**biological species concept**" (BSC). With this view, species are groups of organisms that do not interbreed with other such groups; the groups are reproductively isolated from one another. In molecular population genetic practice, we also often presume that such "biological" species arise through a simple splitting process so that the "evolutionary" or "phylogenetic" species concepts also apply: most gene genealogies coalesce prior to the common ancestor with another species. In reality, however, this genealogical assumption need not be true, as complex patterns of incomplete lineage sorting in real organisms make abundantly clear (Figures 5.9, 5.14, and 5.15, Box 5.4). When analyzing real-world data, it is crucial to keep in mind this distinction between how **reproductive**

(Continued)

Box 5.1 CONTINUED

isolation creates populations that fail to exchange genetic material (biological species) and the timescale over which conveniently shaped genealogical relationships among organisms actually result. Molecular population genetics helps us understand why these views do not always match.

Figure 1 The idea of a phylogeny of species represented as a branching tree stems from the iconic "I think" diagram drawn by Charles Darwin in his notebook "B" in 1837, here reproduced as a tattoo on the arm of an Australian doctoral student. Darwin refined his idea of a branching diagram to illustrate species relationships, which provided the only figure in his magnum opus *On the Origin of Species*. (photo courtesy H. Warland)

Although we will not worry ourselves too much about the intricacies of different **species concepts**, it is nevertheless useful to have a working notion of what is meant by a species. Species can be defined loosely as groups of individuals that interbreed among themselves and that fail to interbreed with other such groups, making them reproductively isolated; this is the biological species concept (Box 5.1). In our population genetic lingo, no interbreeding means no gene flow between their populations. As a consequence, alleles of a given locus will tend to be more closely related and share a more recent common ancestor with other alleles within a species than with alleles of homologous loci in different species (Figure 5.7). Moreover, we should generally expect shorter branch-lengths to separate the gene trees of species that separated more recently from one another. This history of

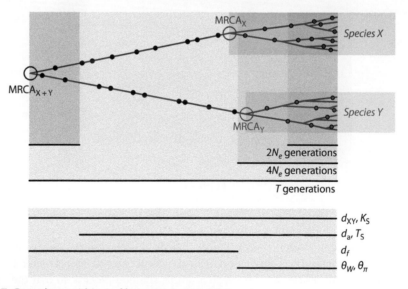

Figure 5.7 Genealogy within and between two species.

This gene tree shows fixed differences between two species (filled black circles) and polymorphisms (filled red and purple circles) that give rise to the eight haplotypes that were sampled from each species. The timescales indicate the times we discuss for species divergence versus coalescence within species to illustrate the connection of within- and between-species analysis of DNA sequences, as well as common metrics used to quantify polymorphism and divergence. Open circles indicate the time to the most recent common ancestor (MRCA) of haplotypes within a species (red, purple) or between species (black). Note that the TMRCA is longer ago than the speciation time (T_S) (Figure 5.9).

population splitting that led to speciation and the present-day collection of species in the world can be summarized with a species tree that describes the phylogenetic relationships among the species. In the ideal case, where all pairs of species in the phylogeny are well-separated with long-but-not-too-long internal and external branches, then most gene trees should have the same pattern of branching, the same **topology**, as the species tree. However, as we will see in section 5.3.2, there are a number of causes for why any given gene tree might not match the species tree.

Box 5.2 HOMOLOGY VS. ORTHOLOGY VS. PARALOGY

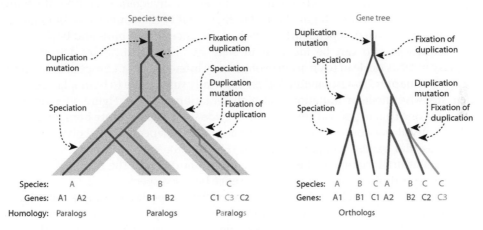

Figure 2

Genes are homologs if they descend from a common ancestor. Homolog is a blanket term that includes both orthologs and paralogs, and so depending on context, homologs could refer to copies of genes in the same species or in different species. So, all of the blue, red, and yellow genes shown in the gene trees in Figure 2 are homologs. Genes that are orthologs represent a special case of homologous loci: they are gene copies from different species that represent each other's closest relative among the gene repertoire of the species being compared. Orthologous genes share a common ancestor gene copy that was present in the genome of the ancestral species prior to the speciation event that led to the descendant species that are being analyzed in the present day. The three blue-colored genes in the diagram represent the group of "one-to-one-to-one" orthologs in this three-species phylogeny. Paralogs represent sets of genes that are related to one another by duplication. Such gene duplication can give rise to multi-gene families. Duplication mutations will initially be polymorphic within a species, termed copy number variants (CNV); duplication mutations that become fixed differences between species usually are what people mean when talking about paralogs. You can refer to the set of paralogous genes within a single species (e.g. genes C1, C2, and C3 in the diagram), or, if the gene duplication event preceded speciation in an ancestral species, you could refer to duplicated gene copies from each of the descendant species as paralogs (e.g. A1 and B2). Paralogy can be confusing in practice in the face of imperfect information about gene content and duplication history, especially when some duplicate copies get deleted from the genomes of some species but not others. Sometimes analyses of genomes include **clusters of orthologous groups**, which includes the set of paralogs that together descend from a single common ancestral copy.

When analyzing features on a phylogeny, it is important to understand the history of ancestry and descent. Homology is the sharing of common derived features that arose through descent from a common ancestor (Box 5.2), and so we can talk about homologous traits or homologous loci or homologous allelic states. Homologous allelic states are also referred to as being identical by descent, an idea that we made extensive use of in deriving the coalescent process in section 5.2. This contrasts with homoplasy, which describes the situation of features being the same, being **identical in state**, without it resulting from shared ancestry. In morphological terms, homoplastic traits are **analogous** rather than homologous, by virtue of **convergent evolution** producing the shared common features from different ancestors or from evolutionary reversals that yield identical phenotypic states. As a familiar morphological example of convergent evolution, the wings of bats and birds both confer flight ability, but flight ability is an analogous, homoplastic trait because it evolved independently in each of the ancestors of birds and bats.

DNA sequences also can be identical in state but not identical by descent. Such homoplasy in DNA can lead to errors in evolutionary inference, if not properly recognized. As we saw in Chapter 2, microsatellite loci are especially susceptible to homoplasy, because of their stepwise mode of mutation and high mutation rate (see Figure 2.8). However, point mutations that happen at the same nucleotide site also can create homoplasy, such that the identical nucleotide variant will have arisen on independent and distinct genetic backgrounds that exhibit unique haplotypes (Figure 5.8). We will explore this issue of repeated mutational change to the same site in further detail in the next section, as we think more technically about sequence divergence between species.

The divergence between homologous loci from different species is a simple byproduct of the accumulation of fixed genetic differences that occur independently in the populations of each species, the outcome of positive selection and genetic drift. This means that new mutations arose in an ancestral population from one or the other species' history and then increased in frequency to fixation. In Figure 5.7, only those mutations between the "species A" common ancestor node and the "species B" common ancestor node contribute to divergence in terms of fixed differences; the other mutations contribute to intraspecific polymorphism.

Common practice for many phylogenetic applications involves taking just a single "representative" sequence from each species and, so, some polymorphic sites would be counted in the sum of sequence differences between the species. If the time to the MRCA T is much larger than $4N_e$ (Figure 5.7), then the within-species polymorphism will not contribute much to the overall amount of sequence difference between the species. Consequently, ignoring within-species polymorphism will not matter much when T is substantially bigger than $4N_e$ (i.e. as $\frac{4N_e\mu}{2\mu T}$ approaches zero). Importantly, and as we will return to in more detail, the common ancestor was itself comprised of a genetically variable population and so we must consider that ancestral polymorphism for its potential to influence both the structure of gene trees and our estimates of the timing of divergence (e.g. time to MRCA$_{A+B}$ in Figure 5.7).

5.3.1 *Substitution rates in divergence*

In thinking about genealogies that represent divergence among species, we will first focus on the simplest scale of multi-species gene trees: two species. When comparing DNA sequences between species, we refer to the single nucleotide changes that fix in different populations and then accumulate as fixed differences between them as **sequence divergence**. In population genetic terms, we would envision these species

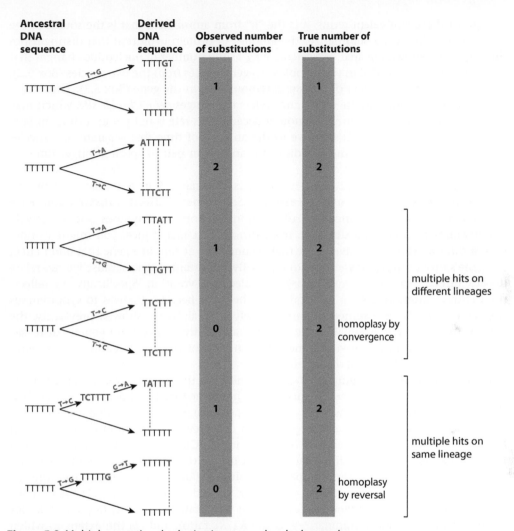

Figure 5.8 Multiple mutational substitutions as molecular homoplasy.
Point mutations to DNA that hit the same position more than once lead to a reduction in the number of observed sequence differences. The top two examples show independent mutational substitutions at distinct sites in two lineages over time, such that the number of observed changes equals the number of actual mutational substitutions. The bottom four examples illustrate different ways in which multiple hits along a single lineage or along a pair of lineages can mask our ability to detect the true number of mutational substitutions. The hidden nature of multiple changes occurs because we only have the derived present-day DNA sequences to work from; the true ancestral sequence is unknown from a sampling of individuals that live in the present day.

simply as distinct populations, groups of individuals that share haplotype lineages within them but that represent non-recombining lineages with respect to the accumulation of the sequence divergence between them; species are population units within which genetic drift and selection operate independently.

When we look at divergence in coding sequences, we typically separate out the sequence divergence according to those substitutions that cause either replacement site or silent site differences. Divergence at replacement sites is summarized as the **non-synonymous site substitution rate**, commonly known by the metric K_A, and with alternate names of the amino acid or replacement site substitution rate, as well as the alternate abbreviation d_N.

In light of these equivalent terms, it is the "a" from amino acid that is the source of the "A" subscript in K_A. K_A summarizes the amount of protein evolution that distinguishes those species being compared. We calculate K_A as the number of nucleotide changes that alter amino acids encoded in the orthologous gene copies from the two species (Box 5.2), divided by the total number of non-synonymous sites in the gene (Box 5.3). Thus, K_A is scaled per nucleotide and the term "rate" refers to changes *per nucleotide site*, which may initially seem a counterintuitive misnomer because this rate is not per unit time. In fact, the numerical value of K_A is sensitive to the amount of time that separates the species being compared and that time will differ for any given pair of species: longer times of separation give higher values of K_A.

Divergence at synonymous sites in genes is summarized in a similar way by the **synonymous site substitution rate, K_S**. Sometimes a **silent substitution rate** is calculated instead that includes substitutions at both synonymous and selectively unconstrained non-coding sites (e.g. from introns). At first glance, you might wonder why should we care about divergence that accumulates at sites in a gene that don't affect the gene's function? K_S, however, is an especially important metric because we can relate it back directly to the Neutral Theory of Molecular Evolution. Specifically, K_S reflects the neutral mutation rate for nucleotides at the locus because changes to synonymous sites can typically be assumed to be selectively neutral: as we showed previously, the neutral mutation rate is equal to the average rate of divergence from genetic drift (see section 4.1.1). Thus, K_S provides a neutral standard for the accumulation of sequence divergence in the absence of direct selection.

Sequence differences accumulate to get summed together to give us K_S in a predictable way. But this predictability does not necessarily hold true for sites under selection, such as replacement sites: replacement sites typically experience purifying selection and so evolve slower. Sometimes, however, positive selection can speed up evolution of proteins and therefore elevate K_A; we will explore this idea more in Chapters 7 and 8. Because K_A and K_S are scaled per site (rather than per locus), it means that we can compare their values directly to each other and also compare the values for different loci that have different sequence lengths. Consequently, by comparing K_A and K_S we can infer (1) the potency of selection on non-synonymous sites, relative to mutation and drift (K_A versus K_S for a given gene, commonly summarized as the K_A/K_S ratio); and (2) the extent to which different loci experience different mutation rates (i.e. K_S differences among genes reflect **mutation rate heterogeneity** across loci).

Measurements of neutral divergence that use K_S have another powerful application, as a molecular clock (see Figure 4.1). Provided that the mutation rate does not change over time or differ among lineages, then the amount of DNA sequence change at selectively unconstrained sites should reflect the time since they shared a common ancestor, between species or between non-recombining haplotypes within a species. This means that we can use a simple calculation to estimate the TMRCA of a pair of species from the observed amount of silent-site divergence (K_S), if we know what the neutral mutation rate is (μ). Specifically, because $K_S = 2\mu T_{\mathrm{MRCA}}$ (see Equation 4.1), we can solve for the TMRCA:

$$T_{\mathrm{MRCA}} = K_S/(2\mu) \tag{5.5}$$

Recall from Chapter 2 that we can estimate μ from mutation accumulation experiments, or infer it indirectly from fossil calibrations. The coefficient 2 enters the equation because fixed differences between the two species occur independently along each lineage (see Box 3.5).

Box 5.3 COUNTING OF NON-SYNONYMOUS AND SYNONYMOUS SITES

	Coding sequence					Protein translation	n_{syn} ($\frac{1}{3}n_2 + n_4$)	n_{rep} ($n_0 + \frac{2}{3}n_2$)	p_{syn}	p_{rep}
Degeneracy	004	004	004	004	004					
Species A	GTT	TCG	CCT	GTG	GCA	VSPVA	5	10		
Species B	GTG	ACG	CCT	GAG	GCT	VTPEA	4.33	10.66		
Degeneracy	004	004	004	002	004					
						Average	4.66	10.33	2 ÷ 4.66 = 0.429	2 ÷ 10.33 = 0.194

Figure 3

As we scan along a string of coding sequence, being aware of its codon structure, we know that mutation of any given site could alter the encoded amino acid (a replacement change) or not (a synonymous change). How many positions represent synonymous versus replacement sites? Usually we use the idea of mutational opportunity to classify how synonymous or how non-synonymous a given codon site is: of all possible mutations to the nucleotide at that site, what fraction would cause synonymous versus replacement codon changes?

For some codon positions, the genetic code makes it easy to classify them. At some sites, any mutation would be a replacement change; these sites are termed 0-fold degenerate, found commonly in the first and second positions of codons. At some other sites, any mutation would be a synonymous change; these sites are termed 4-fold degenerate, found exclusively in the third position of some codons (see Figure 3). But for some sites, it depends on which mutational change happens for you to know whether it would cause a different amino acid to be encoded; these sites are 2-fold degenerate (or, in the case of the third position of an isoleucine codon, 3-fold degenerate). These 2-fold degenerate codon positions count only as $\frac{1}{3}$ of a synonymous site and $\frac{2}{3}$ of a replacement site, because only one of the three possible nucleotide changes is synonymous, whereas the other two possible changes would alter the encoded amino acid (a 3-fold degenerate position counts as $\frac{2}{3}$ synonymous and $\frac{1}{3}$ replacement).

In a pairwise sequence alignment, we can count up the number of synonymous (n_{syn}) and replacement sites (n_{rep}) across the entire coding sequence for each copy of the sequence ($n_{syn} = n_4 + \frac{2}{3}n_3 + \frac{1}{3}n_2$ and $n_{rep} = n_0 + \frac{1}{3}n_3 + \frac{2}{3}n_2$). The values of n include all sites of a given site class, not just those that are identical between species. To compute divergence as the pairwise sequence distance (p_d) for a given site type as the most simple estimate of K_A and K_S, we simply divide the number of changes by the number of sites. Divergence between species in the alignment will lead to slightly different values, so for calculations of K_A and K_S we can use the averages for each site type. In practical applications, one will also ideally incorporate transition:transversion mutation biases into the mutational opportunity of 2-fold and 3-fold degenerate sites. As a result, 2-fold degenerate sites will then count as more than $\frac{1}{3}$ synonymous, given the peculiarities of the genetic code and the fact of transitions being more likely than transversions. We should also adjust divergence estimates to account for the possibility of multiple mutational hits over the course of evolution (section 5.3.2).

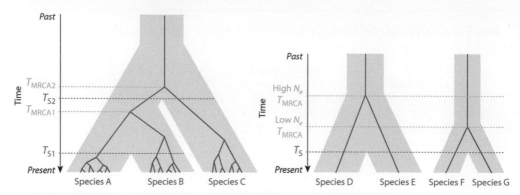

Figure 5.9 Ancestral polymorphism and the estimation of times to the MRCA.

The time at which two populations cease gene flow (T_S) to represent distinct species must necessarily be closer to the present day than is the time at which gene copies share a common ancestor (T_{MRCA}). The reason for this distinction between species split time and coalescent time is that the ancestral species itself was a population, with the alleles that ultimately descended to daughter species also being subject to the coalescent process in that ancestral population. The larger the ancestral population size, the greater the difference between T_S and T_{MRCA} (see also Box 7.2).

Use of the molecular clock in this simple way to date the common ancestry of lineages has a caveat that is especially important for recent speciation events and for species with large population sizes: **ancestral polymorphism**. Let's unpack this idea. Note that when you calculate T_{MRCA} with Equation 5.5, you will not estimate the time of *speciation* but the time that the two orthologous gene copies shared a common ancestor. This time will actually pre-date the speciation event itself. To estimate the **species split time** (T_S) we need to account for the coalescence time within the population of that common ancestral species (Figure 5.9).

So how do we account for the fact that the ancestral species itself was a population with polymorphism so that we can more accurately estimate the species split time? First, remember that in calculating divergence between a pair of species we usually would have sampled a single copy of the locus from each descendant species for analysis. It is this pair of sequences that we are tracing into their common ancestor. It also is this pair of sequences that we need to worry about in terms of their coalescence time in the ancestral population; this is the "extra" divergence between haplotypes that we must factor in to our calculation of T_{MRCA} versus T_S. As John Gillespie and Charles Langley showed in 1979, this adjustment gives the equation:

$$T_{MRCA} = T_S + 2N_{anc} \qquad (5.6)$$

where N_{anc} is the effective population size of the ancestral population and $K_S = 2\mu T_{MRCA}$. We can then substitute and rearrange this equation to solve for T_S. In effect, what this ends up doing in solving for T_S is to subtract the expected time to coalescence for exactly two haplotypes in the ancestral population from the net divergence that we see between the present-day copies (Equation 5.2).

A complication to Equation 5.6 is that we do not actually know N_{anc}. However, it may often be reasonable to assume that the present-day population sizes of the descendant species are similar to their common ancestor. Consequently, we can analyze extant molecular population genetic variation and plug a contemporary measure of diversity like θ_π into the right-hand side of Equation 5.6. This approximation also uses the idea that at equilibrium $\theta_\pi = 4N_e\mu$. So, after some algebraic simplification, empirically we get:

$$T_S = (K_S - \theta_\pi) / (2\mu) \tag{5.7}$$

When K_S is much larger than θ_W, then we can safely ignore polymorphism in estimating T_S, because the relative contribution of ancestral polymorphism becomes negligibly small (Figure 5.9). However, when species are very closely related, then ancestral polymorphism can represent a large fraction of the overall divergence between gene copies that were sampled in the different species. Moreover, many of the differences between the orthologous copies might also overlap with polymorphic sites within one or both of the species in the present day. Because calculation of substitution rates with K_A and K_S presumes, implicitly, that the differences represent mutations that have *fixed* within each species, many applications of divergence using K_A and K_S are not appropriate for very closely related species.

It is important to keep in mind in all of this discussion of divergence that substitution of neutral variants is a stochastic process (mutation *and* genetic drift), so we must expect variation in the substitution rate that we observe at different loci. There will be variation in the accumulation of mutations along each lineage, not a uniform interval between mutational substitution events as you might think of for a literal clock. Early critics of the molecular clock hypothesis argued that this variation was too great for divergence to be sufficiently "clock-like" in practice, but the molecular clock idea from the Neutral Theory of Molecular Evolution has survived the test of time and is in widespread use today. Also remember that the Neutral Theory predicts a constant substitution rate per generation (not per year), so lineages of organisms with longer generation times are expected to accumulate substitutions at a slower rate per year than lineages with rapid generation times. This contrast between time measured in years versus generations is responsible for the **generation time effect** on branch lengths in molecular phylogenies (Figure 5.10).

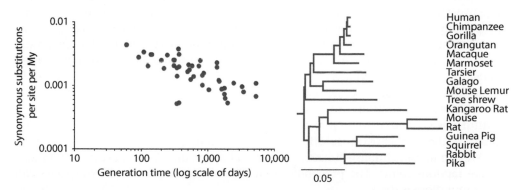

Figure 5.10 Generation time effect on substitutions and phylogenetic branch-lengths.
The Neutral Theory predicts that neutral substitutions accumulate at the mutation rate per site per generation, so species that take longer to pass through a generation will accrue fewer mutations on a per-year basis, as seen among different mammals (data redrawn from Welch et al. 2008). The faster accumulation of substitutions per year in species with shorter generation times will make them have longer branches in molecular phylogenies, as seen clearly for mice and rats compared to primates (data redrawn from Song et al. 2012).

Another problem with the application of molecular clocks is that protein sequence divergence (i.e. replacement site divergence) is often assumed to be clock-like in order to date very ancient common ancestors for which synonymous sites are saturated with mutations (see section 5.3.2). However, the Neutral Theory predicts that only neutral substitutions will accumulate in a clock-like manner, so replacement site divergence might not conform to the assumption of neutrality. Despite this concern, it turns out that protein divergence between anciently related species often relates well with dates estimated from the fossil record.

5.3.2 *Multiple hits in sequence divergence*

The infinite sites model of mutation provides a reasonable approximation of reality at the modest timescales of coalescence within species. However, as the time that separates lineages increases, there is increasing chance that the "infinite" assumption will be violated: some sites will get hit by mutation more than once (Figure 5.8). This issue is called the **multiple hits** problem. Consequently, calculations of K_A and K_S should try to correct for the influence of multiple hits on how much sequence divergence we can observe relative to how much change has truly taken place. In some situations, the observed sequence divergence may be what is important, for example, when testing experimentally for functional differences between orthologous sequences using gene swaps. This raw net divergence, sometimes called **p-distance**, is just the proportion of sites that differ between the two orthologous sequences (p_d), the converse of which is referred to as **percent sequence identity**. But in molecular evolutionary analysis, we are usually interested in estimating the true number of substitutions that have accrued over time, rather than merely counting the number of observed nucleotide differences between an orthologous pair of sequences.

If multiple hits are not properly accounted for, then the number of sequence differences that we observe will underestimate the true number of substitutions that actually occurred in their history (Figure 5.11). This bias becomes more severe as the divergence between the sequences increases: it is more of a problem for DNA sequences from more distantly related species that shared a very ancient common ancestor. Multiple mutational hits are an especially acute concern for deep phylogenetic analysis or for sequence comparisons of gene families that arose from ancient gene duplication events. Molecular population genetic analysis, however, usually focuses on differentiated populations and their divergence from closely related species, making the multiple hits problem less dire in practice than it is for analyses with deeper-time phylogenies.

To illustrate the multiple hits effect, let's think about how often we should see x substitutions given that we expect that K should have occurred. That is, K will be our estimate of the number of true substitutions after accounting for the possibility of multiple hits. We can make a simple estimate of K using a Poisson distribution because mutations are rare random events. The probability that the common ancestor of two species will have x mutations is then:

$$\Pr[x] = \frac{e^{-K}K^x}{x!} \tag{5.8}$$

The probability that zero substitutions will occur is e^{-K}, given that we expect to find K mutations since the common ancestor of the two sequences. From this, it then follows that the probability of one or more substitutions occurring is just one minus the probability that zero substitutions take place: $1 - e^{-K}$. This latter value can be estimated from the

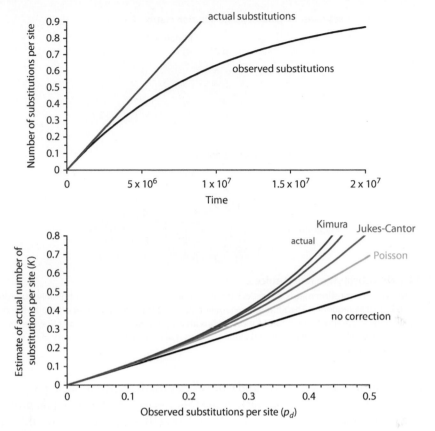

Figure 5.11 Mutational saturation.

As point mutations accumulate over time between orthologous sequences in different species, multiple hits will eventually lead to saturation of divergence. When divergence is saturated, then additional mutational substitutions do not lead to additional observed differences. As a consequence, the number of observed differences underestimates the true number of substitutions that have accumulated (top graph). We can correct the observed number of differences by accounting for the finite number of sites in a sequence that can change (Poisson correction), the number of nucleotides they can change to (Jukes-Cantor correction), and mutational biases inherent to the biochemical structure of DNA (i.e. transition vs. transversion changes; Kimura correction). These corrections that incorporate additional biological detail provide a more accurate estimate of the true number of substitutions that occurred on the lineages that separate the sequences being compared.

proportion of sites that we observe to differ (p_d), such that $p_d = d/n$ as we have talked about previously, where d is the number of sites observed to differ and n is the total number of sites under consideration. When p_d is very small, then it will accurately estimate how many substitutions occurred without any multiple hits correction. But p_d will poorly estimate K as the value of p_d goes up (Figure 5.11). So, solving $p_d = 1 - e^{-K}$ for the expected number of substitutions (K) gives us a way to estimate the true number of mutational changes between the orthologous sequences that accounts for the possibility of multiple mutational hits at the same positions in the sequence:

$$K = -\ln\left(1 - p_d\right) \tag{5.9}$$

A drawback to this "Poisson correction" to address the multiple hits problem is that it does not account for the 4-letter DNA alphabet. To connect better to the molecular basis

Jukes-Cantor 1-parameter substitution matrix

		Mutation to:			
		A	G	C	T
Mutation from:	A	1−3*a*	*a*	*a*	*a*
	G	*a*	1−3*a*	*a*	*a*
	C	*a*	*a*	1−3*a*	*a*
	T	*a*	*a*	*a*	1−3*a*

Kimura 2-parameter substitution matrix

		Mutation to:			
		A	G	C	T
Mutation from:	A	1−*a*−2*b*	*a*	*b*	*b*
	G	*a*	1−*a*−2*b*	*b*	*b*
	C	*b*	*b*	1−*a*−2*b*	*a*
	T	*b*	*b*	*a*	1−*a*−2*b*

Figure 5.12 DNA substitution rate matrices.
The Jukes-Cantor (JC) substitution matrix is simple, with just a single rate among the four possible nucleotides at a given position in a DNA sequence. The Kimura 2-parameter (K2P) substitution matrix incorporates a second parameter to allow separate substitution rates for transitions (*a*) and transversions (*b*). In practice, the ratio of *a:b* is usually close to a value of 2. The rates indicated along the diagonal represent the probability of no substitution (1 − 3*a* for JC; 1 − *a* − 2*b* for K2P).

of sequence evolution, Thomas Jukes and Charles Cantor derived a more appropriate, but still simple, multiple hits correction: each of the four nucleotides has an equal probability of changing to any of the other three nucleotides. Using this approach, we would estimate the true number of substitutions that have accrued as:

$$K = -\frac{3}{4} \ln \left(1 - \frac{4}{3} p_d \right) \tag{5.10}$$

This "JC correction" assumes that all possible nucleotide bases are equally abundant in the sequence and that all possible nucleotide changes are equally likely. Stated more formally in terms of the substitution matrix among nucleotides, there is a single substitution rate parameter (Figure 5.12; see also section 5.3.3). Mutation between different pairs of nucleotides, however, typically occurs at *different* rates. For example, transitions are more prevalent than transversions, which must also be accounted for if we want to make an even more realistic estimate of the substitution rate. More complicated substitution models, like the Kimura 2-parameter model that incorporates different transition and transversion substitution rates, allow even more accurate multiple-hits correction of sequence divergence, although they require that you can provide an appropriate value for the transition: transversion ratio (Figures 5.11 and 5.12). Another popular and sophisticated model of molecular evolution is the 5-parameter HKY model devised in 1985 by Masami Hasegawa, Hirohisa Kishino, and Taka-aki Yano.

Remember also that, for coding sequences, we would want to estimate *K* separately for each particular class of sites (i.e. replacement or synonymous sites, K_A or K_S, respectively), whether we were using a raw p-distance, a Poisson correction, a JC correction, or a Kimura 2-parameter correction for multiple hits. One powerful extension to these ideas is to model the codons of a coding gene sequence more explicitly to derive estimates of substitution rates for non-synonymous sites and synonymous sites. Codon-based estimates of per-site divergence are most commonly referred to as d_N and d_S (analogous to K_A and K_S; and with the ratio $\omega = d_N/d_S$), which can be calculated for each branch on a phylogeny of an arbitrarily large number of species. For most purposes, you can think of d_N and K_A as slightly different ways of calculating the same thing (similarly for d_S and K_S). A detailed study of these issues is beyond our scope here, but is important for studying evolution in a broader phylogenetic context (see Chapter 8). Remember that all of these calculations of divergence presume that the amount of polymorphism within populations is sufficiently small relative to fixed differences that we may safely ignore the polymorphisms in drawing inferences about selection from the values of divergence.

5.3.3 *Phylogeny: topology, discordance, and molecular clocks*

So far, we have emphasized gene trees that display the history of some particular gene, but it is relationships *between* species that provide the most familiar representation of evolutionary tree diagrams. After all, it was species-tree relationships that Darwin printed as the only diagram in his famous book *On the Origin of Species* (Box 5.1). In a **species tree**, or **phylogeny**, external tip nodes represent reproductively isolated species and internal nodes represent the common ancestor species that no longer exist. The **topology** of a **phylogenetic tree** refers to the relationships among the nodes, the *pattern* of branching. Identical tree topologies can be visualized in a variety of ways that might superficially look different; two trees with branching patterns that cannot be made identical by simple rotations of the branches around the internal nodes have different topologies (Figure 5.13). Gene trees follow similar logic regarding their topology, but reflect the branching history at a single locus, which can differ from the overall topology of the species tree (Figure 5.14). The **root** of a phylogenetic tree is defined by including an **outgroup** in the analysis, which is a species that is closely related to the set of species of primary interest (**ingroup** taxa), but that is known from other information to share a more distant common ancestor than observed for any of the ingroup species (Figure 5.2).

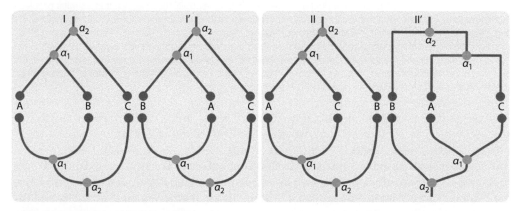

Figure 5.13 Phylogenetic topology.
Trees I and I′ have the same topology, with species A and B sharing a more recent common ancestor than either does with C (rotation of the branches around node a_1 will switch between the visualizations). Think of how the parts of a mobile can rotate while dangling over a baby's crib. Similarly, trees II and II′ share the same topology, but it is different from that of I and I′ because species C shares a more recent common ancestor with A in trees II and II′. Note the distinct orientations (up vs. down) and ways of drawing the lines connecting species to their common ancestors are equivalent and do not alter the topological pattern.

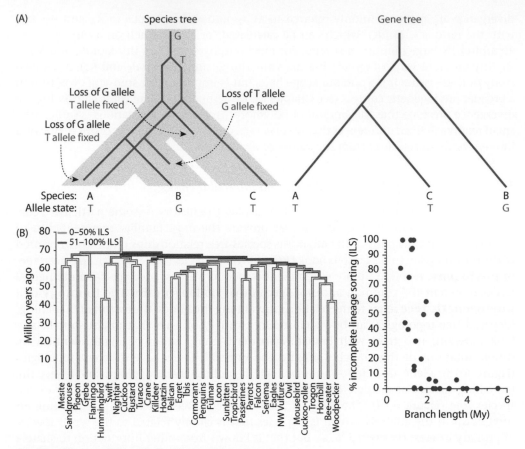

Figure 5.14 Ancestral polymorphism and gene trees.
(A) The incomplete lineage sorting (ILS) of ancestral polymorphisms during speciation events can create discord between the species tree (left) and individual gene trees (right). Gene duplication can result in a similar problem in phylogenetic inference due to the comparison of paralogous rather than orthologous loci. (B) Incomplete lineage sorting was especially important in the adaptive radiation of birds, with most gene trees actually failing to reflect the history of species splitting for those very short branches of the phylogeny (redrawn from Suh et al. 2015). Birds experienced a rapid radiation of species about 60 million years ago, leading to many splitting events in short succession. These short branches in the phylogeny have left an imprint in the genomes of present-day species that indicate incomplete lineage sorting in their history.

Phylogenetic trees generally do not explicitly show any genetic variation within a species, which is something that we should think about a bit more. When might polymorphism within a species matter? Gene trees and species trees will have the same topology for a given locus, provided (1) that homoplasy is absent (i.e. no multiple hits problems), (2) that sufficient change has accrued to estimate the trees with confidence (i.e. no problems of statistical power), (3) there are no methodological artifacts, and (4) that ancestral polymorphism has not been retained. However, ancestral polymorphism in a common ancestor can lead to a gene tree that does not match the history of speciation, due to loss of different allele copies in different lineages (Figure 5.14).

This problem of mismatch between gene trees and species trees, also referred to as **incomplete lineage sorting**, occurs most commonly if the **daughter lineages** speciated recently and if the common ancestor had high polymorphism, for example, because of a large population size in the ancestral species. When the time between species splitting events is less than or similar to the expected coalescence time, then this complication of ancestral polymorphism leading to incomplete lineage sorting will be most severe, and therefore most likely to cause **gene tree–species tree discordance** (Box 5.4). Or, as Wayne Maddison (1997) summarized the issue, "Perhaps it is misleading to view some gene trees as agreeing and other gene trees as disagreeing with the species tree; rather, all of the gene trees are part of the species tree, which can be visualized like a fuzzy statistical distribution, a cloud of gene histories." In any case, it is best to use multiple genes to infer species relationships, just as it is best to use multiple loci to infer population genetic parameters.

The long history of research that aims to work out accurate inference of species trees has a statistically rich literature with a wide diversity of methods. Molecular **distance methods** use similarity of sequence as a proxy for evolutionary distance, involving metrics akin to K_A and K_S. Another common method of estimating phylogenetic relationships uses **parsimony**, which relies on determining **shared derived characters**, not on absolute similarity, and is typically used for morphological traits as the **characters** used in building phylogenies. The principle of parsimony is the idea that the simplest explanation of the pattern in the data is the most plausible.

Unfortunately, the parsimony principle can fail for molecular data, as we actually expect multiple mutational changes to accrue over time; in other words, a "less simple" explanation for the data makes the most biological sense. Methods more commonly applied to DNA sequence data involve sophisticated statistical approaches including **maximum likelihood** and **Bayesian inference** that stipulate explicit mutational models for nucleotides. Maximum likelihood methods differ from parsimony in that they compare different phylogenetic tree topologies with a mutational model in mind to identify the tree that is most likely to give rise to the observed data, rather than starting with the data and inferring the tree or trees that minimize the number of changes needed to explain the data.

Box 5.4 INCOMPLETE LINEAGE SORTING AND GENE TREE-SPECIES TREE DISCORDANCE

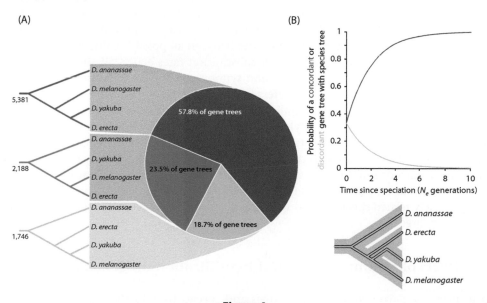

(A)   (B)

Figure 4

(Continued)

Box 5.4 CONTINUED

When gene trees match the splitting relationships of the species that the genes come from (the species tree), then this concordance implies reciprocal monophyly of the gene tree as in Figure 5.7. Approximately one-quarter of the gene trees constructed for human, chimpanzee, and gorilla orthologs, however, differ from the species tree (Ebersberger et al. 2007). Figure 4 (A) In an analysis of 9,315 gene trees for four *Drosophila* species, only 5,381 of them (57.8%) matched the species tree (Pollard et al. 2006). (B) The complication of gene tree–species tree discordance is greatest when the time between speciation events is short relative to the effective population size. The red gene tree conforms to the known species relationships, with yellow and orange colors indicating alternative gene tree topologies that reflect incomplete lineage sorting.

More recently, researchers have applied coalescent theory to multi-locus sequence datasets to infer species trees from a distribution of gene trees, termed **multi-species coalescent** methods. The multi-species coalescent aims to better account for differences among genes in the topologies they produce, termed **gene tree heterogeneity**. In particular, they aim to incorporate the influence that ancestral polymorphism plays in creating incomplete lineage sorting that can induce spurious inferences about species relationships. For our purposes, we shall not delve into any details of phylogenetic inference; however, many resources are available to explore this fascinating topic in greater depth.

To model the sequence substitution process for phylogenetic analysis, we can define a specific transition matrix or **substitution matrix**. It aims to provide a combined summary of mutation origin through to fixation for DNA sequence change between species. A substitution matrix is a table, which also can be written with equations, that gives probabilities of substitution between each possible pair of nucleotides. The matrix may or may not distinguish between different positions within codons, depending on the substitution model. A two-parameter substitution matrix like the Kimura 2-parameter model defines transition and transversion nucleotide changes as occurring at different rates (Figure 5.12). More complicated models commonly employed in phylogenetic substitution matrices can define different rates between all possible substitution pairs, or instead presume a gamma distribution of substitution rates across all sites rather than separating out distinct site classes (e.g. non-synonymous versus synonymous). In this context of a phylogeny, substitution refers implicitly to fixed mutations along a lineage in the history of a species and generally makes no explicit model of selection and drift in their fixation. Most standard phylogenetic analysis treats evolution in this way, with the logic that divergence times between species are much longer than coalescent times within species, and so intraspecies diversity can be safely ignored. In many instances, this can be a satisfactory simplifying assumption. But as discussed above, many factors can undermine this assumption because of ancestral polymorphism and incomplete lineage sorting, closely related species, rapid species radiations with short internal branches, and species with large N_e.

5.4 Effects of recombination and hybridization on genealogies

Up to now, we have considered gene trees and species trees in the absence of recombination. An absence of recombination presumes that the variant sites that make up a

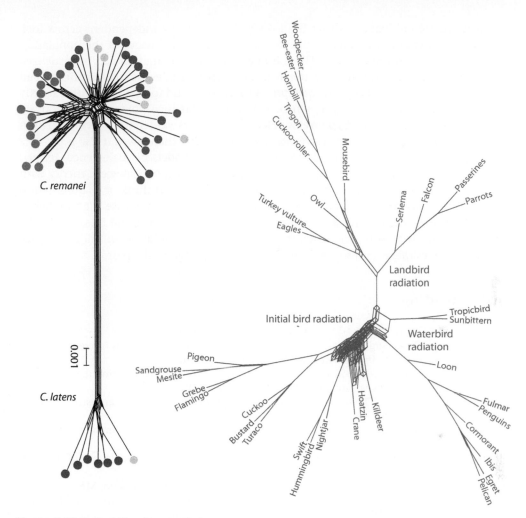

Figure 5.15 Reticulation in genealogy.
Despite commonly depicting gene trees with bifurcating lineages, when long stretches of DNA sequence are considered, recombination can lead to *more* common ancestors rather than fewer for the sequence as a whole. Recombination among haplotypes can be represented as the reticulation of lineages in a genealogical network when visualizing genetic variation within a species like the nematode *C. remanei* (redrawn from Dey et al. 2012). Reticulation also can occur in species phylogenies constructed from multiple loci because incomplete lineage sorting can cause different loci to produce different gene trees, as seen for birds (redrawn from Suh et al. 2015). Suh, Smeds, and Ellegren summarized nicely the implications for birds of recombination and incomplete lineage sorting on genealogies: "a fully bifurcating, universal species tree is an oversimplification of the underlying complexity of speciation."

given haplotype are sufficiently close together that they are perfectly linked. Obviously this is a major simplification of many real-world situations of DNA sequence evolution. In particular, recombination within populations acts to break up linkage groups and, also, can introduce distantly related alleles into a population if hybridization occurs between partially reproductively isolated species. We will go into more detail about recombination in molecular population genetics in Chapter 6, but here I want to touch briefly on the implications of recombination for genealogies, in particular.

A virtue of recombination is that it allows gene trees to be independent across loci, and thus to provide independent replicates of the coalescent history of the population that gets recorded in the genome. These influences of recombination can be visualized with trees by relaxing the unidirectional bifurcation of nodes, so that branches can **reticulate** (Figure 5.15). Typically, we represent gene trees and species trees with a strictly bifurcating topology of branching, whereas diagrams that incorporate reticulation are generally referred to as **haplotype networks**. Looking backward in time, recombination causes a reticulating haplotype to have *more* ancestors rather than fewer. It also is possible to incorporate the complication of recombination into models of coalescence, termed **ancestral recombination graphs**. While a full description of coalescent theory with recombination is beyond our scope here, conceptually, one can think of recombination and coalescence as competing processes: recombination acts to increase the number of ancestral lineages, coalescence acts to decrease the number of ancestral lineages. In the end, coalescence will "win" and reach the MRCA, but recombination will slow the pace at which lineages merge and so result in an MRCA longer ago in the past.

Further reading

Coyne, J. A. and Orr, H. A. (2004). *Speciation*. Sinauer: Sunderland, MA.

Degnan, J. H. and Rosenberg, N. A. (2009). Gene tree discordance, phylogenetic inference and the multispecies coalescent. *Trends in Ecology & Evolution* 24, 332–40.

Dey, A., Jeon, Y., Wang, G.-X. and Cutter, A. D. (2012). Global population genetic structure of *Caenorhabditis remanei* reveals incipient speciation. *Genetics* 191, 1257–69.

Ebersberger, I., Galgoczy, P., Taudien, S. et al. (2007). Mapping human genetic ancestry. *Molecular Biology and Evolution* 24, 2266–76.

Edwards, S. V. (2009). Is a new and general theory of molecular systematics emerging? *Evolution* 63, 1–19.

Gillespie, J. H. and Langley, C. H. (1979). Are evolutionary rates really variable? *Journal of Molecular Evolution* 13, 27–34.

Graur, D. (2016). *Molecular and Genome Evolution*. Sinauer Associates: Sunderland, MA.

Hasegawa, M., Kishino, H. and Yano, T. (1985). Dating of the human-ape splitting by a molecular clock of mitochondrial DNA. *Journal of Molecular Evolution* 22, 160–74.

Maddison, W. P. (1997). Gene trees in species trees. *Systematic Biology* 46, 523–36.

Pollard, D. A., Iyer, V. N., Moses, A. M. and Eisen, M. B. (2006). Widespread discordance of gene trees with species tree in *Drosophila*: evidence for incomplete lineage sorting. *PLoS Genetics* 2, e173.

Song, S., Liu, L., Edwards, S. V. and Wu, S. (2012). Resolving conflict in eutherian mammal phylogeny using phylogenomics and the multispecies coalescent model. *Proceedings of the National Academy of Sciences* 109, 14942–7.

Suh, A., Smeds, L. and Ellegren, H. (2015). The dynamics of incomplete lineage sorting across the ancient adaptive radiation of neoavian birds. *PLoS Biology* 13, e1002224.

Wakeley, J. (2009). *Coalescent Theory: An Introduction*. Roberts and Company Publishers: Greenwood Village, CO.

Wakeley, J. and Takahashi, T. (2003). Gene genealogies when the sample size exceeds the effective size of the population. *Molecular Biology and Evolution* 20, 208–13.

Welch, J. J., Bininda-Emonds, O. R. P. and Bromham, L. (2008). Correlates of substitution rate variation in mammalian protein-coding sequences. *BMC Evolutionary Biology* 8, 53.

CHAPTER 6

Recombination and linkage disequilibrium in evolutionary signatures

Up until now, we have focused primarily on individual loci or loci assumed to be spaced sufficiently far apart in the genome that we can consider them to be independent of one another. Genetic independence means that the loci are **unlinked** and have **free recombination** between them. Unlinked pairs of loci, such as loci on different chromosomes or on opposite ends of the same chromosome, pass along their alleles irrespective of which alleles get passed into gametes at the other locus. This idea goes back to Gregor Mendel's law of **independent assortment**. But, as with many issues in science, things are often more nuanced in reality than in the simplest scenarios. The same is true for thinking about multiple loci in molecular population genetics. Therefore, we should think more deeply and quantitatively about what it means for loci to have partial recombination between them and about what are the consequences for the population as a whole.

At the level of individuals creating recombinant gametes in meiosis, we can quantify the independence between loci on a continuous scale, with the amount of independence controlled by recombination: the crossover **recombination fraction (c)**. The recombination fraction tells us how likely it is that an individual will create gametes that split up locus combinations that came from a single one of their parents. For example, a value of c that equals one-half between a pair of loci means that, on average, an individual will produce half of its gametes to contain their paternal allele for one locus and their maternal allele for the *other* locus (Figure 6.1). This value of $c = 0.5$ is the maximum possible value that c can hold and corresponds to free recombination. The other half of the individual's gametes would have the same allele combinations at the two loci as that individual itself had inherited from one or the other of its parents.

You may notice that I have been using the word recombination to mean, more specifically, **crossover recombination**. Crossover recombination involves the reciprocal exchange of genetic material between homologous chromosomes (Figure 6.1). We will talk about **non-crossover recombination** in section 6.4, sometimes referred to as **gene conversion**. But, for simplicity, I will continue to use the term recombination to refer to crossovers because the effects of crossover are usually the kind of recombination that we most care about.

At the other extreme from independence due to recombination, loci can have **perfect linkage**. Also known as **complete linkage**, perfectly linked loci have a value of $c = 0$

A Primer of Molecular Population Genetics. Asher D. Cutter, Oxford University Press (2019).

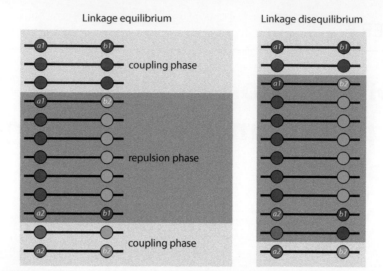

Figure 6.1 Linkage equilibrium and linkage disequilibrium.

Two loci (*a* and *b*) will be in linkage equilibrium when each haplotype combination of alleles occurs at the frequency of the product of the individual allele frequencies at each locus. In this diagram, the a_1 allele (dark red) occurs at frequency $f_{a1} = 75\%$ and the b_1 allele (dark blue) occurs at $f_{b1} = 33\%$. Only the sample of haplotypes on the left occur in the frequencies expected at linkage equilibrium ($f_{a1b1}, f_{a1b2}, f_{a2b1}, f_{a2b2}$; e.g. $f_{a1b1} = 0.75 * 0.33 = 0.25$); there is an overabundance of repulsion-phase haplotypes (f_{a1b2}, f_{a2b1}) in the sample shown on the right. By convention, coupling phase combinations of alleles involve ancestral-ancestral and derived-derived pairs (or common-common and rare-rare pairs; section 6.3).

such that no recombination occurs between the sites. These perfectly linked loci will have identical genealogies. DNA on mitochondrial and Y chromosomes are non-recombining, as is true also for any chromosome of an asexually reproducing organism.

Remember that a locus does not have to be a gene. We can consider each nucleotide site to be a distinct locus—this is the essence of the infinite sites mutation model. Many nucleotide sites make up a given gene. So, we should also think about the potential for recombination to occur between nucleotide sites, whether within or between genes. Importantly, recombination can only be detected between genetically variable loci and its consequences are only relevant to genetically variable loci, so it is the proximity of polymorphic sites to each other that we need to think about. This point is especially salient for distinct polymorphic sites within a gene. Because sites within a gene are in close physical proximity, it will usually be inappropriate to assume that they are genetically unlinked. Instead, closely spaced polymorphic sites will be partially linked. Partially linked loci have intermediate values of the recombination fraction ($0 < c < 0.5$).

This, so far, has been a quick summary of crossover recombination from a classic "transmission genetics" view of inheritance. It operates during meiosis within single individuals to reveal its effects on their progeny. What we next need to do is to expand up to the effects of recombination for populations of many individuals with many polymorphic loci: a population genetic view of inheritance. For recombination to mix up different combinations of alleles for a given pair of loci in a *population*, meiosis has to wait for the right combination of haplotypes to join together in a zygote, rather than relying on an experimenter to put together the relevant cross.

At the level of the population, what does it mean for alleles to be independent across loci? What might lead to predictable patterns of non-independence of allele combinations across loci? What consequences would it have for understanding evolution? We saw a prelude to how recombination can influence genetic variation in section 5.4, in terms of creating reticulation in genealogies. As we will explore in more detail in Chapter 8, such partial linkage can also compromise tests of neutrality that depend on an assumption of historical independence among loci. Our job now is to think more generally about what the impact on patterns of nucleotide polymorphism will be for partially linked variant sites under neutral and non-neutral scenarios of evolution.

6.1 Linkage disequilibrium: what is it?

If reproduction in nature puts together random samples of gametes from a wild population, rather than a controlled cross of specific genotypes, what should we expect to find in nature? Can we make predictions about the combinations of variants, the haplotypes, in the population? What role does recombination have in shaping these allelic combinations across polymorphic loci when we scale up from examining just a single individual to an entire population?

When polymorphic loci are close together, the joint combination of alleles for the two loci might be arranged in non-random associations in a population. Nucleotide polymorphisms within the close confines of a gene are especially likely to have haplotype arrangements of multiple variants that are non-random combinations of those variants. Some haplotype combinations of variants can be more common than others, even if the genotype frequencies of each locus individually are in Hardy-Weinberg equilibrium (see Box 3.2). We call this non-random association of variants of different polymorphic sites **linkage disequilibrium (LD)**, or **gametic phase disequilibrium**.

Given what a mouthful this phrase makes, scientists often stick with the abbreviation "LD." If the phrase "non-random association" also sounds too abstruse, think of it this way: when loci have LD, it means that we can predict the variant at one site if we know the variant at another site. Sites that are physically close together on the same string of DNA are less likely to have recombination between them, thus making those sites more likely to be in LD. But the magnitude of LD depends on the amount of recombination, only indirectly relating to the nucleotide distance between loci by virtue of its association with recombination. It is also possible for loci on different chromosomes to be in LD, though this is less commonly found in practice. The crucial measure of distance for LD is the genetic distance in units of cM (see section 6.3), defined by how much recombination breaks combinations up. The predictability of which variants co-occur together turns out to have quite a lot of important implications for understanding evolution at the molecular level.

Let's work through a simple example to make these ideas about random and non-random allele associations between loci more concrete. Consider two loci, A and B, each with two alleles: the alleles are named *a1* and *a2*, *b1* and *b2*, respectively. These two loci would be in **linkage equilibrium** if each potential haplotype occurs in the population at the frequency expected by random association of alleles (Figure 6.1). For example, if the frequency of *a1* is $f_{a1} = 0.9$ (so $f_{a2} = 1 - 0.9 = 0.1$) and of *b1* is $f_{b1} = 0.2$ (so $f_{b2} = 1 - 0.2 = 0.8$), then we would expect the haplotype *a1b1* to occur at frequency $f_{a1b1} = 0.9 * 0.2 = 0.18$. These calculations use the **product rule**, which lets events with independent probabilities of occurring be multiplied together to determine the joint probability that both events would occur. Correspondingly, the remaining possible haplotypes would occur at frequencies $f_{a1b2} = 0.9*0.8 = 0.72$, $f_{a2b1} = 0.1*0.2 = 0.02$, $f_{a2b2} = 0.1*0.8 = 0.08$ if alleles of each locus get packed into gametes independently and then fuse to combine into zygotes through random mating. If there were, in fact, different haplotype frequencies than these, then the A and B loci would be in LD. For example, if $a2$ always occurred in combination with $b2$, such that the $a2b1$ haplotype was never found in the population (i.e. $f_{a2b1} = 0$), then this would represent one of the many possible ways in which LD could manifest.

As another example of LD, think of $a2$ as a new mutation at the A locus. A new mutation must inevitably occur in just a single genomic context, so imagine that the $a2$ mutation arose on a haplotype that also contained the $b2$ allele at the B locus. Initially, $a2$ would inevitably be associated only with $b2$, causing LD between the A and B loci (Figure 6.3).

LD describes the non-independence of variants, which should be more severe for polymorphic sites that are close together. To see how we might be able to incorporate this non-independence in real-world datasets, we will need (1) to understand the factors that create LD in the first place, (2) to be able to relate LD to ideas that we already know, and (3) a way to quantify LD. The remainder of this chapter will work through these three issues.

6.2 The causes of linkage disequilibrium

LD usually is a transient phenomenon between any pair of loci. It is caused by perturbations from a population's equilibrium state, and eventually it gets eliminated by recombination. When we see LD, it tells us that alleles of different loci have a shared co-ancestry. This partial correlation of their genealogical histories arises by virtue of the loci being genetically linked, either from their close physical association on the same strand of DNA or by being statistically associated due to some extrinsic force (like selection or inbreeding). Now that we have built up a sense of what LD is, we should think about what factors might cause LD to occur in the first place (Figure 6.2).

In working through the causes of LD, let's start with the simpler cases and then move to the more exciting ones. First, new mutation creates LD (Figure 6.2). We already introduced this basic idea in section 6.1: at their origin, new mutations inherently arise in LD with other variants because the new mutation arises on just a single **genetic background**. Until recombination separates that new mutation from all the other variants on the haplotype that it happened to be linked to when it arose, it will be in LD with all those linked variants.

Genetic drift can create LD. Just like the random sampling of gametes causes fluctuations in the frequency of alternative alleles, this process of genetic drift also can cause fluctuations in particular *combinations* of alleles. We can see this influence of genetic drift explicitly in Equation 6.7: the LD between a pair of loci is inversely related to population size, so a given pair of loci will tend to have more LD in smaller populations.

The contribution of drift to LD means that population structure also creates LD in a species. As we know, genetic drift operates independently on each subpopulation and will lead to different allele frequencies in the different subpopulations for any given locus (see Box 3.5). This genetic differentiation will also make variants at different loci associate non-randomly with one another within a subpopulation relative to the entire species. This LD that differentiates subpopulations from one another can confuse our interpretations

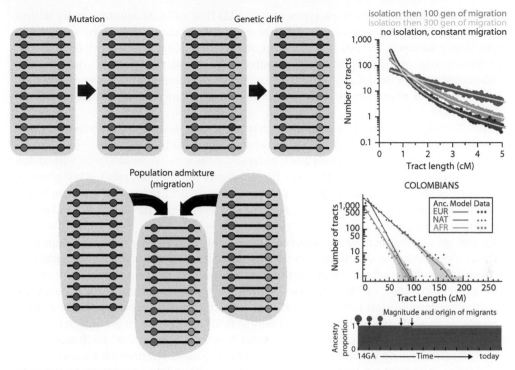

Figure 6.2 Causes of LD.
Both the input of new mutations and the chance fluctuations in allele frequencies by genetic drift can produce linkage disequilibrium. Migration between differentiated populations, in the extreme being complete admixture, also creates non-random associations in allelic states among loci. LD also can be maintained by factors that reduce recombination, like inversions and inbreeding, when combined with mutation, drift, migration, and selection. The size distribution of LD tract lengths from the source populations reflects the time since the admixture events (top right), as LD decays with recombination each generation (redrawn from Pool and Nielsen 2009). The fifteenth-century admixture in Central and South America of human populations that had partial genetic differentiation (Europeans, Africans, Native Americans) has led to characteristic patterns of LD in the genomes of modern people (Moreno-Estrada et al. 2013). Photo published under the CC-BY license. Taken from Moreno-Estrada et al. (2013). Reconstructing the population genetic history of the Caribbean. *PLoS Genetics* 9, e1003925. https://doi.org/10.1371/journal.pgen.1003925

if we pool samples from multiple subpopulations without realizing it. Specifically, a failure to recognize the presence of multiple demes would make us conclude that there is higher than expected LD (Figure 6.2). This accident of not having properly accounted for population subdivision in the analysis of LD is called the two-locus Wahlund effect, echoing the single-locus Wahlund effect on heterozygosity (see sections 3.6.1 and 7.3.1).

While spurious LD can be a headache for analysis if you don't realize that you are analyzing data from multiple subpopulations, population structure can also give biologically interesting signals of LD. **Admixture** in nature between previously isolated subpopulations will induce elevated LD (Figure 6.2), similarly to the Wahlund effect scenario. However, admixture would represent a true biological signature rather than an error on the part of the experimenter. This higher-than-expected LD occurs because allelic states become associated within subpopulations and it takes time for mating and recombination to break up the LD even after subpopulations merge together.

But we can also think of admixture that is less extreme than the complete fusion of previously isolated subpopulations into one: gene flow through normal migration.

Migrants will introduce new haplotypes, creating LD that will subsequently decay within their new destination population as the genomes of their admixed progeny recombine. Interestingly, however, it is possible to use the distribution of haplotype block lengths that are characteristic of the migrant source population to quantify the timing and influence of admixture events or ongoing gene flow. In particular, when a population experiences a recent influx of migration, it will produce an overabundance of long haplotype block lengths (Figure 6.2).

Recombination reduces LD, so anything that suppresses recombination will also make LD greater than would be expected otherwise. Several things might suppress recombination. First, chromosomal inversions generally eliminate recombination within the inverted region in individuals that are heterozygous for the inversion. This physical disruption of recombination results in all the variants within the inverted region being linked to one another, which will mean that the loci overall will have non-zero LD in the population (Figure 6.2).

Second, genomic structures like **centromeres** tend to have little recombination near to them. Consequently, we should typically see stronger LD for loci near centromeres.

Third, inbreeding induces lower effective recombination rates as a byproduct of creating an overabundance of homozygous genotypes. We say that inbreeding reduces the "effective" recombination rate because it does not actually alter the biophysical process of recombination during meiosis, yet that biophysical recombination is made genetically impotent. Remember that recombination is not "genetically effective" for homozygous loci: when there are no polymorphisms, then there are no differences for recombination to rearrange. Although inbreeding, in itself, does not alter allele frequencies in the population, by producing an excess of homozygous genotypes, inbreeding sets the stage for genetic drift or selection to create non-random genotype combinations in the population. Self-fertilization represents the most extreme form of inbreeding, and occurs commonly in many plant species and in some animals and other taxa (Figure 6.5).

Finally, natural selection can cause LD (Figure 6.2). This can happen if the fitness effects of certain combinations of alleles produce disproportionately deleterious or advantageous effects. That is, natural selection can create LD when multiple loci contribute to fitness differences in the population such that they interact in a **non-additive** manner resulting in **epistatic selection**. However, LD induced by selection in this way will only affect those specific selected loci and linked variants, not the entire genome. In the case of advantageous allele combinations, in some circumstances, selection can favor reduced rates of recombination between the loci so as to maintain haplotype identity of the "**coadapted gene complex**" to create a "**supergene**." In the extreme, an inversion mutation could capture such combinations of beneficial loci and preserve their non-random allele associations.

Even if two loci don't interact directly in an epistatic way to influence fitness, selection on one locus nevertheless can influence the response to selection for another locus. This situation happens when the beneficial alleles for the two loci occur on different haplotypes: unless recombination combines them onto the same haplotype, the distinct haplotypes will compete with one another and, in so doing, slow down the fixation process for both of them. You can also think of this scenario as involving a beneficial allele at one locus being linked to a deleterious allele at another locus. This **selective interference** is called the **Hill-Robertson effect**, named after William Hill and Alan Robertson who scrutinized its implications in 1966. Such selective interference creates LD. In particular, as the alternative haplotypes increase in frequency as they are favored by selection, they will each be linked to distinct combinations of variants that will be absent from the opposing genetic background.

Regardless of the cause of LD, remember that LD involves polymorphisms and so is a transient feature of populations. We expect the LD between a given set of loci to change over time and to differ among populations. Once selection, or genetic drift, yields fixation or loss, there will be no more LD for that locus. While neutral variants will be caught in LD with a selected variant during a selective sweep as a result of genetic hitchhiking (see section 7.1.1), it turns out that this linkage to a selected variant has no overall effect on the rate at which neutral variants get fixed. The reason for this indifference of neutral substitution rates to hitchhiking is that the increased probability of fixation of the linked neutral variant is exactly counterbalanced by the decreased probability of fixation of all the other neutral variants that are unlinked to the selected locus; the converse is true for linkage of neutral variants to detrimental mutations. We will talk further in Chapter 7 about stereotypical patterns in genomes caused by the interaction between linkage and different forms of selection, due to the phenomena of genetic hitchhiking and background selection.

LD also has important practical implications. On the academic side, we can exploit LD to quantify population structure or to identify regions of the genome that likely experienced recent positive selection. This idea has been instrumental in understanding the demographic history of human migration across the globe, in particular. It is this kind of application that makes it possible for you to send a tube of spit to a company for DNA extraction and sequencing to receive details about your own genomic history. In medicine, one way that the concept of LD is applied in human population genetics is to use **association mapping** to identify variants that correlate with a disease state in a contrast of **case and control samples** of individuals. Key to analysis of such data is identifying variable molecular markers that are linked to the initially unknown causal disease alleles, so this approach depends on LD between the molecular markers and causal disease alleles. Perhaps unsurprisingly, then, this approach also is referred to as **linkage disequilibrium mapping**. **Genome-wide association studies (GWAS)** have exploded in the last decade as a means of identifying genetic variants across the genome that contribute to heritable human disease. A social health goal of such approaches is to, eventually, adopt "personalized medicine" in which medical diagnosis, drug delivery, and treatment could be attuned to the particular genotypic makeup of an individual person. Commercial agriculture also applies GWAS to plant crops and livestock to improve food production.

6.3 Measuring linkage disequilibrium

Sometimes we just want to answer a simple question about linkage and recombination: do the polymorphisms in our data show evidence of *any* recombination among them at all? One approach to determining the presence of recombination among polymorphic sites is to conduct a 4-gamete test. This test gets its name from the fact that there are a maximum of four possible haplotype configurations that could be packaged into a haploid gamete when we look at two loci that each have two alleles in the population. The logic behind the 4-gamete test is that, if mutation follows the infinite sites model, then recombination will be the only way that some configurations of variants could co-occur on the same haplotype (Figure 6.1). This combinatorial logic gives us a simple way to know that there must have been at least one recombination event in the history of the loci, provided that the infinite sites mutational assumption has not been violated by multiple hits at one or both loci.

Often, however, we also are interested in quantifying the extent to which recombination has rearranged allele combinations between loci: how much LD is there? We can quantify LD between a pair of loci in several ways, each of which is based on contrasting the observed and expected frequencies of alleles and haplotypes in a population.

The first and most classic metric is the **disequilibrium coefficient, D**. We will use some generic locus abbreviations as we did in the example in section 6.1, with loci A and B, each with two alleles ($a1$ and $a2$; $b1$ and $b2$) that could combine into four possible two-locus haplotypes ($a1b1$; $a1b2$; $a2b1$; $a2b2$). Remember that the infinite sites mutation model, as well as most real-world DNA polymorphisms, also are bi-allelic, so this superficially abstract scenario actually does connect to real data. The disequilibrium coefficient D is defined as the difference between the product of **coupling phase** haplotype frequencies (f_{a1b1}, f_{a2b2}) and **repulsion phase** haplotype frequencies (f_{a1b2}, f_{a2b1}). Which alleles we choose to label "1" or "2" is of course arbitrary, as is the labeling of a particular pair of alleles that make up a haplotype as "coupling" or "repulsion" phase. These clunky terms are simply a relic of classical crossing experiments with dominant and recessive visible genetic markers in peas, as William Bateson and Reginald Punnett worked out the logic of genetic linkage in the early 1900s. For the sake of convention with real data, we would usually define the combination made up of the most common allele at each locus as being in coupling phase; another logical alternative is to designate the combination made up of the ancestral allele at both loci as being in coupling phase. The important thing is to be consistent, and so by convention we will refer to the $a1b1$ and $a2b2$ combinations as coupling phase haplotypes as we work through the logic of LD (Figure 6.1).

Based on these ideas about the combinations of alleles of different loci into haplotypes, we can define the disequilibrium coefficient to summarize the amount that the observed haplotype frequencies differ from what we would expect from random assortment of alleles. Remember from section 6.1 that the expected haplotype frequency for each of the four possible combinations of alleles at the two loci is just the product of their individual allele frequencies. For example, the expected frequency of the $a1b1$ haplotype is $f_{a1} \cdot f_{b1}$. Consequently, we relate D to the expected haplotypes as:

$$
\begin{aligned}
f_{a1b1} &= f_{a1} \cdot f_{b1} + D \\
f_{a1b2} &= f_{a1} \cdot f_{b2} - D \\
f_{a2b1} &= f_{a2} \cdot f_{b1} - D \\
f_{a2b2} &= f_{a2} \cdot f_{b2} + D.
\end{aligned}
\tag{6.1}
$$

So, D tells us the amount that each two-locus haplotype differs from what random assortment and free recombination would give the population (Figure 6.3).

By rearranging these relationships in Equations 6.1 to show D as a function of the haplotype frequencies, we get the following mathematical expression for the disequilibrium coefficient:

$$
D = f_{a1b1} \cdot f_{a2b2} - f_{a1b2} \cdot f_{a2b1}.
\tag{6.2}
$$

The disequilibrium coefficient summarizes the degree to which alleles at the A and B loci associate non-randomly. When $D = 0$, the coupling and repulsion phase frequencies cancel out in Equation 6.2 to indicate that the loci are in linkage equilibrium: the alleles for the two loci would *not* have biased haplotype configurations. When $D = 0$, it also means that we would not be able to predict any better than chance which allele occurs at

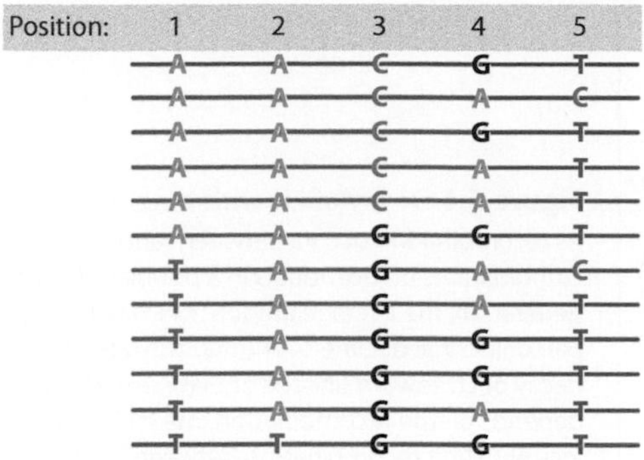

Position:	1	2	3	4	5

Position i	Position j	D	D'	r^2
1	2	0.04	1	0.09
1	3	−0.21	−1	0.71
1	4	0	0	0
1	5	0	0	0
2	3	−0.04	−1	0.07
2	4	−0.04	−1	0.09
2	5	−0.01	−1	0.02
3	4	0.04	0.20	0.03
3	5	0.01	0.14	0.01
4	5	0.08	1	0.20

Figure 6.3 Linkage disequilibrium measurements between sites.
All three standard measures of LD give a value of zero when random associations occur between loci, as for site pairs 1–4 and 1–5 in the diagram, indicating linkage equilibrium. Usually, $|D'|$ is shown instead of D' because the sign of D' is usually arbitrary. Singleton variants, as can arise in a population from new mutation, lead to non-zero LD (e.g. site pair 1–2). Consequently, it is common practice to exclude sites with singleton variants in calculations and analysis of LD among haplotypes.

the B locus if we know what allele occurs at the A locus. For statistics junkies, note that D is the covariance in allelic state between the A and B loci.

Recombination between a pair of loci will act to "destroy linkage disequilibrium" (Figure 6.4). This dramatic phrase means that, when recombination occurs between the A and B loci, the recombination will cause D to approach a value of zero over time. The more often that recombination occurs between the loci, the more rapidly will the LD decay over time (Figure 6.4). We can make this statement quantitative using the recombination fraction c as our measure of recombination between the loci. Specifically, if in generation t the disequilibrium coefficient is D_t, then in the next generation, it will be:

$$D_{t+1} = D_t − c \cdot D_t = (1 − c) \cdot D_t. \tag{6.3}$$

Intuitively, this means that each generation the non-random association between loci will drop by a proportion that depends on the rate of recombination between the two loci. That is, recombination will reduce LD by the amount $c \cdot D_t$.

For example, we know that $c = 0.5$ for loci on different chromosomes, which is equivalent to independent meiotic assortment of the loci with a recombination distance of 50cM. Remember that geneticists measure distance along the "map" of a chromosome either in **physical map units** measured in nucleotide basepairs or in **genetic map units** measured in **centiMorgans** (cM). Plugging this maximal value of $c = 0.5$ for recombination between loci into Equation 6.3, we can see that whatever value of LD was present initially would decline by half each generation. So even though a single generation of random mating will bring the genotype frequencies of each *single* locus to the equilibrium Hardy-Weinberg frequencies (see Box 3.2), the same is not true for the two-locus haplotype frequencies. Even if the two loci are on different chromosomes, one generation of random mating will only decrease the LD between them by half. It takes much more time to achieve linkage equilibrium than Hardy-Weinberg equilibrium. More generally, after T generations, LD initially at D_0 would decline to:

$$D_T = (1 − c)^T \cdot D_0. \tag{6.4}$$

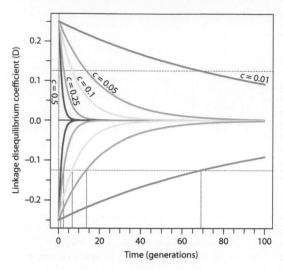

Figure 6.4 Decay of LD with time. As recombination occurs between random combinations of individuals in a population each generation, the average amount of LD between a pair of loci will decline over time. The rate of decay of LD toward linkage equilibrium ($D = 0$) depends on the recombination rate (c) between loci, with free recombination between loci ($c = \frac{1}{2}$) expected to cut LD by half every generation. It will take nearly 70 generations to reduce LD by half of its current value for loci that are separated by a recombination distance of 1 cM ($c = 0.01$). The example in this diagram starts at time zero with the maximum possible LD.

Although the disequilibrium coefficient provides a useful and simple metric, it suffers from the problem that its magnitude and potential range depend on the allele frequencies at the two loci. The maximum possible value of D is 0.25, which you will get when $f_{a1} = 0.5$ and $f_{b1} = 0.5$, which results in the frequencies of both coupling phase haplotypes being equal to 0.5; the minimum of $D = -0.25$ occurs with repulsion phase haplotypes both at 50%. But if $f_{a1} = f_{b1} = 0.3$, then the maximum value that D could take on would be 0.21. Therefore, one cannot make a fair comparison of D values for different pairs of loci that have different allele frequencies, or between populations or species.

To get around this problem, Richard Lewontin devised the **D'** statistic (pronounced "D-prime"). D' is simply D divided by its most extreme possible value, given the allele frequencies of the loci involved:

$$D' = \frac{D}{\min [f_{a1} \cdot f_{b1}, f_{a2} \cdot f_{b2}]} \text{ when } D < 0 \text{ and} \qquad (6.5)$$

$$D' = \frac{D}{\min [f_{a1} \cdot f_{b2}, f_{a2} \cdot f_{b1}]} \text{ when } D < 0.$$

This **normalization** procedure results in D' varying between -1 and $+1$, retaining the sign of D, and thus making D' comparable across different datasets. Often, the sign of D' is unimportant, because we assign the repulsion and coupling states arbitrarily. Consequently, calculation of the absolute value $|D'|$ provides another common summary of LD (Figure 6.3). In all of these permutations of D, the value is usually very close to zero in natural populations of most species *except* when calculated for polymorphisms that are very close together.

In addition to the decay of LD over time, LD also declines as a function of the *distance* between loci (Figure 6.5). Both the passage of time and greater distance between loci give more opportunities for recombination events to accrue between a given pair of variant sites, thus reducing LD. Even though LD between a given pair of loci exists only temporarily in time, the time lag in its decay leads to predictable patterns of LD as a function of the spacing between loci in the genome. We quantify the spacing in genetic map units (cM), also termed **genetic distance** or **recombination distance**, which relate to the recombination fraction (c) that we discussed previously. The recombination

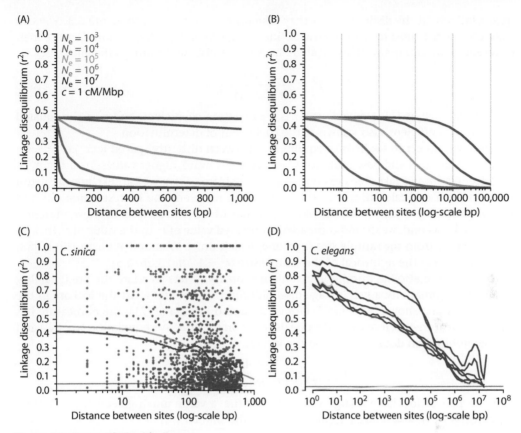

Figure 6.5 Decay of LD with distance.

(A, B) The likelihood of recombination between two loci increases with the distance between those loci. As a result, we expect LD to be lower for pairs of loci that are farther apart than for pairs of loci that are close together. When the effective recombination rate in the population ($4N_ec$) is high, due either to a high likelihood of crossover recombination between a given pair of sites or to a large effective population size, then the decay of LD with distance will be more rapid (showing expected decay in LD from the equation on p. 77 in Hill and Weir (1988) for a sample size of 1,000). To relate c to nucleotide distances, we must use units of cM/Mbp rather than just cM. (C, D) Self-fertilization also reduces the effective rate of recombination in a population by creating excessive homozygosity, leading to higher LD in species with selfing reproduction, as seen for the self-fertilizing nematode *C. elegans* compared to the related species *C. sinica*, which is non-selfing (data redrawn from Thomas et al. 2015 and Wang et al. 2010). For *C. sinica*, the orange line represents the fit of the theoretical LD decay equation and the purple line is a spline fit to the LD between pairs of SNPs; the distinct lines for *C. elegans* each correspond to a different chromosome. Note the huge difference in the distance scale for *C. sinica* versus *C. elegans*.

fraction c is in units of Morgans, a unit of measure named after renowned geneticist Thomas Morgan, so $c = 0.5$ equals 50 cM or 50 "map units" and $c = 0.01$ corresponds to 1 cM. So, how can we relate LD to the distance between loci?

To explore the influence of distance on LD, let's introduce ourselves to another common metric of LD. The $\boldsymbol{r^2}$ ("r-squared") measure of LD conveniently ranges between 0 and 1, because it is the squared correlation coefficient of variant frequencies between a pair of polymorphic loci. When $r^2 = 0$, sites are in linkage equilibrium whereas when $r^2 = 1$, we can predict perfectly the allelic state of one site in a haplotype if we know the state of another (Figure 6.3). You will notice that this way of thinking about LD with r^2 is somewhat different from how we summarize LD with D and D', which you can see quantitatively in the distinct values they take from the same data in Figure 6.3. You can then choose the most relevant metric to link to a given biological question. Despite their differences, we can nevertheless relate r^2 back to the disequilibrium coefficient D as:

$$r^2 = \frac{D^2}{f_{a1} \cdot f_{a2} \cdot f_{b1} \cdot f_{b2}} \tag{6.6}$$

Statistically, just as D is a covariance, r is a correlation coefficient, and r^2 is the squared correlation coefficient, also known as the coefficient of determination.

While we expect to see linkage equilibrium between distantly spaced loci (i.e. $r^2 = 0$), small sample sizes and new mutations can skew r^2 toward higher values. As a result, in practice, we should take steps to counteract these biases. First, we can exclude singleton variants from our calculations, as these often will represent the newest mutations and so by excluding them we can avoid any signal of LD introduced by new mutation (Figure 6.3). Second, we should compare the observed value of r^2 to the value of r^2 that we would expect from the sample size (n) alone, which Bruce Weir and William Hill (1988) showed to equal the reciprocal of the sample size ($r^2 = 1 / n$; Figure 6.5).

The r^2 metric also has another convenient property. It lets us relate the predictability of allele combinations, a.k.a. linkage disequilibrium, to how far apart the loci are in the genome. As first shown by Tomoko Ohta and Motoo Kimura (1971), the r^2 measure has the useful feature that its expected value depends on the effective population size (N_e) and the recombination distance between the loci (c):

$$\hat{r}^2 = \frac{10 + 4N_e c}{22 + 13 \cdot 4N_e c + (4N_e c)^2}. \tag{6.7}$$

When a pair of loci have a large value of $4N_e c$ (i.e. a value much greater than 1), then Equation 6.7 can also be well-approximated by the simpler form of:

$$\hat{r}^2 = \frac{1}{4N_e c}. \tag{6.8}$$

The quantity $4N_e c$ in Equations 6.7 and 6.8 is often referred to as the **population recombination rate, ρ** (pronounced "rho"). You can think of this scaled summary of recombination as the integrated effects of crossover recombination from meiosis within individuals on the overall population, which we can measure in terms of LD. Note the dependence of ρ on N_e, giving it a similar form to the population mutation rate $\theta = 4N_e \mu$. This common feature of population genetic parameters getting scaled by population size is important: it makes explicit the relevance of genetic drift in finite populations. In particular, from the presence of ρ in Equation 6.7, we see that r^2 depends on population size in a way that means we should expect larger populations to have less LD than smaller populations (Figure 6.5), if they are otherwise similar in terms of genome-wide profiles of recombination rate, mutation, selection, and demography. The reason for bigger populations having less LD is simply that there are more chromosome copies overall that can recombine with one another each generation to bring variants at different loci closer to linkage equilibrium, offsetting the tendency of genetic drift to skew the relative abundance of alternative haplotypes. Although not often implemented in practice, Equation 6.7 could allow us to estimate N_e by using LD information, analogous to how we might estimate N_e from θ_W (see Box 3.1). To estimate N_e from r^2, we would need to assume neutral equilibrium and know the genetic distance between loci.

These equations show how LD is inversely related to the genetic distance between pairs of loci: remember that c does double-duty by representing the recombination *rate* and recombination *distance*. The farther apart two loci are, the closer to linkage equilibrium we expect them to be (Figures 6.5 and 6.6). While the recombination distance for loci on different chromosomes is $c = 0.5$, polymorphic sites within a gene will be more likely to have a recombination distance of $c = 0.00005$. Equivalently, the higher the rate of recombination between loci, the more rapidly will LD decay with distance. The amount of LD that we expect to see between loci also will depend inversely on physical distance, which we also could describe with Equation 6.7 by adding a scaling factor, as long as genetic and physical distances are linearly related. While we might be willing to accept the assumption of a linear relationship between genetic and physical distances at small genomic scales, the reasonableness of this assumption will break down at larger genomic scales because some parts of the genome have higher recombination rates than do others. The decay of LD with distance is simply a consequence of recombination having been more likely to break up non-random allele associations between loci that are farther apart.

Can LD occur only between those polymorphic sites that are physically linked by being connected on the same strand of DNA on a single chromosome? The general answer is "no," although it takes special circumstances to create LD between chromosomes, such as strong inbreeding, population structure, or epistatic selection (section 6.2). LD is usually strongest between sites that are in close spatial proximity along a stretch of DNA. Conveniently, we can incorporate inbreeding into our predictions about LD in a simple way. Specifically, we just have to replace the value of c in Equations 6.3, 6.4, and 6.6 with $c \cdot (1 - F)$ where F is Wright's inbreeding coefficient (see Box 3.3); for random mating, we can ignore this generalization of those equations because $F = 0$ in that case. For inbreeding that takes the form of self-fertilization, we can replace F with $S_f/(2 - S_f)$, where S_f is the rate of self-fertilization ranging from 0 to 100%.

Given that a gene or other long stretch of DNA will contain many polymorphic sites, how should you summarize the overall amount of LD across them all? One option at small spatial scales is to simply take the average value of $|D'|$ or r^2 across all pairwise comparisons of polymorphic sites. However, the dependence of LD on distance between sites makes this a poor approach when comparing long segments of chromosomes. A second option is to calculate the distance at which LD decreases by half. A short LD decay distance indicates that the genomic interval being analyzed has a large value of ρ. A third option is to calculate a multi-locus metric of LD. The index of association (I_A) provides one example of a multi-locus LD measure, which Eviatar Nevo and his colleagues came up with to quantify the degree to which individuals that have the same variant at one locus also have matching variants at other loci, relative to what would be expected from linkage equilibrium across loci.

While we can think of recombination in a probabilistic way across the population, crossovers between homologous chromosomes are discrete events that happen at specific locations in the DNA. This means that discrete chunks of a chromosome might have one pattern of LD, with the boundaries defined by recombination events in the history of the chromosome. These chunks of co-inherited DNA with distinctive LD are termed **haplotype blocks** (Figure 6.6). Haplotype blocks are especially pronounced in the human genome, in which recombination tends to recur at the same "hotspot" locations again and again. Variants in strong LD will share the same genealogy. This shared genealogy makes sense: variants that are close together genetically in the same region of the genome will be more likely to be passed along together and not have been separated

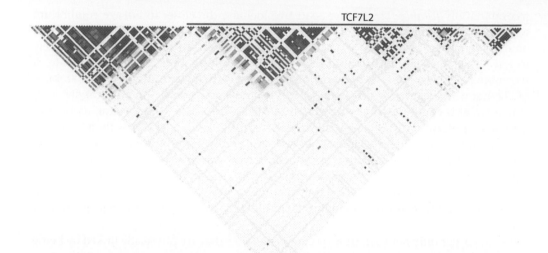

TCF7L2

Figure 6.6 Triangle plot representation of LD.
Another way to visualize LD among variant sites is with a triangle plot. It is half of a pairwise matrix of LD measures showing only the polymorphic sites. Points along the hypotenuse of the triangle represent neighboring polymorphic sites, the point at the vertex opposite the hypotenuse represents the two most distantly spaced sites, and points embedded within the triangle correspond to pairs of sites at intermediate distances from one another. These data show a portion of human chromosome 10 that contains *TCF7L2*, a gene implicated in type 2 diabetes. Darker red points indicate higher values of $|D'|$ for a sample of Caucasian-American individuals, and depict five haplotype blocks of linkage disequilibrium demarked by the regions within the dark red triangles (redrawn from Lehman et al. 2007, courtesy D. Lehman).

by recombination. The distribution of lengths of these haplotype blocks in a genome provides another way to visualize the extent of LD (Figure 6.2).

In all of this talk about measuring LD, we have presumed that we can, in fact, tell which allele at one locus is linked to which other allele at another locus. In other words, it presumes we know the **phase** of the haplotypes. However, it can be technically challenging to know the phase of different loci from DNA sequence data for diploid organisms, especially when the loci are far apart. Before analyzing polymorphisms for LD, it is crucial that you are confident in the haplotype phasing of the underlying data. Depending on whether it is possible to analyze phased haplotypes, one can then choose an approach that does or does not require phasing.

6.4 Gene conversion: non-crossover recombination

In this chapter, I have emphasized recombination in terms of crossovers, the reciprocal exchange of DNA between homologous chromosomes. Crossover recombination encapsulates that classic way that everyone learns about for the genetic rearranging that takes place along a chromosome to make different allele combinations. This rearranging has easily detectable and important genetic effects and a crossover event is usually required on each chromosome for successful meiosis. But the recombination process is not as simple as this, and part of our goal in molecular population genetics is to think about how the details of molecular biology can be important for evolution. What else do we need to think about? As Bret Payseur wrote, "Part of meiosis involves a bizarre series of events in which cells deliberately damage their DNA and then repair it. Of the many double-strand breaks that are generated, a minority are resolved as reciprocal exchanges between homologous chromosomes." So what happens to the *majority* of double-strand breaks? How do they get repaired? The answer: non-crossovers.

Non-crossover recombination is a form of recombination that involves the copying of a short stretch of DNA sequence between chromosomes. This copying of DNA from a donor strand to a recipient strand, **gene conversion**, does not rearrange distant variants along the chromosome like we normally think about for crossover recombination, and so leads to **non-reciprocal exchange** of DNA (Figure 6.7). Sometimes people use the term gene conversion as a shorthand to refer to the more general process of non-crossover recombination. In fact, however, even crossover recombination involves short gene conversion tracts of sequence nearby the crossover point.

For our purposes, we are mostly interested in classic gene conversion tracts between allelic sequences at the same locus on homologous chromosomes (Figure 6.7). Classic **allelic gene conversion** affects the composition of haplotypes only in a small zone

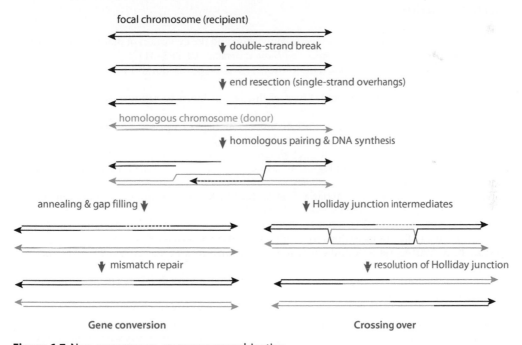

Figure 6.7 Non-crossover vs. crossover recombination.
When double-strand breaks (DSB) in DNA are made as part of meiosis, the cellular machinery sometimes repairs the break by undergoing crossover recombination. Crossovers result in the reciprocal exchange of genetic material between the homologous chromosome pair to create new haplotype combinations relative to either of the original chromosome haplotypes. This is usually what we think about when we think about recombination. Non-crossover recombination provides a common alternate mechanism to repair the DSB. Non-crossover recombination does not result in reciprocal genetic exchange, but instead essentially yields the replacement of some DNA from one chromosome to its homolog through gene conversion; genetically, it looks like a tiny double-crossover. When one chromosome type is used preferentially as the template, then it can lead to biased gene conversion; in many organisms, the GC-rich copy will tend to be used as the template. Gene conversion tracts usually are short, just a few hundred nucleotides long. It is important to keep in mind also that short tracts of gene conversion also occur at sites of crossover recombination, so recombination is never completely free of gene conversion.

because conversion tract lengths usually are only a few hundred nucleotides long. Nevertheless, the exchange of DNA can create new haplotypes. In some organisms, the identity of the donor versus recipient DNA tends to be biased toward copying over A/T nucleotides with G/C nucleotides at polymorphic sites, termed **GC-biased gene conversion**. This bias can mimic selection by leading to fixation of the G/C variant faster than would be expected from genetic drift.

Despite our emphasis on homologous recombination, gene conversion sometimes affects non-allelic DNA between duplicate copies of loci that are located in different regions of the genome, even non-homologous chromosomes. This **non-allelic gene conversion**, also known as interlocus or ectopic conversion, tends to homogenize the duplicate copies. This promotion of sequence similarity among paralogous gene copies is termed **concerted evolution**. In some cases, non-allelic conversion can create novel alleles at one of the loci, so in that sense we can think of ectopic conversion as a source of mutational input.

Further reading

Duret, L. and Galtier, N. (2009). Biased gene conversion and the evolution of mammalian genomic landscapes. *Annual Review of Genomics and Human Genetics* 10, 285–311.

Gaut, B. S. and Long, A. D. (2003). The lowdown on linkage disequilibrium. *The Plant Cell* 15, 1502–6.

Hill, W. G. and Robertson, A. (1966). Effect of linkage on limits to artificial selection. *Genetical Research* 8, 269–94.

Hill, W. G. and Weir, B. S. (1988). Variances and covariances of squared linkage disequilibria in finite populations. *Theoretical Population Biology* 33, 54–78.

Lehman, D. M., Hunt, K. J., Leach, R. J. et al. (2007). Haplotypes of transcription factor 7-like 2 (*TCF7L2*) gene and its upstream region are associated with type 2 diabetes and age of onset in Mexican Americans. *Diabetes* 56, 389–93.

Moreno-Estrada, A., Gravel, S., Zakharia, F. et al. (2013). Reconstructing the population genetic history of the Caribbean. *PLoS Genetics* 9, e1003925.

Ohta, T. and Kimura, M. (1971). Linkage disequilibrium between two segregating nucleotide sites under the steady flux of mutations in a finite population. *Genetics* 68, 571–80.

Payseur, B. A. (2016). Genetic links between recombination and speciation. *PLoS Genetics* 12, e1006066.

Pool, J. E. and Nielsen, R. (2009). Inference of historical changes in migration rate from the lengths of migrant tracts. *Genetics* 181, 711–19.

Thomas, C. G., Wang, W., Jovelin, R. et al. (2015). Full-genome evolutionary histories of selfing, splitting and selection in *Caenorhabditis*. *Genome Research* 25, 667–78.

Wang, G.-X., Ren, S., Ren, Y. et al. (2010). Extremely high molecular diversity within the East Asian nematode *Caenorhabditis* sp. 5. *Molecular Ecology* 19, 5022–9.

CHAPTER 7

Natural selection and demography as causes of molecular non-randomness

Evolution at its most basic has a simple recipe: heritable change over time. But to study evolution at the molecular level, we need to reverse engineer the blend of ingredients that give a species its special flavor. You now have some of the raw ingredients to studying evolution at the molecular level. The instructions call for: (1) genetic variation, having originated from mutation, with (2) quantitative means of assessing that variation and (3) integrating those assessments into a conceptual framework, in the form of the Neutral Theory and Nearly Neutral Theory. We have built an understanding of each of these components over the past few chapters. But what happens in the real world to our predictions about genetic variation, the shape of genealogies, and the accumulation of divergence between lineages? That is, how do we accommodate violations of the standard neutral model (SNM) to predict how factors other than mutation and genetic drift will influence molecular evolution? What do molecular data look like when and if Neutral Theory is an *inaccurate* simplification of reality? In particular, what is the effect of natural selection and of non-standard demographic scenarios, each of which might be revealed in molecular population genetic data?

In this chapter, we will first think through the general, qualitative impact of selection and demography. We will talk about all of the general modes of selection from the standpoint of the genome, and the basic ways that demography can be "non-standard": increases and decreases in population size as well as multiple subdivided populations connected by gene flow. This overview will build up our intuition about selection and demography as molecular population genetic processes (Table 7.1). Then, in Chapter 8, we will dive into explicit quantitative *tests* to evaluate these qualitative distinctions from neutrality.

7.1 Natural selection

Just as the flow of the Colorado River cuts deeper into the Grand Canyon over time, we can think of natural selection pushing as a persistent force in evolution. It pushes to maximize fitness, with changes to DNA as a record of its handiwork. In this respect, selection contrasts with the purely stochastic evolutionary forces: mutation and genetic drift. At the molecular level, we can most simply view selection as coming in three flavors: **positive selection** favoring new or rare beneficial mutations, **balancing selection**

A Primer of Molecular Population Genetics. Asher D. Cutter, Oxford University Press (2019).
© Asher D. Cutter 2019. DOI: 10.1093/oso/9780198838944.001.0001

Table 7.1 Summary of the effects of selection and demography[*] on patterns of polymorphism and divergence.

Effect	Positive selection	Purifying selection	Balancing selection	Population expansion	Population contraction
Polymorphism	Lower	Lower	Higher	Lower	Lower
Genealogies	Shorter TMRCA	Shorter TMRCA	Longer TMRCA	Shorter TMRCA	Longer TMRCA
Site frequency spectrum	Excess of low-frequency variants, excess of high-frequency derived variants	Excess of low-frequency variants	Excess of intermediate-frequency variants	Excess of low-frequency variants	Excess of intermediate-frequency variants
Divergence	Higher K_A	Lower K_A	—	—	—

[*] relative to expectations from the standard neutral model for a single population of a historically constant size equal to the present-day size; TMRCA = Time to the Most Recent Common Ancestor

that acts to maintain multiple alleles at a locus, and purifying selection that eliminates detrimental mutations from populations. Let's now evaluate, in turn, how each of these general modes of selective pressure at the molecular level leaves its imprint in genomes. We will talk about the direct effects of selection and about the indirect effects of selection on linked portions of DNA. With these ideas about selection in mind, we will then make clear the distinctions between viewing selection at the molecular level versus the phenotypic level.

7.1.1 *Positive selection*

Positive selection favors the increase in frequency of a beneficial mutation as a byproduct of the superior reproductive output of its carriers (see Box 1.2 and Figure 4.5). Sometimes referred to as directional selection, positive selection represents a directional pressure to increase one allele at the expense of all others at that locus. Usually we think about positive selection favoring an allele that is relatively rare in the population, not an allele that already is the most common variant. When positive selection acts successfully on a *new* beneficial mutation, then we expect it to lead to fixation of the haplotype on which the mutation arose in what is termed a **hard selective sweep**. That is, after the beneficial mutation has swept to fixation in the population, all copies of the locus in the population will have that new mutation, of course, as a result of the **direct selection** on the variant that confers a fitness advantage.

Immediately following a hard selective sweep, everyone in the population *also* will have whichever variants happened to occur on the same genetic background in which the beneficial mutation arose (Figure 7.1). These other variants feel **indirect selection** simply by virtue of their linkage to the direct **target of selection**. If the selective sweep occurs sufficiently rapidly, then all haplotypes bearing the adaptive mutation will be identical (Figure 7.1). Consequently, the population will have no genetic variation in the chromosomal vicinity of the selected locus. This extreme case shows us a major effect of positive selection: to reduce genetic variation (Table 7.1). Even those sites in the genome that are not direct targets of selection will have reduced genetic variation simply by virtue of being genetically linked.

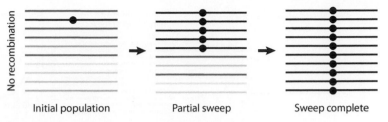

Figure 7.1 Hard selective sweep from a new beneficial mutation.
Each distinctly colored horizontal line represents a unique haplotype sampled from the population, comprised of a unique combination of polymorphic variants at different sites along its length. A new beneficial mutation arises in the initial population, indicated by the black circle linked to the black haplotype. The beneficial allele experiences positive selection and increases in frequency in the population, represented by the "partial sweep" of the invading black allele at intermediate frequency and the loss of some of the haplotypes that were present in the initial population and lacked the beneficial mutation. When the beneficial mutation reaches fixation, the selective sweep is complete. In the absence of recombination among haplotypes, as depicted in the diagram, the population will be comprised solely of the black haplotype on which the beneficial mutation arose, thus resulting in a reduction in genetic diversity for DNA near the selected target mutation. This influence of selection on loci that are linked to the selected target is called genetic hitchhiking.

Let's come back to the issue of **linked selection** for a little more detail. For loci with tightly packed polymorphic sites over a short span of DNA sequence, it is often safe to assume that we can ignore the potential for recombination to rearrange the combinations of polymorphic variants among haplotypes. However, as we consider longer and longer sequences, it becomes more and more likely that recombination will, in fact, have occurred between some variant sites in the history of the locus. Positive selection will drive a beneficial variant to fixation, as we previously described for a hard sweep (Figure 7.1). We call that specific beneficial variant the target of selection. All loci that are closely linked to the target of selection, i.e. those variants on the same haplotype as the beneficial variant, will also become more common, rising in frequency toward fixation, despite the fact that they may have no effect on fitness. Such a phenomenon was termed **genetic hitchhiking** by John Maynard Smith and John Haigh in 1974. With genetic hitchhiking, variants linked to a beneficial allele are dragged along as part of the haplotype that the beneficial variant happens to be associated with.

Following a selective sweep, new unique haplotypes will arise by mutation. Most of the polymorphisms that we observe among haplotypes at the locus will be rare, low-frequency variants because they will have arisen so recently by mutation. We will see this pattern in the site frequency spectrum (SFS) as a skew compared to the SNM (Figure 7.2). We can describe this particular kind of skew as an **excess of rare variants**. When we say that new rare mutations are unusually common among haplotypes in this way, it means that the haplotypes at such a selected locus will also have a shorter time to the most recent common ancestor than expected for a neutral locus unaffected by selection (Figures 7.2 and 7.3). Hard selective sweeps of this kind provide the canonical model of positive selection in molecular population genetics, so we will emphasize the consequences of single sweeps before worrying about the implications of alternative aspects of positive selection at the molecular level. In Chapter 9, we will go into the details of exciting case studies of positive selection in nature.

A second feature that is even more distinctive of a selective sweep in a recombining region of the genome is an excess of high-frequency derived variants. We can observe this skew only in an unfolded SFS, which distinguishes ancestral versus derived variants. The

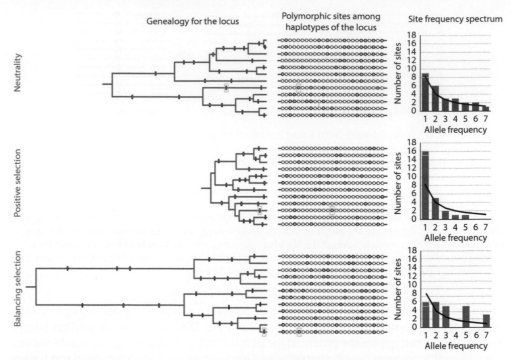

Figure 7.2 Effects of selection on genealogies and the pattern of polymorphisms.

Relative to the expected shape of neutrally evolving gene trees in a population (purple), loci subject to recent positive selection will have a shorter time to the most recent common ancestor (pink). Consequently, a greater proportion of new neutral mutations, that arose since the origin of the beneficial allele, will occur on terminal branches of the genealogy than expected under neutrality. Such mutations on terminal branches of the tree will be represented as singletons when comparing among haplotypes, and will sum up to yield an excess of rare variants in the site frequency spectrum for a locus that experienced a hard selective sweep (note the taller bar for singletons with minor variant frequency equal to one). The neutral frequency spectrum is indicated with black lines for reference. By contrast, a locus subjected to long-term balancing selection will have a more ancient time to the most recent common ancestor (brown), with a correspondingly greater fraction of neutral variants having arisen on the long internal branches of the genealogy. These factors produce an abundance of shared variants among haplotypes and an excess of intermediate-frequency variants for a locus experiencing balancing selection. One example singleton variant (blue) is highlighted to show its represented position on the tree, on a haplotype, and in the site frequency spectrum.

overabundance of derived variants at high frequency reflects genetic hitchhiking of sites that have partial linkage to the target of selection: neutral variants that were themselves rare when the new beneficial mutation arose on their haplotype, or new mutations on the haplotype containing the beneficial mutation when it was rare in the population, which then increased in frequency through genetic hitchhiking but recombined with other haplotypes during the sweep. Those recombination events during the sweep keep some of the neutral variants from getting fixed, leaving them polymorphic and at high frequency when the beneficial variant fixes.

Despite a beneficial variant *eventually* becoming fixed, it takes time for it to spread through the population to complete the selective sweep (see Box 1.2). During that time, recombination might separate the beneficial variant from other linked variants that happened to occur on the original haplotype. Recombination will separate variants more readily if those sites are farther away from one another along the DNA sequence. As

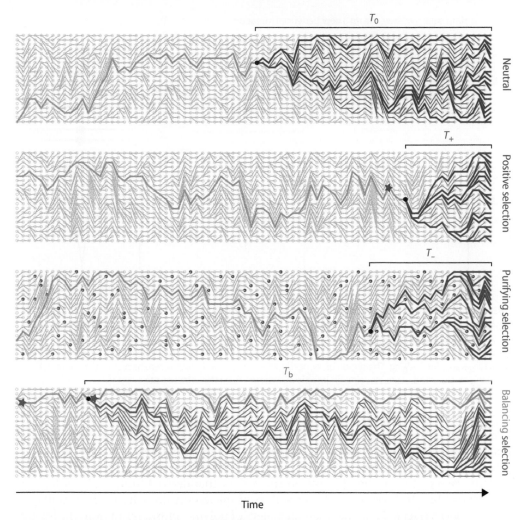

Figure 7.3 Genealogies for loci evolving neutrally or under different modes of selection.
Dark-colored lines trace the coalescent genealogy, representing lineages that occur in the present day that share the same most recent common ancestor (MRCA, black circles). The light-colored lineages represent all descendants of that common ancestral haplotype that have since gone extinct and so are not part of the coalescent genealogy. The expected time to the MRCA for the neutral scenario is $T_0 = 2N_e\mu$ generations for this haploid asexual example of a constant-size population (vs. $4N_e\mu$ generations for a diploid). The coalescence times for genealogies subject to positive or purifying selection are shorter than the case of neutrality (i.e. $T_+ < T_0$ and $T_- < T_0$), whereas the coalescent time is longer for balancing selection (i.e. $T_b > T_0$). The pink stars represent the new beneficial mutation(s) favored by positive or balancing selection, which pre-dates the MRCA in the positive selection scenario depicted. The red circles in the purifying selection scenario represent deleterious mutations, which are responsible for the more rapid loss of individual lineages. The dark gray lines indicate the lineage extending further back in time from the MRCA.

a result, positive selection should produce a "valley" of low genetic variation in the close vicinity of the positively selected site, centered at the target of selection, with polymorphism gradually rising to normal levels with greater distance from the target of selection (Figure 7.4). In this way, selection interacts with recombination over time to leave an imprint on space, the spatial location of polymorphisms along a chromosome.

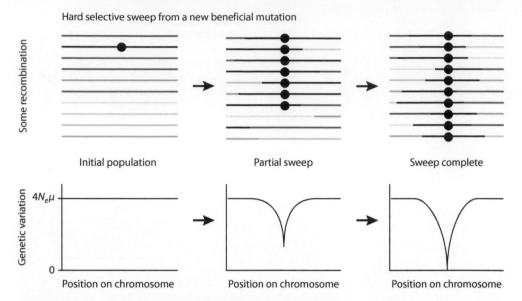

Figure 7.4 Hard selective sweep in the presence of recombination.
Unique haplotypes sampled from a population as depicted in Figure 7.1, except that recombination among them can occur in the intervening time between beneficial mutational origin and fixation (black circle). Consequently, only a portion of the initial haplotype gets dragged to fixation with the beneficial mutation owing to genetic hitchhiking. The region of reduced genetic diversity is most severe near the selected target, recovering to expected neutral levels with increasing distance from the selected site.

The width of the region of reduced genetic diversity will be narrower in regions of the genome with a high recombination rate because the more frequent recombination will make loci more independent of the change in frequency at the selected locus (Figure 7.5). Similarly, sweeps that occurred longer in the past or that derived from weaker selection will also exhibit shallower or narrower valleys of nucleotide diversity in the vicinity of the target locus (Figure 7.5). These subtler valleys of polymorphism result from the longer timespans: more time allows recombination and mutation to erode the sweep signals from those sites that are partially linked to the selected target.

In talking about these indirect effects of linked selection, keep in mind that the signals of genetic variation that we actually look at come from selectively neutral polymorphisms. By looking at linked neutral polymorphisms, we can compare the patterns to what we would predict from the SNM (see section 4.1).

Molecular views of positive selection do go beyond single hard sweeps, including recurrent hard sweeps in which multiple selective sweeps happen one after another, partial sweeps in which the favored allele is in the process of invading but has not yet fixed, and **soft selective sweeps** in which selection acts on variants that pre-existed in the population or on identical-in-state beneficial variants that arose convergently by independent mutations on distinct genetic backgrounds (see section 9.6). When sweeps occur repeatedly in a genome for a population, the frequency changes for alleles that are linked to the targets of selection can mimic the stochastic fluctuations caused by genetic drift. In 2000, John Gillespie called these fluctuations of allele frequencies induced by selection at linked loci **genetic draft**. A consequence of a genome being strongly influenced by genetic draft is that its evolution will behave somewhat like

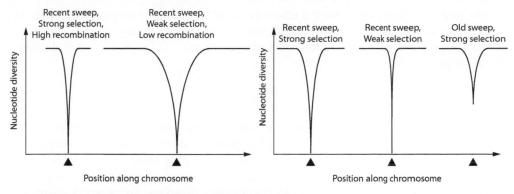

Figure 7.5 Effect of a selective sweep on genetic diversity.
When a selective sweep occurs in a region of lower recombination, the signature of perturbed genetic variation around the target of selection (black triangle) spans a broader portion of the genomic region than when rates of recombination are higher. Stronger selection will affect a broader region than will weak selection because the sweep occurs faster with less opportunity for recombination among haplotypes. Over time, new mutational input will eliminate the signature of reduced genetic variation.

what you would otherwise expect to see for a population with a smaller effective population size.

There are other footprints left by selection in the genome that we can look for. We can also examine the distributions of "haplotype block lengths" (see Chapter 6). Remember that recombination between different polymorphic sites will create distinct haplotypes that comprise unique combinations of variants in a given stretch of DNA. Over time, recombination will generally cause the span of sequence that has variants in linkage disequilibrium (LD) to be very short, so recombination results in short "blocks" of linked sites that comprise haplotypes (see Chapter 6). Selection, however, will disrupt this expectation. Because the rapid spread of a positively selected variant constrains the amount of time available for recombination to separate closely linked polymorphisms, the selectively favored variant ought to occur within a long stretch of perfectly linked variants. In particular, we expect the selectively favored variant to occur on longer haplotypes than would be predicted by neutral evolution alone, as a result of the more restricted timeframe for recombination to whittle away at the haplotype block that contains the selectively beneficial variant. This pattern can be especially helpful in identifying a selective sweep that is ongoing in the population that has not yet fixed the beneficial variant, called a partial sweep. Such a partial sweep can induce extended tracts of identical haplotypes in individuals from the population as a result of such individuals having higher fitness (see section 8.1.2).

Here is a quick summary of the characteristics that should be typical of positively selected loci after a hard sweep relative to neutral evolution (Table 7.1). We should expect selected loci to have: (1) low overall variation among haplotypes (reduced polymorphism), (2) an excess of low-frequency variants and an excess of high-frequency derived variants (shifted SFS), (3) a short coalescence time (shallow genealogies), and (4) long haplotype block lengths. The direct and indirect effects of positive selection will lead to all four of these features, meaning that we will observe them both at the targets of selection and in selectively neutral polymorphism linked to the direct targets of selection. If positive selection occurs repeatedly at a locus, then it will also result in (5) greater divergence (more fixed differences) at the selected set of sites than expected under neutrality. For genes, we assume that replacement sites are the selected class of sites as direct targets of selection. By contrast, we can usually assume that changes to synonymous sites are neutral. Therefore, we expect recurrent positive selection on a gene to increase K_A but not K_S, potentially to the point that $K_A > K_S$ (Table 7.1).

7.1.2 *Balancing selection*

Balancing selection is a general term that refers to any kind of selection that acts to maintain genetic variation at a locus. Just like positive selection, balancing selection is an adaptive force. However, unlike the variation-eliminating effect of positive selection that drives divergence between populations, balancing selection generally operates in the opposite fashion to retain alleles and to restrict their divergence between populations. The key types of balancing selection include **overdominance** (heterozygote advantage), **negative frequency-dependent selection** (rare allele advantage), or selection that varies over space or time **(spatially varying selection** or **temporally varying selection**; Box 7.1). This is quite the panoply of distinct modes of selection that get encapsulated under the umbrella term "balancing selection." These different modes are incredibly interesting, and so a careful investigation of any given case should eventually aim to understand which specific balancing selective regime is responsible. However, this can be tricky. To distinguish one from another will often require integrating many forms of biological information: it will help to have information about population structure, a time series of population genetic data, individual-level heterozygosity, and a phenotypic understanding of the selection pressure. But the key first step, from the perspective of molecular population genetics, is to focus on that one singular thing that all the modes have in common: an overabundance of polymorphisms.

Box 7.1 TYPES OF BALANCING SELECTION

The simplicity of balancing selection as a generic term is a double-edged sword: it encapsulates the idea of natural selection acting to maintain polymorphisms within a species by whatever mechanism, but neglects the distinct selective processes that lead to persistence of genetic variation. Molecular population genetics provides the tool to determine whether *some* type of balancing selection impacts a portion of the genome, with subsequent analysis generally required to figure out which specific type. For example, host–pathogen co-evolution or sexual selection can drive arms-race evolution that favors novel rare alleles in a population (negative frequency-dependent selection), with the resulting diversifying selection producing many haplotype classes. In these cases, most individuals might be heterozygous at the locus because of the many alleles, but not necessarily due to heterozygote advantage of fitness effects. If newly favored alleles are actually detrimental when homozygous, then their rise to intermediate frequency will appear like a partial selective sweep in the short term. When balancing selection persists over the long term, then trans-specific polymorphisms can produce alleles with greater similarity to alleles in different species than to the alternate alleles segregating within a species, as in Figure 1. Alleles of the MHC immunity loci and the ABO blood groups in primates show such trans-specific polymorphism. Spatial variation in selection provides a simpler way to test for molecular signals of balancing selection across a species range, using metrics related to population differentiation like F_{ST} (section 7.3.2; also see section 8.2.2). Variation in selective pressures over time also can lead to long-term retention of polymorphisms in populations. However, long time series of allelic frequencies are still rare in molecular population genetics, making tests of this type of balancing selection unusual.

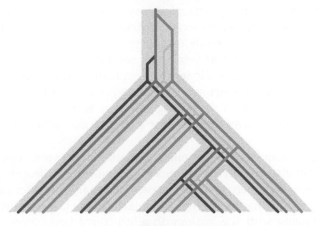

Figure 1

Balancing selection that favors two variants over the long term is the simplest way to think about the selective maintenance of polymorphisms at the molecular level. In this scenario of long-term balancing selection, two ancestral haplotypes will have persisted for much longer than their expected neutral coalescence time (Figure 7.3). Mutational differences will have accumulated on the two kinds of haplotypes, producing more polymorphic sites in the vicinity of the locus than expected under neutrality. That is, neutral variants will show LD with one selectively favored variant or with another and this reflects the balancing selection version of genetic hitchhiking (section 7.1.1). This linkage of neutral polymorphisms to distinct selectively favored variants is termed **associative overdominance** when it is heterozygous genotypes that have a fitness advantage over homozygotes: **heterozygote advantage** is also termed overdominance for the direct target of selection, and the neutral variants *associate* with the advantage because of linkage.

In the extreme, polymorphisms can be selectively maintained for so long that they persist through speciation events, termed **trans-specific polymorphisms**: members of different species actually share a more recent common ancestor than do the alleles within either species! Some classic examples of trans-specific polymorphisms include the sex chromosomes of mammals, major histocompatibility complex (MHC) alleles involved in innate immunity in primates, and alleles conferring self-incompatibility (SI) in hermaphrodite plant reproduction (Box 7.1).

We can also take a genealogical view of a locus that experiences balancing selection. When we look at the gene tree, we will see that the long, deep branches in the gene tree will contain an overabundance of the mutations linked to the selectively maintained variants (Figure 7.2). Many descendant haplotypes will necessarily share those mutations that occur on deep branches. As a byproduct, those numerous shared variants will also be present at intermediate frequencies in the population. I hope that you see that I have used this logic about the genealogy of a locus under balancing selection to lead us to a prediction about the SFS. Specifically, we should expect that long-term balancing selection will cause a skewed SFS toward an excess of intermediate-frequency variants or, stated the other way around, a deficit of low-frequency variants relative to the SNM (Figure 7.2).

Another way to think about the distinct alleles that are favored by balancing selection is to view them as structured haplotypes. From this view, we can think of there being separate "subpopulations" in the genealogical history of the locus (see section 3.6). Just as physically subdivided populations have restricted genetic exchange between them, causing deeper genealogies as a result (see section 7.3.1), we can also think of the selectively maintained haplotypes as being kept separate and having restricted genetic exchange. Rather than being physically separate populations that differentiate the entire genome, however, they are distinct sets of haplotypes defined by mutations that are linked to the target of selection. Balancing selection keeps those haplotype groups separated and genetically differentiated at just a single locus rather than genome-wide (Figure 7.3).

Analogous to the accumulation of genetic differentiation between populations, haplotypes that contain the distinct selected variants will accumulate nucleotide differences and LD. However, the long-term persistence of favored alleles also provides recombination a long time to mix variants that were originally linked to one favored allele or the other. In this way, continuing with the analogy to a structured population, recombination acts like migration between those haplotype classes that experience balancing selection. This recombination will erode the span of LD in the DNA sequence that surrounds the selectively maintained haplotypes (see Figure 6.5). Consequently, when recombination takes place over a long period of time, only a very narrow interval around the target of balancing selection will remain to show the skewed population genetic signals indicative of long-term balancing selection (Table 7.1). This decay of skewed signals with the distance from the target of selection means that recombination will make it hard to detect loci affected by balancing selection.

In summary, we should expect three key signatures of molecular evolution for loci associated with balancing selection, as compared to neutral evolution. We should observe (1) high overall sequence variation among haplotypes, (2) an excess of intermediate-frequency variants, and (3) deep coalescent times for the genealogy overall compared to the neutral expectation (Table 7.1). You may have noticed, however, that I have given no prediction about divergence per se. Balancing selection preserves variation within a species, and so makes no strong predictions about unusual patterns of accumulated divergence between species.

7.1.3 Purifying selection

Purifying selection favors the elimination of new deleterious mutations and the persistence of the prevalent high-fitness ancestral allele. Usually we think about purifying selection as favoring the most common allele in the population, with any new mutations to the locus being deleterious or neutral. In this way, purifying selection is a kind of directional selection. It also is termed negative selection due to its role in weeding out harmful mutations. Everyone has the first inclination to think of purifying selection as a bit dull, nagging and unsexy, diligently cleaning up the garbage left by the exciting party next door. But purifying selection is, hands-down, *the* most pervasive form of natural selection in genomes. Most new mutations that affect fitness will be deleterious (see section 2.2), which gives good reason for why we should anticipate that purifying selection will represent the dominant mode of selection in populations.

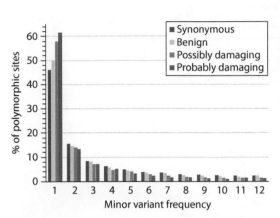

Figure 7.6 Purifying selection and the SFS. Neutral Theory predicts the relative incidence of neutral variants that occur at different frequencies in a population. When a portion of sites are direct targets of selection, however, the distribution of variant frequencies will be skewed compared to the distribution predicted for sites that only contain neutral variants. In coding sequences of genes in the human genome, many changes to non-synonymous sites will be detrimental whereas changes to synonymous sites will nearly all be selectively neutral. As a result, mutations tend to be especially rare at those non-synonymous sites that are more likely to be damaging to protein function and fitness, causing a skewed SFS toward an excess of rare variants relative to synonymous sites as a neutral reference (data redrawn from Ng et al. 2009).

By removing deleterious mutations from the population, the most obvious thing that purifying selection does is reduce genetic variation at the target locus. In this way, it is similar to positive selection. In contrast to positive selection, however, purifying selection promotes sequence **conservation** rather than sequence divergence between species. This conservation enforced by purifying selection results from the selective constraint imposed on the relevant pieces of DNA by their functional intolerance to mutations that exert detrimental effects. Because deleterious mutations get eliminated from the population by selection, they are especially unlikely to reach intermediate or high population frequencies. Consequently, the SFS for sites under purifying selection will be skewed such that rare variants are overrepresented relative to the distribution expected under neutrality (Figure 7.6). Unlike the case for positive selection (see section 7.1.1), derived variants are especially unlikely to reach high frequency with purifying selection.

We can also view the influence of purifying selection in terms of genealogies. In this case, we are imagining that we would be analyzing the haplotypes defined by neutral polymorphisms, but that are linked to other sites that are direct targets of purifying selection. Again, this will let us compare observations to predictions based on Neutral Theory. Haplotypes that experience purifying selection will have a more recent coalescent time than under neutrality because those lineages with deleterious mutations get eliminated faster by purifying selection than by drift (Figure 7.3). This shorter genealogical history can therefore be used to think about the reduced polymorphism of the locus in its own way: because the depth of the coalescent genealogy is shorter than for the neutral case, less time will have elapsed from tip to common ancestor. Less time means that fewer new mutations will have accumulated on the branches of the genealogical tree, thus resulting in less standing genetic variation in the present day.

Just like a positively selected locus will drag linked neutral variants with it to fixation, a detrimental variant will drag linked neutral variants to extinction (Figure 7.7). This process was termed **background selection** by Brian Charlesworth, Martin Morgan, and Deborah Charlesworth in 1993. Background selection describes the indirect effects of purifying selection, analogous to genetic hitchhiking but as applied to linkage to deleterious mutations. All of these indirect effects are different forms of linked selection when alleles get driven to fixation or extinction, also called **selection at linked sites**, and represent special cases of the Hill-Robertson effect. William Hill and Alan Robertson

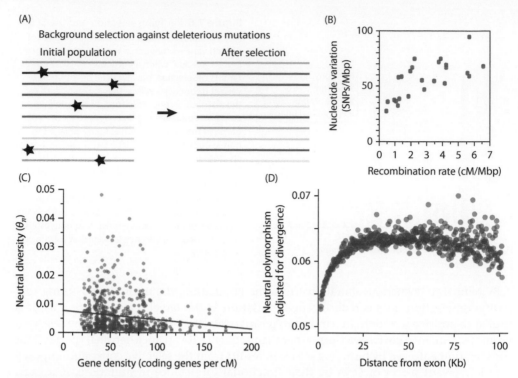

Figure 7.7 Background selection against deleterious mutations.
(A) Unique haplotypes sampled from a population, as depicted in Figure 7.1. Deleterious mutations (black stars) arise randomly on haplotypes and, in the absence of recombination, doom those haplotypes to extinction owing to their inferior fitness relative to the remaining haplotypes. After selection has weeded out the deleterious mutations, the corresponding region of the genome contains less genetic diversity owing to the elimination, via "background selection", of polymorphisms that were linked to the deleterious mutations. (B) Background selection will more strongly reduce genetic variation in regions of the genome that experience less frequent recombination, contributing to the pattern seen in the nematode *C. elegans* (data redrawn from Cutter and Payseur 2003). (C) Genomic regions with a greater density of selected targets will also experience stronger background selection and have lower polymorphism at selectively unconstrained but linked sites, as seen in the plant *Arabidopsis thaliana* (data redrawn from Nordborg et al. 2005). (D) Background selection has a stronger influence on loci closer to coding exons than for loci far from exons, as seen in the mouse genome (data redrawn from Halligan et al. 2013). Note that recurrent selective sweeps can lead to similar patterns due to genetic hitchhiking, with genomic patterns generally reflecting a combination of both forms of linked selection.

laid the theoretical groundwork for these fundamental pieces of molecular population genetic thinking back in 1966 when they described how a direct target of selection will influence the evolution of other sites that happen to be linked to it.

So what does background selection do to genomes? Background selection reduces genetic variation nearby to the loci that are targets of purifying selection: it reduces linked neutral polymorphism. The swath of sequence that gets affected by background selection will be greater in regions of low recombination than in regions of high recombination, just like it is with genetic hitchhiking (Figure 7.7). When we talk about these effects of selection at linked sites (genetic hitchhiking, background selection), in practice, we quantify their effects on neutral polymorphism. By seeing how they perturb polymorphism for sites that are selectively unconstrained, not the direct selective targets themselves, it lets us compare things to the SNM in a fair way.

Box 7.2 LINKED SELECTION INFLUENCES CALCULATIONS OF DIVERGENCE

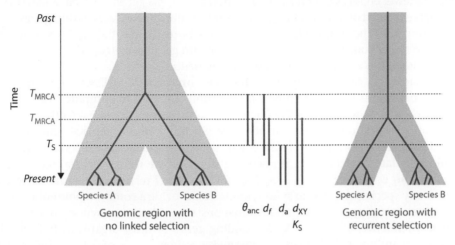

Figure 2

Figure 2 The time at which two populations cease gene flow (T_S), which defines the formation of distinct species, must necessarily be closer to the present day than is the time at which gene copies share a common ancestor (T_{MRCA}; see Figure 5.9). When species separate by simple isolation without gene flow, then all loci across the genome will share the same T_S. However, coalescence times will differ between some loci. Genomic regions having lower polymorphism because of especially strong effects of linked selection will make those loci have more recent T_{MRCA} than elsewhere in the genome. Remember that longer genomic spans in regions of low recombination will be subject to background selection and genetic hitchhiking (Figures 7.5 and 7.7), and the ancestral population would likely have experienced the same purifying selection pressure as present-day populations of the descendant species. As a result, we should expect that some measures of divergence between species will differ consistently for loci found in high versus low recombination regions of genomes (see Table 3.1). As species become more distantly related, the fraction of divergence between them that reflects ancestral polymorphism will become smaller. However, the residual influence of ancestral polymorphism on the remaining variability among loci in divergence can be substantial, even for species that diverged long ago like humans and mice: "intuition has understated the importance of even small amounts of ancestral polymorphism on the variability of genome-wide patterns of divergence between species" (Phung et al. 2016).

As a first approximation, we can think of this reduced polymorphism caused by background selection in terms of effective population size (N_e). Given the same rate of input of new mutations, genomic regions with less polymorphism have a smaller N_e (see Chapter 4). That means that we can think of background selection as causing a reduced N_e in a localized portion of the genome. This analogy is not perfect, however, because background selection also can subtly affect the shape of genealogies and the SFS. It also is important to remember that purifying selection will tend to be consistent over time, which means that selection in ancestral populations can leave a persistent imprint in genomic patterns into the present day (Box 7.2; see Figure 5.9).

Directional purifying selection also lets us make predictions about how fixed differences will accumulate as substitutions between species. The direct targets of purifying selection accumulate fewer changes over time because new mutations get eliminated, and so evolve slower than expected of neutral changes. In terms of substitution rates within genes, as we saw in section 5.3.1, purifying selection causes K_A to be much less than K_S. Most mutations that change the encoded protein are detrimental and get eliminated, whereas mutations to synonymous sites are all, or mostly, selectively neutral.

More broadly, we can quantify the **selective constraint** on DNA sequence as the fraction of mutations that get removed by selection (C). We can compute C for any site class i as $C_i = 1 - f_0$ where f_0 is the fraction of substitutions that are effectively neutral and we now are back to thinking with the SNM. So, if i represents replacement sites then we could calculate constraint on replacement sites; if i represents non-coding sites then we could calculate constraint on non-coding sites. But we need a neutral reference. In general, we can estimate f_0 as $f_0 = K_i/K_{neu}$ for site class i relative to neutral divergence K_{neu}. In many species, changes to synonymous sites provide a common point of reference for selective neutrality. So, for non-synonymous sites relative to synonymous sites as our neutral reference, $C_A = 1 - K_A/K_S$. Coding genes of most organisms indicate that constraint for amino acid changes is high, giving values for C_A of 70% or higher. This definition of C_A makes the implicit assumption that all of those substitutions that we actually observe *both* at non-synonymous sites and at synonymous sites have fixed as a byproduct of genetic drift because they were effectively neutral changes. Consequently, C_A ignores the possible contribution of adaptive substitutions contributing to K_A, an issue that we will return to in section 8.3.3.

Moreover, purifying selection causes the K_A/K_S ratio to be less than θ_A/θ_S, so long as recurrent positive selection on replacement-site changes is not too frequent. Why the difference in these ratios of divergence versus polymorphism? We tend to find that $K_A/K_S < \theta_A/\theta_S$ because some of the variants contributing to replacement-site polymorphisms (θ_A) are slightly deleterious. These slightly deleterious variants will *eventually* get weeded out by selection on the longer timescale of divergence that gets captured in the K_A metric, but in the meantime they inflate the value of θ_A. I hope you notice that this explanation comes from the Nearly Neutral Theory (see section 4.2). This logic will become important for interpreting McDonald-Kreitman tests of neutrality, as we will see in section 8.3.2.

To sum up, purifying selection results in four key effects on molecular evolution relative to neutrality. Purifying selection causes (1) lower overall variation among haplotypes (reduced polymorphism), (2) an excess of low-frequency variants (skewed SFS), and (3) shorter coalescent times (shallow genealogies). As you can see, these first three influences are similar to the effects caused by positive selection. All three of these first three effects should also be visible both at the direct targets of selection as well as at other sites nearby to the direct targets, at sites that are unconstrained by direct effects of selection. In contrast to positive selection, however, (4) purifying selection *reduces* divergence (fewer fixed differences) at the targets of selection (Table 7.1).

7.1.4 *Connecting notions of selection at phenotypic and molecular levels*

In this chapter, we have thought about selection as it is experienced by DNA sequences, but, in fact, selection does not directly "see" DNA sequence differences. Instead, selection acts on phenotypic differences that affect fitness, and the mapping of genotype to phenotype provides the crucial arbiter of how selection drives allele frequency change in the genomes of populations. Even for the simplest genetic mapping of genetic variation at a single locus controlling phenotypic variation, it is the diploid genotype that underlies a given trait value, and there is no unique predetermined match of selection on phenotype and genotype.

This distinction between the effects of selection on phenotypes and on genotypes can be verbally confusing, as the terminology is very similar for both phenomena. But there are key conceptual differences in meaning. For example, directional selection favoring larger phenotypic trait values could be responsible for a multitude of possible molecular selection regimes: positive selection favoring a new allele that confers a larger phenotype when heterozygous or homozygous, purifying selection against new deleterious mutations that induce smaller trait values when homozygous, balancing selection via heterozygote advantage owing to individuals with heterozygous genotypes exhibiting the largest trait values, among other scenarios. All of these molecular incarnations of the selection process *can* be, but are *not required* to be, associated with directional phenotypic selection. The molecular manifestation of other modes of phenotypic selection will similarly depend on the underlying genetic causes of phenotypic differences, often referred to as the **genotype-phenotype map**, and the pattern of selection across time and space. For example, **heterozygote disadvantage** provides one, but not the only, molecular means of realizing phenotypic **disruptive selection**. Of course, most polymorphic traits have still more complicated **genetic architectures**, involving multiple loci and potentially **epistasis** among them.

So, is it impossible to interrelate these distinct ways of thinking about selection? Despite the challenge to connect phenotypic selection to selection at the molecular level, some situations are easier than others. Stabilizing selection represents, perhaps, the most pervasive form of selection at the phenotypic level and, conveniently, we can connect it to DNA in a fairly straightforward manner. In particular, stabilizing selection at the phenotypic level generally corresponds to purifying selection at the molecular level. This may seem counterintuitive, given that you were taught that stabilizing selection favors an *intermediate* phenotype, with mutations pushing trait values above or below the selective optimum, whereas purifying selection is "directional" at the molecular level. However, viewed for a given locus, most haplotypes in the population will have the favored allele that will produce a phenotype near its fitness optimum and so fitness-affecting mutations *all* will be deleterious, regardless of whether they induce an increase or decrease in trait value. New deleterious mutations are rare at their inception, of course, and will stay rare until they are eliminated by selection. Consequently, the variant frequency spectrum for fitness-affecting alleles at the locus will be skewed toward an excess of rare variants. Despite this general view of stabilizing selection back-translated as purifying selection at the molecular level, in some cases, it also is possible to conceive of and model stabilizing selection on genomic features directly; for example, the size of a microsatellite in a coding sequence could be subject to stabilizing selection on its length.

7.2 Population size changes

Populations that are stable in size are a convenient reference point. But in nature, of course, populations commonly undergo changes in population size over time. Consequently, if we assume that population dynamics are at equilibrium (i.e. a stable population size) then we will often be wrong, another kind of departure from the SNM. Because changes in population size do occur in real natural populations, it is important that we understand how such changes influence polymorphism and the genealogical relationships among haplotypes.

Demographic changes in a population, such as an increase or decrease in size, will affect the *entire* genome. In contrast, we typically expect positive or balancing selection to target just individual loci, and therefore differences from the SNM caused by selection should occur only at particular loci in a genome. Even for a polygenic trait, it still will be just a subset of loci in the genome that are direct targets of selection. Therefore, loci that are outliers with respect to most other loci in the genome provide likely candidates for especially recent or pervasive positive or balancing selection having been important in shaping their history. This targeting of individual loci contrasts with the effects of demography, which can leave its mark on *all* loci in the genome. While it is true that the entire genome experiences the effects of demographic change, the coalescent properties of different loci will be more or less consistent with one another depending on the mode of demographic changes.

7.2.1 *Population expansions*

First imagine what happens to the SFS during a **population expansion**, in which case the population size increases (Figure 7.8). Under this scenario, the number of new mutations that enter the population ($2N_e\mu$ for a diploid locus) will go up: the increased N_e makes more mutational targets. This influx of new, and consequently rare, mutations into the population generates a skew of the allele frequency distribution to reflect an excess of rare variants compared to the distribution of variant frequencies in a stable population (Figure 7.8). We expect to see this happen at all loci across the genome. A consequence of having an excess of rare variants due to new mutations, from a genealogical perspective, is that alleles at a given locus will have a recent common ancestor relative to what we would otherwise expect for a population at its present-day size.

Viewed in terms of coalescence, looking backward in time, there will be disproportionate lineage coalescence at the point of population size change because coalescence will be more rapid in that ancestrally small population (Figure 7.8). Sometimes this feature of the gene tree, the clustering of branch coalescence events near the time of expansion, is summarized graphically for non-recombining loci with the **mismatch distribution**. Expansions lead to a unimodal distribution of pairwise differences among haplotypes that reflects the clustering of coalescences in the genealogy, as opposed to the more "ragged" multimodal mismatch distribution seen for a stable population.

Looking across loci, the time to the most recent common ancestor should be more consistent across loci than expected given the current population size. Relative to a stable population of a size equal to the large present-day size of our expanded population, the expanded population would contain less genetic variation because most of its history was spent experiencing a smaller N_e than it now enjoys. Relative to that smaller ancestral population, however, the expanded population would have more genetic variation that derives from the influx of new mutations. Thus, it is important to be clear in your thinking as to what the reference population is: the smaller ancestral N_e or a stable population with a large N_e of the same size as the current population.

When we quantify molecular variation at selectively unconstrained sites, we expect that a recent population expansion will result in four key effects. A population expansion should cause most loci in the genome to have (1) an excess of low-frequency variants, (2) short times to the MRCA, and (3) lower polymorphism and (4) more uniformity in genealogies among loci, as compared to what the SNM would predict for a stable population as big as the current expanded population size (Table 7.1). Note that this pattern is similar to what we expect to see for targets of positive or purifying selection. The key difference between the patterns predicted by population expansion and by

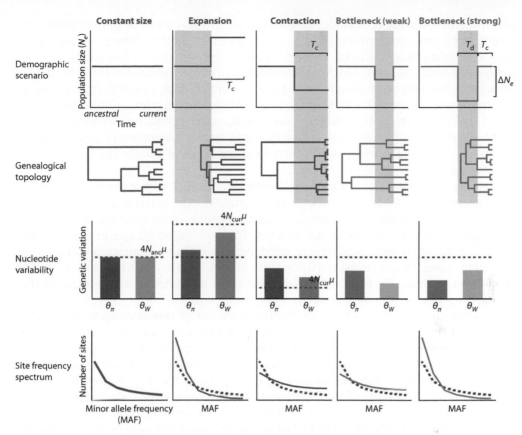

Figure 7.8 Population size changes perturb patterns of molecular variation.

For simplicity, we can represent population growth or decline as an instantaneous expansion or contraction of the effective population size (N_e) and compare how we expect such demographic shifts to differ from neutral expectations for a population of constant size (purple). We expect coalescence to be faster when N_e is small (gray regions; see section 5.2), and small populations to have less genetic variability (ancestral N_{anc}, current N_{cur}). The SFS becomes skewed with demographic perturbations, as well, which results in different ways of quantifying nucleotide polymorphism giving different values (see section 3.3). These scenarios depict just a single "representative" locus with information derived from selectively unconstrained sites, but appreciate that one must evaluate many loci to make general conclusions about demography because of stochasticity among loci. The coalescent times across loci in expanding populations, however, will be more consistent than for a constant population. For contractions and bottlenecks, we should instead anticipate much greater heterogeneity in gene tree depth among loci than seen in a constant-size population, depending on whether or not lineages manage to escape coalescence during the period of low population size. The severity of demographic effects will be exacerbated by larger size changes (ΔN_e), shorter times since the size change (T_c), and longer bottleneck durations (T_d).

directional selection is that demographic change should influence the entire genome, whereas selection should affect only a small subset of loci.

7.2.2 Population contractions and bottlenecks

A **population contraction** involves shrinking of the population size. This decline in size will cause more rapid change of neutral allele frequencies because drift will have become a stronger force now that the population size is reduced. As a consequence, variants that occurred at low frequencies in the ancestral larger population will tend to get lost in the new smaller population. This, of course, involves a loss of genetic variability overall. But some variants will manage to remain in the population after the contraction. These remaining variants will then tend to occur at *higher* relative abundance than would be expected under the SNM, which we term a deficit of rare variants with respect to the SFS (Figure 7.8). In terms of genealogies, some lineages will fully coalesce within the contracted size time period, whereas other lineages will extend long enough into the past to have their rate of coalescence depend on the larger ancestral N_e. As a consequence, we should expect to see a deeper average total coalescence time and greater heterogeneity among loci in the times to the most recent common ancestor as compared to a stable population of the current size.

In general, a population contraction will induce (1) a deficit of rare variants, (2) longer times to the MRCA, (3) lower genetic variation, and (4) more heterogeneity among loci, *as compared to the neutral expectation for a stable population of the same size as the current population* (Table 7.1). Like other size changes, this will impact all loci in the genome, which distinguishes this demographic effect on the variant frequency spectrum and coalescent times from that expected for particular loci subject to balancing selection. In addition to these effects on the average pattern in the genome, it also increases the stochasticity from one locus to another.

Biologists are often concerned with the consequences of **population bottlenecks** (Figure 7.8), in which a population shrank transiently and then rebounded. Because bottlenecks involve phases both of contraction and of growth, their implications for molecular diversity are necessarily more complex. In particular, while bottlenecks result in reduced diversity overall, the shape of genealogies and the type of skew in the SFS will depend on the severity of the contraction, the duration of the contraction, and the time since the population size recovery (Figure 7.8). As a rule of thumb, strong, long, and relatively old bottlenecks tend to produce genealogical coalescence almost entirely within the contracted period, leading to an excess of low-frequency variants, which mostly reflects the signal of the post-bottleneck expansion phase. In contrast, coalescence tends to extend back prior to the contracted period for weak, short, and relatively recent bottlenecks to show an excess of intermediate-frequency variants, which mostly reflects the consequences of the transient drop in population size (Figure 7.8). In practice, given such conceptual complications—not to mention the organismal details of a particular focal study system—researchers employ coalescent computer simulations of plausible bottleneck scenarios to compare with empirical data (see Box 8.2).

So far in talking about population size changes, I have presumed that we would measure polymorphism at selectively unconstrained sites. However, different DNA site classes will respond on different timescales to demographic perturbations like a bottleneck (or even to a nearby selective sweep). Specifically, the time it takes for non-synonymous sites to recover to equilibrium levels of polymorphism (i.e. mutation-selection balance) will be quicker than it takes for synonymous sites (i.e. mutation-drift balance). This difference comes about because, despite their equivalent rates of mutational input per site each generation, the amount of polymorphism at equilibrium is expected to be lower for non-synonymous sites than for synonymous sites.

7.3 Population structure and migration

Up to now, we have discussed genetic variation almost entirely in the context of species that are made up of a single well-mixed group of individuals. Real-world distributions of

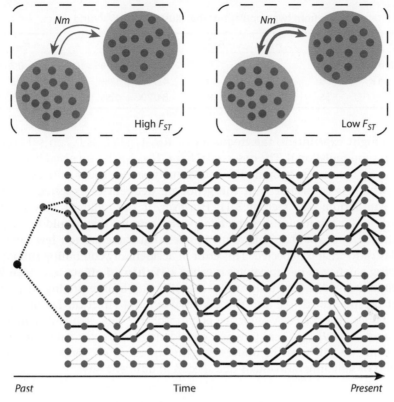

Figure 7.9 Subdivided populations with partial genetic connectivity.
When the individuals of a species are structured into two or more subpopulations, with less exchange of genetic material than would be expected for random mating across the entire species range (i.e. restricted gene flow), then genetic drift and selection can operate partially independently in each subpopulation. F_{ST} provides a relative measure of genetic differentiation (see section 3.6). The number of migrants moving between the populations each generation is Nm, with higher rates of migration making the subpopulations genetically more similar (thicker arrows indicate more migration).

individuals in a species, however, often do not fit neatly into the notion that they comprise a single population in which any given individual is equally likely to mate with any other. Oftentimes, instead, populations are structured. Population structure results from restricted genetic exchange between one or more subpopulations (Figure 7.9), creating genetic differentiation between them (non-zero F_{ST}; see section 3.6). Individuals within each subpopulation might be thought of as randomly mating, but the total collection of individuals in the species as a whole cannot. Consequently, how much the total population differs from the expectations of random mating depends on how much gene flow connects the various subpopulations, in addition to the various forms of natural selection and demographic size change.

As you might have anticipated, population structure influences patterns of genetic variation of a species. That is what we will be working through in this section. As an example of how population structure can affect our interpretations of genetic variation, consider the following simple thought experiment. Imagine a species that is subdivided into two subpopulations, and a biologist samples a locus for many individuals and quantifies variation to observe data like in Table 7.2. What should we make of such data?

Table 7.2 Heterozygosity from two populations that illustrate the Wahlund effect.

Genotype	A_1A_1	A_1A_2	A_2A_2	H_{obs}	H_{exp}
Subpopulation J	49	42	9	$42/100 = 42\%$	$2pq = 2 \cdot 0.7 \cdot 0.3 = 42\%$
Subpopulation K	9	42	49	$42/100 = 42\%$	$2pq = 2 \cdot 0.3 \cdot 0.7 = 42\%$
Total	58	84	58	$84/200 = 42\%$	$2pq = 2 \cdot 0.5 \cdot 0.5 = 50\%$

In this thought experiment, the frequency of the A_1 allele is 0.7 in subpopulation J versus 0.3 in subpopulation K. The observed (H_{obs_J}, H_{obs_K}) and expected heterozygosities (H_{exp_J}, H_{exp_K}) for each subpopulation individually are all equal to 0.42, as shown in Table 7.2, which is consistent with the genotype frequencies expected under Hardy-Weinberg equilibrium (see section 3.1). However, if the biologist naively calculated heterozygosities for this locus using the *total* pooled sample, they would find $H_{obs\_total} = 0.42$ but that $H_{exp\_total} = 0.5$! This result implies that there are too few heterozygotes present relative to what is expected. This error of accidentally combining subpopulations and observing a deficit of heterozygotes is called a **Wahlund effect** after Sten Wahlund who described it in 1928, and it illustrates the difficulty in interpreting data for species that might be subdivided (see also the 2-locus Wahlund effect on LD in section 6.2). Analogous issues arise for DNA sequence polymorphism data. Remember that θ_π is the DNA counterpart to H_{exp}. Clearly, when considering molecular data and patterns of polymorphism, we will want to understand the influence on genomes of population structure and of sampling individuals of a species in different ways.

7.3.1 *Population structure and genealogies*

In talking about population subdivision so far, I have implicitly alluded to the effects of population structure on a single locus. But it is important to keep in mind that we should expect the structuring of populations to have similar effects on *all* loci in the genome—just like how demographic size change in a single population affects all loci in the genome. In contrast, adaptive evolution by positive selection is generally thought to "act locally" within the genome, only on particular genes. Remember, of course, that these effects layer on top of the intrinsic stochasticity among loci that is expected for the coalescent process. If this sounds like it is getting complicated, then we should pause and try to think about what general trends we should be able to see. To make general inferences about demographic effects (structure or population size change), one needs to look at many loci to observe the patterns that emerge from the genetic data overall. With this in mind, we can then ask: if population structure is present, what effect should we typically see on genealogies?

The partial genetic isolation between subpopulations means that the inhabitants of each deme will generally share a more recent common ancestor with each other than with members of a different deme. We should see this uneven degree of relatedness across the species in genealogical relationships. Specifically, we should observe deeper coalescent events on average between individuals that derive from different subpopulations than would be expected for a single panmictic population (Figure 7.10).

We saw with the Wahlund effect what the consequences are for heterozygosity if we have structured populations for which we mistakenly presumed the data to come from only a single population. But what are the consequences for genealogies and nucleotide

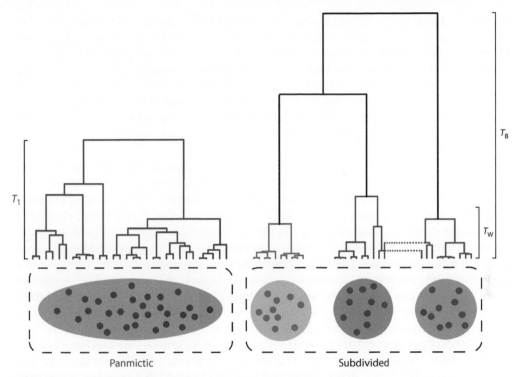

Figure 7.10 Coalescent genealogies for panmictic versus subdivided populations.
Haplotypes tend to coalesce within subpopulations before they coalesce with haplotypes from other
subpopulations ($T_W < T_B$), reflecting the genetic differentiation that separates the demes. The restricted
migration among subpopulations pushes further into the past the time to common ancestry relative to a single
population of equivalent total size ($T_B > T_1$). Because here we presume the total size of the species to be the
same for panmictic and subdivided scenarios, each subpopulation is smaller than the panmictic population and
so has a more recent expected coalescence time within demes ($T_W < T_1$). Migration between populations
(dashed lines), however, can perturb the SFS for local samples.

polymorphism? It is useful to think about what the expected effects are for such a "pooled
sample" on a neutral genealogy, and how that translates into patterns of genetic variation.

First, the TMRCA for the haplotypes in the total population will be older than
you would expect for a single population with the same number of individuals. The
restricted gene flow impedes coalescence of haplotypes found in different subpopulations,
pushing between-population coalescence further into the past. This resulting deeper
age of common ancestry across the species as a whole is a byproduct of the genetic
differentiation created by restricted gene flow among subpopulations. This means that
you would calculate a higher value for metrics of polymorphism using a pooled sample; remember that θ_π is analogous to the expected heterozygosity in Table 7.2 (see
section 3.3.1).

Second, the longer branch-lengths of the multi-population genealogies of the pooled
sample of haplotypes will also affect the variant frequency spectrum. In effect, it will
resemble a deficit of rare variants. Because drift operates independently within each deme,
each with smaller effective size than a single panmictic population with the same number
of individuals, rare variants will be less common. The long branches separating demes

will also lead to an excess of intermediate-frequency variants because more mutations will have accumulated along them, which will be represented among all descendant lineages (these are fixed differences between subpopulations). In this way, population structure can mimic the patterns that we described for a population contraction or for balancing selection (sections 7.2 and 7.1.2). These predictions for pooled samples presume equal sample sizes from each deme and so the predictions can differ if the subpopulations are unevenly represented in the pool (Table 7.3). Therefore, it is important to try to determine whether or not a sample of haplotypes derives from a single population, and whether the pattern is consistent among many loci.

We can intentionally construct "pooled samples" of individuals from multiple subpopulations to compare with "local samples" of individuals from a single subpopulation (Table 7.3). We should expect local samples to often "behave" like a single panmictic population to compare more easily to the SNM. However, even if you collect data from just a single deme in a structured population, the molecular population genetic signals can still be perturbed by haplotypes that get introduced through migration. Migration can result in local samples having distorted genealogies and site frequency spectra relative to a single panmictic population. How drastic these distortions are will depend on the details of gene flow. We can also analyze "**scattered samples**" of just a single individual from each of many subpopulations. John Wakeley (1999) showed that scattered samples can have the convenient property that, under neutrality, the genealogy will match the neutral coalescent process, except that one must rescale time in N_e generations to additionally be proportional to the number of demes. This feature of scattered samples presumes that the metapopulation has many, many demes and that we sampled DNA from just a subset of them. By comparing measures of nucleotide polymorphism and the SFS for each of these three **sampling schemes** (i.e. pooled vs. local vs. scattered samples), it may be possible to improve one's inference of global versus local population size changes and the extent of population structure (Table 7.3). In practice, however, it may often be hard to identify many distinct, discrete subpopulations with which to construct a scattered sample.

In contrast to species that have had multiple subpopulations for a long time as stable population genetic entities, some species are comprised of more ephemeral subpopulations. Frequent extinction and recolonization of habitat patches, often termed **metapopulation dynamics**, tends to (1) reduce polymorphism within any given subpopulation, (2) reduce species-wide levels of polymorphism, and (3) skew the SFS toward having an excess of low-frequency variants. These effects occur because of the bottlenecks

Table 7.3 Summary of the effects of subpopulation sampling scheme on patterns of polymorphism and divergence relative to a single panmictic population of equal total size.

Effect	Local sample	Pooled sample	Scattered sample
Polymorphism	Lower	Higher[*]	Higher[**]
Genealogies	Shorter TMRCA	Longer TMRCA	Longer TMRCA
Site frequency spectrum	Neutral or excess of intermediate-frequency variants (migration)	Excess of intermediate-frequency variants (even sampling among demes), excess of low-frequency variants (uneven sampling among demes)	Neutral or excess of low-frequency variants

[*] lower if there are strong extinction-recolonization dynamics
[**] if scaled by deme number, then lower than for a panmictic population

associated with the recolonization process and the loss of diversity from subpopulation extinction.

I have told you about the general effects of population structure and population size change, but for any given real-world species, the situation can be quite complex (Figure 7.11). Researchers have developed a wide variety of advanced analytical techniques for deciphering complex demography. For example, some methods rely on coalescent theory to conduct coalescent computer simulations for a range of plausible historical scenarios to test for fit to observed patterns (see Box 8.2). Researchers also use **approximate Bayesian computation** (ABC) to search vast ranges of parameter space to expand

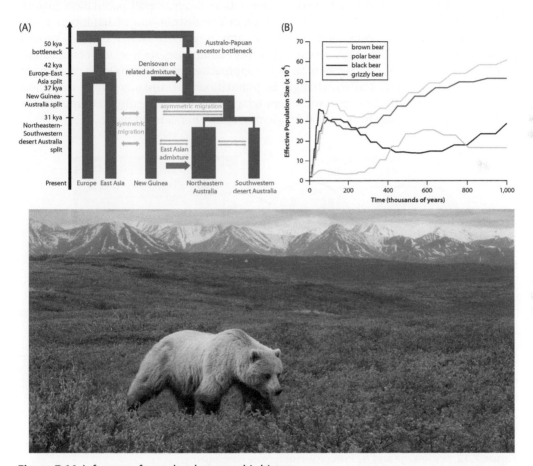

Figure 7.11 Inference of complex demographic history.
With genome-scale data, it is possible to construct complex models of population demography that build on the basic principles of how population growth, contraction, subdivision, and migration influence patterns of polymorphism. (A) The most intensive efforts to understand species demographic history from molecular population genetic information are for our own species, with the peopling of Southeast Asia and Australia requiring an especially sophisticated demographic model to account for patterns in the data (redrawn from Malaspinas et al. 2016). (B) Using the distribution of polymorphism tracts in genomes, the pairwise sequential Markovian coalescent (PSMC) approach allows one to reconstruct a time series into the past of population size. Several different species of bears, including grizzly bears, exhibit declines in effective population size in the late Pleistocene and Holocene (graph reprinted from Miller et al. 2012. Polar and brown bear genomes reveal ancient admixture and demographic footprints of past climate change. *Proceedings of the National Academy of Sciences USA* 109 (36), E2382–90). Yellow line, brown bear; blue, polar bear; black, black bear; red, grizzly bear. Photo by Gregory "Slobirdr" Smith, reproduced under the CC BY-SA 2.0 license (cropped from original).

the scope of possible demographic scenarios to consider. A basic aim of ABC is to estimate parameter values for models of demography for one or more demes, doing so by evaluating the goodness of fit of population genetic summary statistics from stochastic simulations for millions of model parameter combinations. Other methods are based on **diffusion theory**, although they are limited to considering just a small number of subpopulations. Another sophisticated procedure uses **pairwise-sequential Markovian coalescent** model-fitting of heterozygous tracts of sequence in long segments of chromosomes to deduce the dynamics of N_e across a population's history (Figure 7.11). The density of polymorphisms and the length of tracts from just the single pair of chromosome copies in one diploid individual provide information about overall population history by integrating patterns across the whole genome. The distribution of haplotype tracts that get introduced to populations following large migration events also has proved powerful in understanding very recent population admixture, such as the colonization history of the Americas (see Figure 6.2). These approaches allow one to infer the current N_e of populations, ancestral N_e, changes in population size, the amount of migration between demes, and the timing of these demographic events. These disparate approaches collectively provide tools for what we call demographic model building.

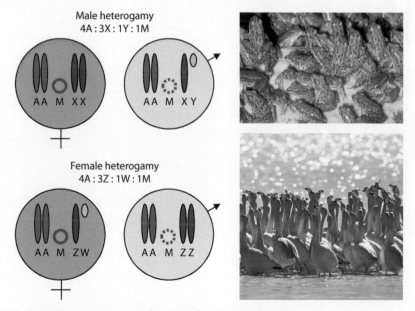

Figure 7.12 Sex linkage and N_e in male versus female heterogamy.

In a population of diploid organisms with an even sex ratio (50% males and 50% females), there will be three copies of each X-chromosome (or Z-chromosome) for every four autosomal copies ("A" chromosomes in the diagram). Similarly, there will be just one copy of each Y-chromosome (or W-chromosome) for every four autosomal copies (e.g. male heterogametic leopard frogs, female heterogametic flamingos). Despite the fact that both males and females have a haploid copy of the mitochondrial genome ("M"), only females transmit it to offspring, also giving mitochondrial loci a 1:4 representation relative to autosomal loci in the population. Sex chromosomes in some species are not easily distinguishable from one another, and are termed homomorphic sex chromosomes. Homomorphic sex chromosomes are more likely to have a portion of their length undergo normal recombination, and so are "pseudo-autosomal" regions with respect to the expected effective population size. Flamingos photo by Pedro Szekely, reproduced under the CC BY-SA 2.0 license (cropped from original). Northern leopard frogs image courtesy of the Canadian Museum of Natural History.

Demographic models are valuable to learn about the history of a given species, but also are important for studying natural selection. Once we build a reasonable demographic model that captures the principal features of a species' demographic history, then we should use it as a new baseline null model to supplant the SNM for that species. By using a more realistic demographic model, we can more accurately and more confidently test for selection or evaluate more subtle evolutionary phenomena. With huge amounts of data in the form of a genomic scale of polymorphisms from hundreds of individuals representing scores of populations, including **ancient DNA**, it is possible to infer elaborate details in a model of demographic history, as is the case for humans (Figure 7.11; also see Figure 6.2).

7.3.2 *Local adaptation in subpopulations*

When you assess population subdivision to understand demography, it is critical to use genetic variants that we can presume to evolve neutrally. As we have learned, however, not all evolution is neutral. In fact, population structure in a species might occur partly as a consequence of important ecological and environmental differences across the range of habitats occupied by the species. As a consequence, population structure can facilitate **local adaptation**, in which specific phenotypes, and the alleles that confer those phenotypes, are favored in some subpopulations but not in others. Similarly, the phenotypes associated with local adaptation may impede gene flow and thus reinforce population subdivision. Molecular population genetic methods can be used to identify loci associated with such adaptation by finding unusual patterns of polymorphism in one subpopulation as compared to another, provided that the subpopulations can be defined clearly. The basic idea is that loci that contribute to local adaptation should show more extreme genetic differentiation between the populations, for example as quantified by higher F_{ST} values, than do most other loci in the genome. The loci with the most extreme values of F_{ST} will provide the most likely candidates for adaptive differentiation. On the other hand, when loci experience purifying selection in all subpopulations, then those loci should exhibit lower F_{ST} than unconstrained, neutral loci. We will discuss these issues in detail in Chapter 8.

7.4 Genomic compartments of chromosomal linkage

We have focused almost entirely on autosomal loci of diploid individuals, either implicitly or explicitly, as we have gone about discussing polymorphism and divergence. However, eukaryotic organisms often have genes partitioned into compartments that are transmitted by just one sex or more commonly by one of the sexes. Does this matter? What are the molecular population genetic consequences of a gene being linked to a **sex chromosome** or to mitochondria or chloroplasts?

Genes encoded on **plastids**, the mitochondrial and chloroplast genomes of eukaryotic cells, are haploid and get transmitted only through the maternal germline in most species. In some interesting exceptions, plastids are inherited bi-parentally or through male germ cells, as for mussels, bananas, and redwood trees. Haploidy and sex-specificity each individually cause a twofold reduction in the effective number of plastid genome copies in the population. Consequently, the effective population size of such loci is expected to be just ¼ that of a typical autosomal locus that is diploid and transmitted by both sexes (Figure 7.12). Similarly, Y-chromosomes occur at only ¼ the frequency of a diploid

Box 7.3 OPERATIONAL SEX RATIO AND EVOLUTION

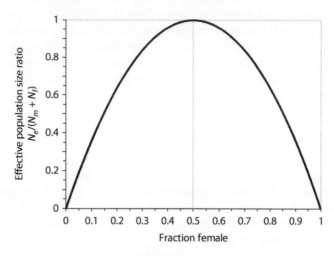

Figure 3

Figure 3 Most species with both males and females have a 1:1 primary sex ratio at fertilization, but the proportion of males versus females can differ from 50% at the time of breeding. Such biases in the operational sex ratio at the time of reproduction can result from many sources, including sex-biased dispersal, sex differences in mortality or development time, lek or territorial mating systems, and meiotic drive and gamete competition. For a given total number of individuals, the effective population size is highest when there are equal numbers of males and females (Equation 7.1). Coalescence theory also can be applied to arrive at the inbreeding effective size for unequal sex ratios shown in Equation 7.1 (see Box 3.1). The overall probability of coalescence, $1/(2N_e)$, is the sum of the probabilities of coalescence for males and females separately, where we can abstractly think of males and females as representing separate populations. Within each sex class, the probability of picking a pair of alleles from males is $\frac{1}{2} \times \frac{1}{2} = \frac{1}{4}$ and for picking a pair of alleles from females likewise is $\frac{1}{4}$. Consequently, the overall probability of coalescence is $\frac{1}{2N_e} = \frac{1}{4} \cdot \frac{1}{2N_m} + \frac{1}{4} \cdot \frac{1}{2N_f}$. If you solve this equation for N_e, you will arrive at the same equation as that shown in Equation 7.1!

autosome (A) within a population because they occur only in the haploid state and only in one of the sexes (males, the heterogametic sex; analogously for the W-chromosome in females as the **heterogametic sex** in species with ZW sex-determination systems like birds).

These details of genomic compartmentalization have important population genetic implications. Given a population with an equal 1:1 sex ratio of males:females, the effective population size of autosomal loci should be 4 times greater than for loci linked to a plastid or to a Y- or W-chromosome (Figure 7.12). This difference in N_e means that loci in different genomic compartments have different expected values for nucleotide polymorphism. If male variance in reproductive success is higher than a Poisson-distributed number of offspring, then the Y:A ratio of N_e will be even lower, approaching 1/8 in the extreme. Loci that are sex-limited also can't recombine. The perfect linkage among genes encoded on non-recombining plastid genomes like mitochondria and on degenerated sex chromosomes, like Y chromosomes in mammals, can lead to even further consequences for their molecular evolution. The perfect co-inheritance of loci means that linked selection from genetic hitchhiking and background selection will be especially strong, potentially decreasing polymorphism to an even greater degree than the 1:4 reduction expected from N_e differences alone.

In a similar way, X- or Z-chromosomes occur in just a single **hemizygous**, haploid copy in one sex despite being diploid in the other sex. As a result, in a population with a 1:1 sex ratio, loci that occur on X- or Z-chromosomes will have ¾ the effective size of loci linked to a diploid autosome (A) (Figure 7.12). However, if the **operational sex ratio** of a population differs from 1:1 or if females have non-Poisson variation in reproductive success (Box 7.3), then this can alter the X:A ratio above or below ¾. You can see this influence of skewed sex ratios reducing N_e explicitly in the classic formula that defines the contribution of male and female reproduction to N_e (Box 7.3):

$$N_e = 4 \cdot \frac{N_m N_f}{N_m + N_f} \tag{7.1}$$

Higher variability in the number of offspring per parent also will reduce the effective population size: $N_e = \frac{4N-2}{V+2}$, where V is the variance in offspring number among parents and the "2" corresponds to the average number of offspring produced per pair of parents in a stable population.

The smaller effective population size for **sex-linked loci** implies that genetic drift should be more important for loci encoded on sex chromosomes and plastid genomes than for loci on the autosomes. In the absence of selection, these N_e differences translate directly into our molecular population genetic expectations: sex-linked loci should have proportionately less genetic diversity and shorter times to the most recent common ancestor in genealogies than will autosomal loci. Sex-linkage also affects the amount of genetic differentiation (F_{ST}) that is expected for a given amount of migration between populations: higher F_{ST} for mitochondria, Y-chromosomes, and X-chromosomes than for autosomes. You can see the influence of different N_e on F_{ST} from Equation 3.23. In practice, mitochondria often experience a different average mutation rate than the nuclear genome, which also affects polymorphism: a higher mutation rate in animals, lower in plants. More generally, these distinct expectations for N_e among loci that are found on different genomic compartments illustrate the abstraction of the "effective population size" idea (see Box 3.1). Remember that N_e is not really a statement about the number of individuals in a population that one can count, but instead applies to a gene-centric view of evolution at the molecular level.

In addition to the relative abundances of distinct chromosome types in a population, the hemizygosity of sex-linked loci in the heterogametic sex means that only a single allele will be expressed in such individuals. Consequently, any new recessive deleterious and beneficial mutations that arise on sex-linked loci will be exposed to natural selection in the heterogametic sex. By contrast, the initial fate for recessive fitness-affecting new mutations that occur on autosomes is to be "masked" in the heterozygous state of a diploid chromosome. Consequently, the ploidy differences due to sex-linkage could cause selection to be relatively more important for sex-linked loci, depending on the dominance of most new deleterious and beneficial mutations. Importantly, this stronger effect of selection on sex chromosomes would work to oppose the influence of smaller effective population size. Because most new mutations are deleterious and partially recessive, we might expect purifying selection on X (or Z) chromosomes to be more efficient than for autosomes, resulting in a lower load of deleterious mutations on the X (or Z) within the population. Perhaps counter-intuitively, such a lower mutational load would induce higher levels of neutral polymorphism, due to weaker background selection on the X (or Z). Despite these theoretical insights, the combined influences of differences among genomic regions in effective population size, ploidy, variance in sex-specific reproductive success, dominance, and selection have not yet been well-established empirically, and this is an active ongoing area of population genetics research.

We have now seen how linkage to distinct genomic compartments has fundamental population genetic implications. However, molecular evolution of the loci linked to sex chromosomes and plastid genomes is intriguing for other reasons as well. The sex-biased transmission of these portions of the genome suggests that the molecular determinants of traits involved in **sexual selection** and **sexual conflict** might occur more readily on them. Because traits subject to sexual selection, and the genes underlying them, can evolve very rapidly, they form an enticing subject for studying the molecular basis to evolutionary change (Box 7.4). Even beyond sexual selection, adaptation in general could occur more readily on sex chromosomes because of more effective selection, depending on the dominance of beneficial mutations. One way to explore these ideas is by analyzing species with **neo-sex chromosomes** that originated in relatively recent evolutionary history. Unlike the ancient origins and degeneration of mammalian Y-chromosomes, the evolutionary signatures on "new" sex chromosomes should more easily be able to tell us about what shifts in molecular evolution coincide with shifts in the sex-linkage of genes.

Box 7.4 RAPID EVOLUTION OF SEXUALLY SELECTED GENES

Sexual selection is a special form of natural selection in which members of one sex compete with one another for reproductive access to the other sex. Evolutionary responses to sexual selection can lead to exaggerated male traits like the horns of moose and dung beetles, the fanciful coloration of cardinals and guppies, or the elaborate courtship displays of jumping spiders and bowerbirds. These sex-biased traits can be triggered in development by sex-biased gene expression. Male competition over fertilization can even take place after mating, between the sperm of different males that inseminated the same female. Proteins found in sperm and seminal fluid also must interact with the female reproductive tract, which also can promote a form of sexual selection termed sexual conflict that can lead to co-evolutionary arms-races between male and female traits. Because of the direct connection of sperm and seminal proteins to reproduction, such selection can be especially strong and recurrent at the molecular level to produce rapid divergence. Figure 4 (A) For example, proteins expressed in the testes and seminal vesicles (SV) of house mice show accelerated molecular evolution with unusually high d_N/d_S ratios that are higher than other tissues of the male reproductive tract of mice, being nearly three times higher than most genes in the genome (data redrawn from Dean et al. 2009). (B) The accessory gland proteins (*Acp*) that male fruit flies transfer to females upon mating also show evidence of adaptive evolution (*a*), based on molecular evidence from both polymorphism and divergence along with significant McDonald-Kreitman tests (see section 8.3.2). (C) This selection in flies also leads to skewed site frequency spectra (negative values of Tajima's *D*, see section 8.1; data redrawn from Proschel et al. 2006).

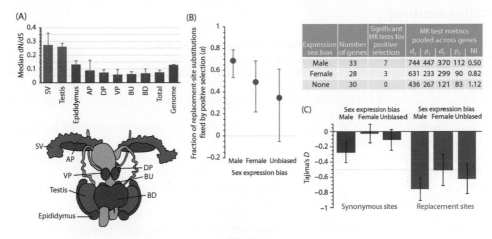

Figure 4

Mitochondrial loci are darlings for molecular analysis of phylogeographic inference to determine historical causes of present-day species ranges, and have been since the dawn of DNA sequencing in molecular population genetics. It was with the analysis of mitochondrial DNA in mind that John Avise first coined the term "phylogeography" in 1987. There are some good reasons for using mitochondrial loci to understand the history of species. The 12–13 mitochondrial protein-coding genes are nearly universal across species, they usually follow the simple logic of maternal transmission across generations, and they experience high mutation rates in animals to provide an ample source of sequence differences.

But limiting one's molecular analysis to mitochondrial loci comes with some major costs. Despite mitochondrial genomes containing multiple distinct genes, their perfect co-inheritance means that they represent just a single, linked genetic locus in genealogical terms. Selection at linked sites, sex-biased gene flow, and multiple mutational hits from the higher mutation rate of animal mitochondria all also introduce complications for analysis and interpretation. In their analysis of the sources of differences among 3,000 species in how much population genetic diversity was present in mitochondria, Eric Bazin, Sylvain Glemin, and Nicolas Galtier (2006) went so far as to say that "mitochondrial diversity of a given animal species does not reflect its population size . . . and will, in many instances, reflect the time since the last event of selective sweep, rather than population history and demography."

I have emphasized again and again the stochasticity of gene trees among loci and the importance of looking at many independent realizations of genealogies, so I hope you appreciate these limitations of studies that only examine mitochondrial genes. Nevertheless, thousands of studies have quantified polymorphism and divergence of the mitochondrially encoded "DNA barcode" cytochrome oxidase I gene (COI) in diverse species. While often providing valuable preliminary information, unfortunately, such data do just represent a single locus. Broader generalizations about demographic and selective processes require integration with many independent unlinked loci from the nuclear genome.

Further reading

Avise, J. C. (2009). Phylogeography: retrospect and prospect. *Journal of Biogeography* 36, 3–15.

Azevedo, L., Serrano, C., Amorim, A. and Cooper, D. N. (2015). Trans-species polymorphism in humans and the great apes is generally maintained by balancing selection that modulates the host immune response. *Human Genomics* 9, 21.

Bazin, E., Glemin, S. and Galtier, N. (2006). Population size does not influence mitochondrial genetic diversity in animals. *Science* 312, 570–2.

Charlesworth, B., Morgan, M. T. and Charlesworth, D. (1993). The effect of deleterious mutations on neutral molecular variation. *Genetics* 134, 1289–303.

Cruickshank, T. E. and Hahn, M. W. (2014). Reanalysis suggests that genomic islands of speciation are due to reduced diversity, not reduced gene flow. *Molecular Ecology* 23, 3133–57.

Cutter, A. D. and Payseur, B. A. (2003). Selection at linked sites in the partial selfer *Caenorhabditis elegans*. *Molecular Biology and Evolution* 20, 665–73.

Dean, M. D., Clark, N. L., Findlay, G. D. et al. (2009). Proteomics and comparative genomic investigations reveal heterogeneity in evolutionary rate of male reproductive proteins in mice (*Mus domesticus*). *Molecular Biology and Evolution* 26, 1733–43.

Eyre-Walker, A., Keightley, P. D., Smith, N. G. C. and Gaffney, D. (2002). Quantifying the slightly deleterious mutation model of molecular evolution. *Molecular Biology and Evolution* 19, 2142–9.

Gillespie, J. H. (2000). Genetic drift in an infinite population: the pseudohitchhiking model. *Genetics* 155, 909–19.

Halligan, D. L., Kousathanas, A., Ness, R. W. et al. (2013). Contributions of protein-coding and regulatory change to adaptive molecular evolution in murid rodents. *PLoS Genetics* 9, e1003995.

Li, H. and Durbin, R. (2011). Inference of human population history from individual whole-genome sequences. *Nature* 475, 493–6.

Malaspinas, A. S., Westaway, M. C., Muller, C. et al. (2016). A genomic history of Aboriginal Australia. *Nature* 538, 207–14.

Maynard Smith, J. and Haigh, J. (1974). Hitch-hiking effect of a favorable gene. *Genetical Research* 23, 23–35.

Miller, W., Schuster, S. C., Welch, A. J. et al. (2012). Polar and brown bear genomes reveal ancient admixture and demographic footprints of past climate change. *Proceedings of the National Academy of Sciences USA* 109, E2382–90.

Ng, S. B., Turner, E. H., Robertson, P. D. et al. (2009). Targeted capture and massively parallel sequencing of 12 human exomes. *Nature* 461, 272-U153.

Nordborg, M., Hu, T. T., Ishino, Y. et al. (2005). The pattern of polymorphism in *Arabidopsis thaliana*. *PLoS Biology* 3, e196.

Phung, T. N., Huber, C. D. and Lohmueller, K. E. (2016). Determining the effect of natural selection on linked neutral divergence across species. *PLoS Genetics* 12, e1006199.

Proschel, M., Zhang, Z. and Parsch, J. (2006). Widespread adaptive evolution of *Drosophila* genes with sex-biased expression. *Genetics* 174, 893–900.

Stadler, T., Haubold, B., Merino, C. et al. (2009). The impact of sampling schemes on the site frequency spectrum in nonequilibrium subdivided populations. *Genetics* 182, 205–16.

Wakeley, J. (1999). Nonequilibrium migration in human history. *Genetics* 153, 1863–71.

CHAPTER 8

Molecular deviants

Sequence signatures of selection and demography

Our foray into molecular population genetics has focused on describing some quantitative predictions of the standard neutral model (SNM) and some qualitative predictions for when the real world violates its assumptions. But how do we integrate the two? How can we test quantitatively whether selection or historical demographic changes affect the genome? In the jargon of molecular population genetics: can we determine whether a dataset *deviates from neutrality* and, if so, what is the cause? Yes, we can! We will now dive into some specific ways that we can analyze molecular data to see whether or not they differ from the predictions of the SNM, and how to tell why they differ.

When we look at DNA polymorphism and divergence, remember that deviations from neutrality can arise from demographic causes (historical population size changes, non-random mating, population structure) and from natural selection (positive, negative, balancing). Some tests aim to isolate selection as the driver of non-neutral **signatures** in data, whereas others merely identify departures from equilibrium irrespective of the cause. Some tests are sensitive only to very recent timescales and only cover the timeframe of the coalescent within a species or less, whereas other methods only detect more ancient selection that accumulates over the long term in divergence between species.

In this chapter, we will walk through the logic and mechanics of some of the most common approaches to quantifying molecular deviations from neutral evolution. In many cases, these tests will use neutral polymorphism or neutral divergence as a reference point. In reality, changes to the corresponding sites will be "effectively neutral," based on sites in the genome that are as selectively unconstrained as is possible for us to examine. Our goal with these methods is to read the story of evolutionary history from the genome, the words of that story having been recorded in the DNA, and the story written by a bickering committee of authors: natural selection and genetic drift and mutation and the demography of populations.

8.1 Skewed patterns of polymorphism

Although we can see evolution happen by monitoring changes in allele frequency over a few generations in the wild, our lifetimes are too short, with some beautiful and sad exceptions (Box 8.1), to watch the kind of adaptations evolve that most capture our imaginations from a walk in a forest, from snorkeling on a reef, or from driving

A Primer of Molecular Population Genetics. Asher D. Cutter, Oxford University Press (2019).
© Asher D. Cutter 2019. DOI: 10.1093/oso/9780198838944.001.0001

Box 8.1 CONTEMPORARY EVOLUTION IN ACTION: EXCEPTIONS TO THE PERCEPTION OF SLOW ADAPTATION

Evolution by genetic drift takes a long time, but responses to selection can drive rapid heritable change that is visible within our lifetimes. Figure 1 (A) Evolution of early flowering by purple loose-strife (*Lythrum salicaria*) allowed it to advance its range north, where it now represents an iconic invasive species in Canada (Colautti and Barrett 2013). (B) Alleles resistant to the antibiotic erythromycin have risen rapidly in frequency in the bacterium *Streptococcus agalactiae* over just decades among human infections (Lamagni et al. 2013), as has resistance to antibiotics in many other bacteria. (C) Rapid evolution in peppered moths (*Biston betularia*) has led to rarity of dark-colored genotypes, controlled by the *cortex* locus (see Figure 2.3), since the decline of smoke pollution in England over the past 50 years (data redrawn from van't Hof et al. 2016). Data replotted for *Streptococcus* resistance courtesy T. Lamagni and Public Health England (©Crown copyright 2018, contains public sector information licensed under the Open Government Licence v3.0). (photo of *Lythrum* courtesy R. Colautti and S. C. H. Barrett). Peppered moth © IanRedding/Shutterstock.com.

Figure 1

along a straight desert highway. It is natural to want to know what selection pressures or population changes have been pushing species in the relatively recent past to leave molecular scars in their DNA. To understand this "recent" timescale of evolutionary change or of demographic history, nucleotide polymorphism is our friend. The sites in the genome that were perturbed by recent events will often still be polymorphic. We can use non-neutral signals in the patterns of polymorphism as molecular "canaries in the coal mine" of the genome to indicate the locations where selection might have been most important and how potent non-equilibrium demography might have been.

8.1.1 *Too little or too much polymorphism*

One of the simplest predictions about how directional selection will affect the genome is that it will reduce the amount of genetic variation. This prediction holds true both for the locus that is the direct target of selection itself, as well as for nearby loci because of linkage through genetic hitchhiking or background selection (see section 7.1.1). A simple validation of the direct effects of selection is that, when we measure DNA sequence variability separately for replacement sites and synonymous sites, purifying selection on replacement sites causes them to have much lower polymorphism (see Figure 3.4). In this case, we can think of changes to synonymous sites as being selectively neutral and so they provide us a baseline neutral hypothesis to compare to the pattern at replacement sites.

We can take polymorphism data a step further in searching for non-neutral patterns by incorporating the linear order of DNA along chromosomes. Let's restrict ourselves to only measuring synonymous-site polymorphism in genes, so we have the straightforward neutral prediction that all genes should have similar values of θ_W. More accurately, we should expect that genes will have a distribution of θ_W values around some average, and we can predict the variability of that distribution from neutral coalescent simulations (Box 8.2). With this backdrop, put in your mind's eye the idea of visually scanning along the length of a chromosome from left to right, plotting the value of θ_W for genes as we slide along the chromosome from one gene to the next, with the aim of finding loci with pronounced reductions in the amount of synonymous-site polymorphism (Figure 8.1; see also Figure 7.5). Any spots along the chromosome with values of θ_W that are substantially below the range that we would expect to see, given variability in the coalescent process, thus reveal to us a signature of positive selection.

This kind of "fishing expedition" analysis that identifies parts of the genome that have unusual patterns of polymorphism, in this case a **valley of polymorphism**, is one type of **genomic scan** or **scan for selection**. Such scans are a bottom-up way of letting the genome tell us what are important targets of selection, rather than an anthropocentric top-down test of selection of a particular feature of an organism. One could also seek out locations on the chromosome that contain an excess of polymorphism at synonymous sites, relative to the baseline neutral expectation; this pattern would represent a signature of balancing selection (Figure 8.1). In this kind of scan, we measure neutral polymorphism, so, these sites themselves will not represent direct targets of selection. Instead, the selected targets would be linked to the loci with unusual patterns of polymorphism as a byproduct of genetic hitchhiking. Consequently, such genomic scans for selection are sometimes referred to as **hitchhiking mapping** of selected targets.

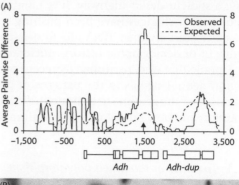

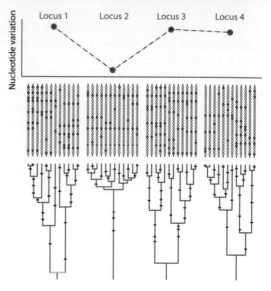

Figure 8.1 Sliding window scan for selection. Regions of the genome with molecular population genetic patterns more unusual than expected from neutral evolution help to pinpoint the targets of natural selection. (A) Analysis of the region surrounding *Drosophila*'s *Adh* locus by calculating nucleotide diversity within 100-bp windows along its length demonstrates how nucleotide polymorphism is especially high near position 1,490 within exon 4 of the *Adh* gene sequence, the position of the non-synonymous polymorphism that is subject to balancing selection (Kreitman and Hudson 1991). This idea of sliding windows of polymorphism can be generalized to scan across a whole chromosome, for example calculating θ_W on a per-gene basis or in bins of nucleotides of a constant length (see Figures 9.7 and 9.8). (B) Loci or genomic regions with especially low polymorphism, like locus 2 in the diagram, would provide candidates for deeper study as potential targets of positive selection. (photo of *D. melanogaster* courtesy E. Ho)

In addition, this approach uses the overall amount of synonymous-site polymorphism in the genome as the reference point and, as a result, any deviations from neutrality inherently rule out demographic changes as causing the skew in polymorphism. But one caveat is that any distribution will have a tail with extreme values. Another caveat to ruling out demography in genomic scans, however, is that contractions and bottlenecks can induce greater heterogeneity in coalescent times between loci than would occur under the SNM with a stable population size. Ideally, we would use a demographic scenario as part of our null model to incorporate such demographic non-equilibrium, rather than just relying on the SNM (Box 8.2). In practice, we would also want to incorporate the possibility that mutation rates could vary along the chromosome in such a way that the valley of polymorphism might simply be caused by an exceptionally low rate of mutational input. Recall that neutral divergence accumulates at a rate equal to the neutral mutation rate (see Chapter 4). Consequently, sequence divergence from a related species could help us deal with this issue of **mutation rate heterogeneity** across loci. We will return to this issue of needing to account for mutation rate heterogeneity in section 8.3.

Box 8.2 COMPUTER SIMULATIONS OF NEUTRALITY

Molecular population genetics theory lets us specify rules that define null models that we can then use to compare to data. These comparisons let us test whether those rules are sufficient to explain real-world observations. But, as I have emphasized, those basic rules have stochastic parts, meaning that they will give rise to a distribution of possible outcomes. Computer simulations let us reconstruct the plausible distribution of outcomes in a rigorous way.

Reverse-time coalescent simulation provides an efficient way to construct null distributions. These simulated gene trees (Figure 5.3) let you calculate expected distributions for measurable metrics like θ_W and D_{Taj} based on millions of simulated genealogies. Clever add-ons to the basic coalescent simulation algorithm allow you to incorporate recombination, population size changes, and multiple subpopulations with or without gene flow. This flexibility lets you define detailed hypotheses about population history that might differ from the SNM so as to test for selection or for the plausibility of alternative demographic scenarios.

Despite the efficiency of coalescent simulations for individual loci, genome-scale polymorphism data have actually made them too computationally burdensome to apply effectively to genomes. New and cunning approaches are being devised to perform forward-time simulations of the evolutionary process, exploiting modern computing power and the enormous scale of genomic data for population variation. A benefit of forward simulators is that they can more easily incorporate selection and recombination along genomes to provide powerful general tests of evolutionary hypotheses.

An algorithm for coalescent simulation:

(1) Simulate a coalescent gene tree.
   (a) For an initial sample of k haplotypes, set the times of these tip nodes to zero.
   (b) Draw a coalescent time t from an exponential distribution with rate parameter θ from the active set of lineages.
   (c) Randomly choose two lineages to coalesce.
   (d) Make a new internal node, record its time t and the identity of its descendant nodes.
   (e) Decrement the number of lineages that are actively being considered.
   (f) If just a single node remains, then stop, otherwise iterate back to step (1b).
(2) Randomly place mutations on the coalescent gene tree.
   (a) For each branch of length b, the number of mutations is drawn from a Poisson distribution with mean $\frac{1}{2} * \theta * b$.
   (b) The timing for each mutation along the length of a branch is drawn from a uniform distribution.
(3) Calculate polymorphism metrics for the mutations (e.g. θ_W, D_{Taj}).

An important statistical consideration in any genome scan or sliding window analysis is to acknowledge the fact that many statistical tests get performed simultaneously, which introduces the danger of producing false positive results. To deal with the potential problem of false positives, one should apply a **multiple-test correction**. The simplest multiple-test correction is the **Bonferroni correction**, in which the P-value threshold (α) for concluding a significant difference from the null model is simply divided by the number of tests being conducted (α/n). So, if a sliding window analysis were to include 1,000 independent statistical tests, then the usual P-value threshold of 0.05 would be adjusted to a value of $\alpha/n = 0.05/1,000 = 0.00005$. A drawback to this adjustment is that it is overly conservative, meaning that while it is effective in avoiding false positives, it does so at the expense of missing many true positives and therefore having a high false negative rate. Other common multiple-test corrections that help avoid this limitation include the **false discovery rate (FDR)** procedure or use of appropriate permutation testing methods.

8.1.2 *Skewed site frequency spectra*

When natural selection or demographic changes occur in a population, both would represent violations of the SNM. In sum, both of these factors can drive a shifted variant frequency spectrum: directional selection (positive or negative) and population expansion can lead to an excess of rare variants, whereas balancing selection, population contraction, and population structure can lead to a deficit of rare variants (see Chapter 7). How can we quantify such skews in the site frequency spectrum (SFS)?

In some ways, the SFS is cumbersome to work with and to think about because it is a vector of numbers (see section 3.3.2). Wouldn't it be nice if, instead, we could have just a single summary metric that could tell us how well an SFS from data matches the theoretical neutral SFS? Thankfully, in 1989, Fumio Tajima devised a convenient way to do just that, to calculate a single number that summarizes how much the SFS deviates from neutrality. Tajima's test presumes the infinite sites model of mutation, and therefore is well-suited to polymorphisms in DNA sequences. This metric, known as **Tajima's D** (or D_{Taj}), is used widely in molecular population genetics as a simple test of neutrality based on skews in the SFS.

The logic that goes into calculating Tajima's D is deceptively simple, which we can think about in terms of two quantities that we already know very well. Specifically, remember that we can quantify SNP variation in different ways (θ_π and θ_W) and that if the population size is stable and the locus is unaffected by selection, then both of these measures of DNA sequence variability will equal each other and also equal the scaled mutation rate parameter θ ($= 4N_e\mu$ at equilibrium in diploids; see section 3.3.1). But, importantly, we calculate θ_π differently from θ_W because θ_π incorporates both the haplotype frequencies and the count of differences between them, whereas Watterson's estimator θ_W uses only the count of segregating sites and ignores frequency information. This distinction in the use of frequency information means that selection and demography will influence θ_π and θ_W in different ways, which we can quantify by comparing their values to one another.

Rare variants and common variants both contribute the same amount to the value of θ_W: no matter the frequency, any variant counts as a segregating site. However, the value of θ_π is affected more by common variants because common variants create more pairwise differences between haplotypes. Consequently, the value of θ_π is more strongly inflated by balancing selection than is θ_W, because this type of selection will cause more segregating sites to occur at intermediate frequencies, yet the number of segregating sites will not be strongly altered. By contrast, directional selection will lead to fewer segregating sites than under neutrality and they will tend to occur at only very low frequencies; consequently, θ_W will be more affected by purifying selection than will θ_π, as compared to sites unaffected by selection.

Based on these ideas, we calculate Tajima's D as the difference between θ_π and θ_W, which we then scale by the variance in this difference:

$$D_{Taj} = \frac{\theta_\pi - \theta_W}{\sqrt{Var\,[\theta_\pi - \theta_W]}}. \tag{8.1}$$

To avoid confusion with other population genetic metrics that also use the letter "D," like the disequilibrium coefficient, I will refer to this metric as "Tajima's D" or as D_{Taj}. **Under neutrality**, according to the SNM, D_{Taj} is approximately equal to zero because, under neutrality, $\theta_\pi = \theta_W$. As a rule of thumb, whenever $|D_{Taj}| > 1.8$ it is likely that the locus deviates significantly from the SNM. However, the exact magnitude that indicates statistical significance will depend on the sample size, the number of polymorphic sites, and the amount of recombination in the locus.

The sign of D_{Taj} is determined by the difference between θ_π and θ_W because mathematically the denominator must be positive. So, just comparing the values of θ_π and θ_W gives us a quick way to gauge whether the data will suggest that the SFS might be skewed one way or the other. Purifying selection, positive selection, and population expansion all will cause $D_{Taj} < 0$ (i.e. $\theta_\pi < \theta_W$), reflecting an excess of rare variants. An excess of rare variants could also occur if the data are analyzed from a pool of individuals derived from very unequal sampling of two or more subpopulations because nearly all of the variants that are "**private**" to the less-well-sampled subpopulation will appear as low-frequency variants in the dataset (see Table 7.3). By contrast, balancing selection, population contraction, admixture, and pooled data from evenly sampled subpopulations all will make $D_{Taj} > 0$ (i.e. $\theta_\pi > \theta_W$), indicating an excess of intermediate-frequency variants.

Clearly, a wide variety of evolutionary perturbations will affect Tajima's D. By calculating D_{Taj} for many, many genes in the genome, the distribution of values for D_{Taj} provides a useful yardstick to gauge the overall degree to which a population conforms to the SNM (Figure 8.2). Any genome-wide trend will reflect the influence of demography on the SFS, because population size changes affect the whole genome. Note that, for genes, we would usually calculate D_{Taj} for synonymous sites, which we can presume to *not* be the direct targets of any selection. This is a roundabout way of saying that we would use D_{Taj} in this way to detect signatures of natural selection at linked loci. Similarly, we could look

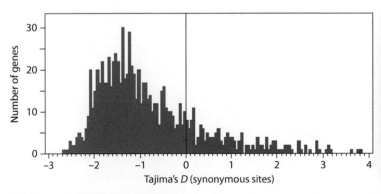

Figure 8.2 Distribution of Tajima's D.

The values for Tajima's D calculated for 876 *Arabidopsis thaliana* gene sequences from a collection of 96 copies of each locus show that the distribution is skewed toward negative values, compared to the standard neutral model expectation (black line at $D_{Taj} = 0$; data redrawn from Nordborg et al. 2005). Photo courtesy of https://www.minnesotawildflowers.info/© Peter M. Dziuk, 2008.

for evidence of genetic hitchhiking by calculating D_{Taj} for selectively unconstrained silent sites in non-coding regions of the genome. Because purifying selection will be common for replacement sites, which will contain direct targets of selection, generally we would observe a lower value of D_{Taj} for replacement sites than for synonymous sites.

Strong demographic effects make it challenging to infer selection from unusual signals on any given locus using Tajima's D alone (Figure 8.2). One way to gain some confidence that positive selection was the force that led a particular locus to have a negative value of D_{Taj} is to also evaluate the amount of polymorphism that it has. The combination of $D_{Taj} < 0$ and exceptionally low values of θ_π and θ_W would provide separate pieces of evidence that would be consistent with a selective sweep.

In part, the limitation of D_{Taj} in distinguishing selection from demography reflects the fact that it only uses information about polymorphism from a single species. That is, it is a summary of the folded frequency spectrum (see section 3.3.2). Given an appropriately close outgroup, however, you can **polarize** mutational changes to infer which variant is ancestral and which is derived, with derived variants having arisen uniquely in the species of interest. Doing so allows one to analyze the unfolded SFS, where we consider explicitly the relative abundance of derived variants (see section 3.3.2). Despite requiring an outgroup to identify the derived variants, the unfolded SFS still focuses on polymorphic sites and so does not incorporate divergence between species. We will go into detail in sections 8.2 and 8.3 about how we can use information from divergence per se to understand selection.

Those variants that are both derived and present at high frequency in the population are a hallmark of recent positive selection because positive selection drives the increase in frequency of the favored haplotype. Consequently, the distinction between the folded and unfolded SFS can be useful for inferring positive selection with metrics like **Fay and Wu's H_{FW}**, which is an unfolded SFS analog of Tajima's D introduced by Justin Fay and Chung-I Wu (2000). A similar distinction between folded and unfolded SFS can be applied to the **joint site frequency spectrum** of two or more subpopulations (Figure 8.3). The

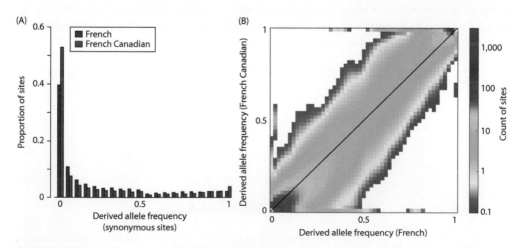

Figure 8.3 Unfolded and joint site frequency spectra.
SNPs from exome samples of French-speaking people from France and Canada can show the (A) unfolded SFS for each subpopulation individually or (B) together as a joint SFS (redrawn from Casals et al. 2013. Whole-exome sequencing reveals a rapid change in the frequency of rare functional variants in a founding population of humans. *PLoS Genetics* 9, e1003815). Site counts <1 in the joint SFS reflect the data analysis procedure for projecting data to 50 samples per population.

joint SFS is just a slightly more complicated version of the standard SFS that is extended to multiple populations. You can then draw inferences about selection, migration, and time to common ancestry between populations by comparing the observed joint SFS to what would be expected under neutrality.

The **Composite Likelihood Ratio** (**CLR**) approach to testing neutrality also uses information about the frequencies of derived variants. In particular, Yuseob Kim and Wolfgang Stephan showed in 2002 that you can combine the probabilities of observing a given number of derived variants for a particular sample size for all sites along a sequence. This procedure is known statistically as a "composite likelihood." These probabilities are partially correlated, however, because of the shared genealogical histories of linked sites. Consequently, this calculation of non-independent factors does not produce a conventional statistical likelihood, so it can't use a standard likelihood ratio test; instead you use simulations to create an appropriate test distribution. A big advantage of the CLR approach is that it includes information from the full frequency spectrum, unlike the simplistic distillation of the SFS that gets captured by Tajima's D and related metrics. Expansions on the CLR idea allow the approach to account for non-equilibrium demography and to implement it in a genomic context.

A major caveat to using the unfolded SFS, however, is that inferring ancestral states is itself prone to uncertainty (Chapter 5), and so in real data the cost to the uncertainty can sometimes outweigh the benefit. Part of the trick in knowing when the benefit might outweigh the cost is having that "appropriately close" outgroup: not so close as to create problems of incomplete lineage sorting, not so distant as to create problems with multiple hits. Some species are so polymorphic that multiple hits become an issue even for within-species polymorphism (Figure 3.2), and this violation of the assumption that mutation follows the infinite sites model also affects the SFS (Box 8.3). Specifically, multiple hits in polymorphism data will tend to produce a skew toward an excess of intermediate-frequency variants ($D_{Taj} > 0$).

Box 8.3 FINITE SITES MODELS OF MOLECULAR EVOLUTION

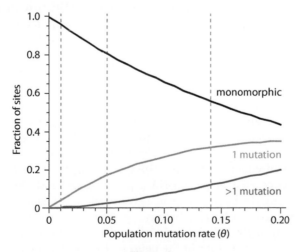

Figure 2

(Continued)

Box 8.3 CONTINUED

Figure 2 We have emphasized the elegant simplicity and real-world relevance of the infinite sites model of mutation: it is easy to think about there being just two alternate variants at a polymorphic site. This works pretty well for species with $\theta_\pi < 1\%$. For some species, however, this presumption of only two variants will fail. As the population mutation rate (θ) gets higher and higher, a greater and greater fraction of sites will be polymorphic from the input of more than one mutation (Tavare 1984). This concern applies most obviously to species that have very high levels of population polymorphism, like some viruses, bacteria, and very abundant eukaryotes that have values of $\theta_\pi > 5\%$ per site. For example, the highest nucleotide polymorphism among animals averages $\theta_\pi = 14\%$ in the genome of the nematode *Caenorhabditis brenneri* (see Figure 3.2), a species named after Nobel laureate Sydney Brenner, the geneticist and founder of *C. elegans* as a model organism. With $\theta_\pi > 5\%$, the proportion of polymorphic sites that are expected to have tri-allelic or tetra-allelic SNPs becomes high enough that we should no longer ignore it. Unfortunately, this means abandoning the infinite sites mutation model in favor of a finite sites model of mutation that allows the possibility of all four nucleotides being present as segregating SNPs. As modern genome sequencing technologies shift molecular population genetics away from being a "theory-rich data-poor" discipline, more realistic and more sophisticated finite sites models of mutation are becoming essential. By incorporating this added realism, we can better model evolution in terms of diversity, the variant frequency spectrum, and divergence.

There are a whole slew of other frequency spectrum methods to test for deviation from neutrality. These alternate metrics typically focus on a particular subset of the SFS. For example, it might be useful to look specifically at singletons because, after all, they are the most abundant kind of polymorphic site and they will tend to be made up of the most recent mutations by virtue of being rare. This emphasis on singletons is exactly the approach taken by Yun-Xin Fu and Wen-Hsiung Li (1993) with their metric D_{FL}. The idea behind **Fu and Li's D_{FL}** harkens back to Watterson's θ_W and the fact that we expect the total number of segregating sites S in a locus that is L nucleotides long to be $S = La\theta_W$, where a is that sample size correction factor of Watterson's (see section 3.3.1). However, the number of *singletons* that we should expect under neutrality is just $S_1 = L\theta_W$. Consequently, under the SNM, we would expect that $S_1 = S/a$, and if observed data differ from this prediction then it would provide evidence of deviation from the SNM. We calculate the D_{FL} metric analogously to D_{Taj}, except that D_{FL} uses this "singleton" neutral expectation, presuming that the singletons are all derived variants. The qualitative interpretation of D_{FL} also is similar to D_{Taj} in terms of what can cause an excess or deficit of low-frequency variants. The special emphasis on singletons, however, comes with some catches. First, errors in DNA sequencing are most likely to masquerade as singletons, thus inflating their abundance in a given real-world dataset. Second, we expect deleterious mutations also to be rare because of selection against them, so singletons presumed to be neutral might be confused with rare deleterious polymorphisms. Thirdly, the relative abundance of singletons in the neutral SFS also depends on sample size, with a greater proportion of polymorphic sites expected to be singletons in smaller samples, which must be accounted for in statistical testing (Figure 8.4).

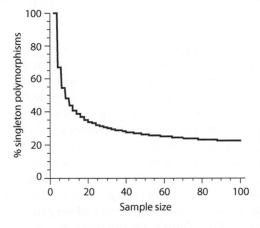

Figure 8.4 Dependence of the proportion of sites that are singletons on sample size.

The exact shape of the SFS depends on the number of copies of the locus that have been sampled. Statistical sampling theory predicts the rarest category of variants, singletons, to be most prevalent among polymorphic sites when sample sizes are small (see Equation 4.5).

In some situations, it is reasonable to assume that each new mutation creates a new unique allele, as for haplotypes of non-recombining DNA. The classic **Ewens-Watterson test** of neutrality is based on this infinite alleles model of mutation. It works by comparing the expected homozygosity (G_{exp}) that you calculate from the data to the value expected under a neutral allele frequency distribution. You will recall from section 3.1 that we calculate the expected homozygosity from the sum of squared frequencies of all of the alleles at a locus, $G_{exp} = \Sigma p_i^2$. To compare this value to the neutral prediction, it is easiest to use coalescent computer simulations (Box 8.2). The basic idea is that if the value of G_{exp} from the observed data is more extreme than the range of values that the neutral coalescent process would predict, then the locus deviates from neutrality. To interpret the outcome of the Ewens-Watterson test, we can use the following logic. The magnitude of the value we calculate for expected homozygosity from data is most strongly controlled by the frequencies of common alleles, whereas rare alleles contribute little to G_{exp}. This is just a simple byproduct of the fact that the square of a small number, like a rare allele frequency, is a *very* small number. When G_{exp} is greater than predicted under neutrality, then it indicates that the most common allele is *even more common* than expected and that, consequently, there are also more rare alleles than expected. This pattern is an excess of rare alleles, which can be caused by purifying selection, positive selection, or by population expansion if we see the pattern at most loci in the genome.

Even in the parts of genomes that recombine, haplotype information can be powerful to give insight into the history of selection at the molecular level. When a selective sweep is in process but not yet complete, termed a "partial sweep," we should expect to find the beneficial mutation on a long haplotype block that is at high frequency in the population (>10% but less than 100%). We should see such **extended haplotypes** during an ongoing selective sweep because the short timeframe will mean that recombination won't yet have had time to shorten the haplotype around the beneficial target of selection. Diploid individuals will also tend to be homozygous for these haplotypes, if selection favors a beneficial variant within them. This logic forms the basis of a collection of metrics, like the **EHH** (extended haplotype homozygosity) approach devised by Parvati Sabeti and her colleagues (Vitti et al. 2013). For example, EHH has been applied in genome-wide scans for evidence of which parts of genomes have been subjected to very recent adaptive evolution in humans. Other versions of EHH include comparisons between populations (the "cross-population" **XP-EHH** metric) and the integrated haplotype score (**iHS**) that polarizes ancestral and derived changes.

8.2 Sequence divergence and differentiation

Evolution is defined by the accumulation of heritable changes, so a natural way to look for how molecular evolution proceeds is to compare DNA from different species or different populations. This comparative approach to testing for non-neutral evolution focuses on a longer timescale than do approaches that focus on polymorphism alone. Even with this longer timescale, we need to be clear about how we quantify sequence differences. Saying that distinct populations or species will accumulate genetic differences sounds simple enough. But there are alternative ways to conceive of what one means by "genetic differences." Do you mean "fixed differences" such that each population has a unique variant? Or do you mean "allele frequency differences" such that the relative abundance of a given variant is distinct in each population, but all populations might contain some copies of all variants? These alternate views define molecular evolutionary **divergence** as the *absolute* sequence distance between groups of individuals, on the one hand, and, on the other hand, genetic differentiation as the *relative* distinctiveness of groups (see section 3.6.1). Characterizing divergence and differentiation, each comes with its own benefits and assumptions, and both provide powerful means to understand the evolution of populations and their genomes.

Box 8.4 SELECTION ON SYNONYMOUS SITES

We often assume that synonymous sites evolve neutrally because changes to such sites do not alter the encoded amino acid. The nearly neutral amendment of the Neutral Theory tells us, however, that selection can distinguish even subtle fitness differences if N_e is large enough. In fact, alternative synonymous codons are not selectively equivalent in highly expressed genes for species with very large population sizes (N_e of 10^6 and higher, e.g. bacteria, yeast, fruit flies, nematode worms). Nematode worms illustrate how impressive natural selection can be by distinguishing even tiny fitness differences, with selection coefficients on the order of 1 in 10 million. This means that mutation of a synonymous site might be slightly deleterious or slightly beneficial, rather than effectively neutral ($2N_e|s| < 1$; see Chapter 4). Over the long term, such selection produces a genomic pattern of **codon usage bias**, in which one of the synonymous codons for a given amino acid generally is observed in highly expressed genes more frequently than expected at mutation-drift equilibrium (so-called, optimal or **preferred codons**).

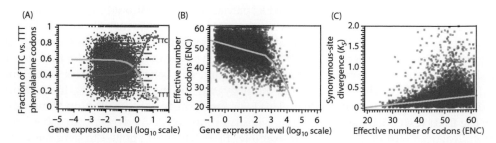

Figure 3

Figure 3 (A) Selection for codon bias is most pronounced among genes with high expression, as for the TTC preferred codon of phenylalanine in yeast that comprises 90% of phenylalanine codons among the most highly expressed genes (data redrawn from

Shah and Gilchrist 2011). Metrics like ENC (effective number of codons), F_{op} (frequency of optimal codons), and RSCU (relative synonymous codon usage) quantify such non-random codon usage in genes. (B) Highly expressed genes tend to have the most skewed codon usage patterns, as reflected by a small effective number of codons among genes with high expression in the nematode *C. remanei*. It is important to recognize this weak selection and to account for it appropriately in any analyses that assume selective neutrality of synonymous sites. (C) For example, mistakenly ignoring selection on synonymous sites will lead to underestimation of K_S (and overestimation of K_A/K_S), as for genes with biased codon usage that have low values for **ENC** in *C. remanei*.

8.2.1 K_A/K_S *ratios*

Analyzing polymorphisms in a population helps us to learn about recent and ongoing selective pressures, but oftentimes we are also interested in understanding long-term evolutionary trends. For this goal, we should look at those changes that distinguish different species or populations: mutations that represent fixed differences that we can measure as DNA sequence divergence for comparison with a neutral model of evolution. Protein-coding genes are a good place to look for non-neutral evolution; genes are plentiful in genomes and the genetic code imposes a convenient organizational structure to separate selected and unselected changes. Specifically, we can test for deviation from neutrality by comparing divergence at synonymous sites (K_S) as a neutral reference for the divergence at replacement sites (K_A). Remember that K_A measures the differences between species that alter the encoded amino acid sequence of orthologous genes, the number of non-synonymous substitutions per non-synonymous site; K_S is the number of synonymous

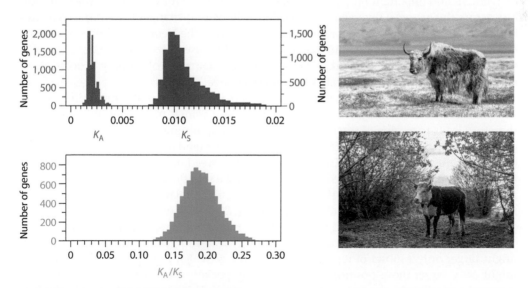

Figure 8.5 Distribution of divergence values.
Substitutions accumulate more slowly at non-synonymous sites than at synonymous sites, so that most genes will have a K_A/K_S ratio less than 1. Divergence between orthologous coding sequences aligned between yak and cow genomes illustrates how purifying selection leads to $K_A < K_S$ in these 10,000 genes (data redrawn from Qiu et al. 2012). Yak © Nowak Lukasz/Shutterstock.com. Hereford cow © Darko Zivlakovic/Shutterstock.com.

Figure 8.6 Rates of protein evolution and population size.

The rate of evolution of protein-coding sequences is slower, on average, in species estimated to have larger populations. A greater fraction of non-synonymous mutations will be non-neutral in larger populations, and thus get eliminated by purifying selection. Synonymous-site polymorphism (θ_W) provides a proxy for effective population size, when we assume that polymorphisms at synonymous sites are selectively neutral, point mutation rates are similar, and populations are at equilibrium ($\theta_W = 4N_e\mu$). The graph shows average values for 44 vertebrate and invertebrate taxa recalculated with Jukes-Cantor correction to K_S and θ_W, redrawn from data reported by Galtier (2016).

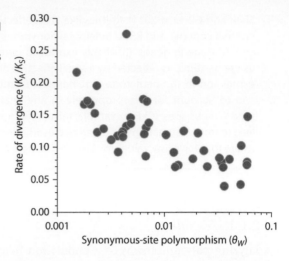

substitutions per synonymous site (see section 5.3). These two metrics summarize the amount of sequence divergence across the entire length of a gene. Normally, we would have applied an appropriate multiple hits correction in our calculation of K_A and K_S, rather than just relying on observed sequence differences (see section 5.3.2).

To get a simple metric that summarizes the rate of protein evolution relative to the rate of change that we would predict to occur solely from mutation and drift, we compute the **K_A/ K_S** ratio. A key assumption to "K_A/K_S ratio methods" is that changes to synonymous sites are effectively neutral because they do not alter the peptide sequence of the encoded protein. Usually, but not always, this assumption is reasonable (Box 8.4). As a result, we can use changes to synonymous sites as an internal neutral control. This "internal control" is a convenient and important property that allows us to account for variation in the mutation rate across the genome: divergence from effectively neutral changes to synonymous sites is a proxy for the local mutation rate, as we learned from the Neutral Theory (see sections 4.1.1 and 5.3.1). Mutation rate differences alone could cause differences in rates of protein sequence change, but this would reflect a selectively neutral evolutionary process when our goal is to identify non-neutral evolutionary change.

Purifying selection leads to fewer substitutions than expected under neutrality, and so induces K_A to be less than K_S. When summarized by just a single metric, purifying selection therefore leads to a value of the ratio $K_A/K_S < 1$ (with a minimum possible value of zero) (Figure 8.5). Species with larger effective population sizes tend to have lower average values of K_A/K_S, reflecting selection being more effective in weeding out deleterious mutations to replacement sites (Galtier 2016; Figure 8.6). Positive selection can lead to an accelerated rate of replacement-site substitution, which will result in relatively higher values of K_A. Finding a value of $K_A > K_S$, which means that $K_A/K_S > 1$, is strong evidence of repeated positive selection having favored the fixation of mutations to replacement sites.

However, positive selection does not necessarily lead to $K_A/K_S > 1$. Positive selection might target only a subset of replacement sites in a gene. For example, positive selection might only target those positions in the binding pocket of an enzyme, with purifying selection operating on the rest of the coding sequence. This means that, even in the face of positive selection, we will still often observe $K_A/K_S < 1$ for a gene overall. One way to test for statistical significance is to construct a 2×2 matrix of the counts of replacement and synonymous sites and differences, then to perform a contingency table statistical analysis such as **Fisher's exact test**. It turns out that it takes a lot of molecular evolution to detect

K_A/K_S significantly greater than 1: as a rule of thumb, Adam Eyre-Walker (2006) suggested that positive selection would need to drive about 70% of replacement-site substitutions to be detected as adaptive evolution. Consequently, the K_A/K_S test is **conservative** with respect to identifying positive selection on a gene despite being very powerful at detecting purifying selection.

Box 8.5 IDENTIFYING SITES OF ADAPTIVE EVOLUTION IN A PHYLOGENETIC FRAMEWORK

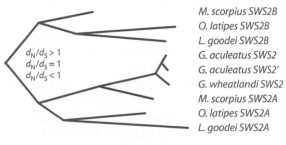

Figure 4

Some of the most compelling examples of evolutionary adaptation of phenotypes come by comparing across species: the phylogenetic comparative method. We can learn about adaptation in DNA sequence change by applying this same logic that different species take independent evolutionary trajectories from a common ancestor.

First, we take orthologous coding sequences from many species and align them, being sure to indicate the phylogenetic tree topology that relates the species to one another. Next, we apply explicit models of molecular evolution using Neutral Theory and Nearly Neutral Theory (i.e. d_N/d_S along the phylogeny). This requires sophisticated maximum likelihood or Bayesian statistical methods, but allows us to determine that particular codons or particular branches on the tree likely experienced adaptive evolution. This kind of inference is possible because each branch in a phylogeny represents a distinct evolutionary path with each codon represented on every one of those branches—each branch is a replicated unit of evolution for that codon. The phylogenetic replication of unusual patterns can allow us to distinguish neutral from non-random processes to say that selection must have driven the evolution of a given codon. Researchers may then perform sophisticated biochemical assays to determine whether the sites putatively responsible for adaptive evolution have important functional consequences.

Figure 4 Opsin proteins, which are involved intimately in the perception of light, illustrate how colonization of new environments can lead to molecular adaptation. In particular, *Gasterosteus aculeatus* stickleback fish that colonized "blackwater" lakes from clearwater ancestors experienced rapid evolution of the SWS2 opsin gene such that $d_N/d_S > 1$, which also conferred an adaptive shift in the wavelengths of light detectable by the fish (data redrawn from Marques et al. 2017).

The molecular changes detected with the macroevolutionary phylogenetic approach actually originated as microevolutionary DNA sequence changes within species, built on assumptions about how polymorphism within species leads to divergence. Therefore, it is important to understand how real-world differences from those assumptions might or might not alter how you interpret the output of a phylogenetic analysis of molecular evolution.

With a third, **outgroup** lineage included in a DNA sequence alignment, you can calculate branch-specific rates of molecular evolution and perform a **relative rate test** to see if the divergence along the two ingroup lineages are the same or if one ingroup lineage has accumulated significantly more substitutions than the other (see Figure 5.2). The idea is that an unrooted phylogeny for a locus with a simple molecular clock will have an identical number of substitutions between ingroup species A and its common ancestor with species B to the number species B has with that same common ancestor. Given outgroup species C that is equally distant from both species A and B in the accumulation of sequence divergence, we can write that predicted equality as $K_{AC} = K_{BC}$ (where K_{ij} is the sequence divergence between lineages i and j) and test whether or not it is true with a χ^2-test.

When replacement and silent substitution rates are equal ($K_A/K_S = 1$), then replacement substitutions have accumulated at the rate expected under neutrality. This could reflect the relaxation of purifying selection, more often abbreviated with the term **relaxed selection**. For example, **pseudogenes** experience relaxed selection, due to the elimination of purifying selection on them as a byproduct of prior mutations that have rendered the original gene non-functional. Alternatively, $K_A/K_S = 1$ might reflect a subset of replacement sites being subjected to positive selection with the remaining sites experiencing purifying selection. It is crucial that the sequence alignment is accurate when used for calculations of K_A and K_S, because errors in the codon frame caused by indels, sequence repeat motifs, or incorrect exon boundaries can also lead to spuriously high values of K_A.

Like all approaches, the K_A/K_S ratio has its own set of limitations for understanding molecular evolution. First, the K_A/K_S ratio method cannot easily detect balancing selection. This limitation is a simple byproduct of presuming adaptation to be mediated by fixed sequence differences between species, whereas balancing selection acts to maintain polymorphisms and preclude fixation. For very closely related species, the K_A/K_S method also encounters the challenges of ancestral polymorphism inducing incomplete lineage sorting and of a large fraction of differences between closely related species representing extant intra-specific polymorphism, rather than fixed differences (see section 5.3). In the face of this shared variation between the species that conflates polymorphism with divergence, it may not be possible to accurately infer the selection pressure from K_A/K_S that acts on genes. In such circumstances, polymorphism- or differentiation-based tests of neutrality should be applied instead of methods that presume strict divergence.

Another issue with closely related species is that there may be very few sites that differ between the orthologous sequences in calculating K_A and K_S. With few differences, there will be a lot of uncertainty in the accuracy of the divergence estimation, thus making the K_A/K_S ratio especially noisy. In general, methods that incorporate polymorphism are often better suited for detecting recent positive selection than the K_A/K_S ratio method. At the other end of the evolutionary timescale, involving very distantly related species, the multiple hits problem for homologous sequences presents another limitation of K_A/K_S analyses (see section 5.3). The multiple hits problem results in inaccurate estimation of the neutral substitution rate, so that K_S no longer acts as a reliable neutral standard for very divergent sequences. This is a problem for values of $K_S > 1$, meaning that, on average, each silent site has experienced more than one substitution since the common ancestor of the pair of species being compared (see Figure 5.11). Finally, when species differ in N_e or experience fluctuating population sizes over time, then the fraction of effectively neutral mutations (f_0) can change in ways that may be unanticipated from a naïve application of

K_A/K_S methods: smaller or repeatedly bottlenecked populations ought to exhibit higher K_A/K_S on average.

In talking about divergence, we have presumed a simple two-species alignment of orthologous genes to calculate K_A/K_S. In this situation, K_A and K_S are pairwise genetic distances between the two sequences in the alignment, with divergence being the combined divergence between each species and their common ancestor. This means we can't distinguish whether a high value of K_A/K_S is due to rapid evolution in one species' lineage, or the other's, or both, when we only compare the sequences of two species. To overcome some of the limitations of analyzing divergence in this way, it also is possible to analyze molecular evolution across many species simultaneously. This is the tack taken by phylogenetic approaches to quantifying divergence that use the repeated, independent evolutionary trajectories in each of a whole group of species to look for distinctive features of amino acid substitutions among them (Box 8.5).

To start a many-species analysis, we would construct a **multiple sequence alignment** of orthologous gene copies from as many species as possible from within some taxonomic group of interest. Using sophisticated maximum likelihood statistical methods, it is then possible to test for particular branches in the phylogeny that have unusually high rates of amino acid substitution. Divergence in this phylogenetic context is often termed ω or d_N/d_S, which are analogs of K_A/K_S, with the "N" subscript referring to "non-synonymous" sites. This multi-species approach uses outgroups to obtain lineage-specific rates of divergence along each branch, rather than the combined divergence between species. Alternatively, given enough species in the analysis, you can tweak this approach to test for particular codons in the multiple sequence alignment that have unusually high rates of substitution. Either of these scenarios would signal that positive selection has driven divergence of the protein sequence encoded by the gene, either in specific portions of the phylogeny or in specific portions of the gene (Box 8.5).

8.2.2 *Population differentiation and* F_{ST} *outliers*

The individuals of many species spread themselves over large ranges, forming subdivided populations that each can experience unique ecological conditions. So far, we have primarily thought of such population structure as a feature of demography. But those unique environments experienced by different subpopulations create the opportunity for population-specific selection pressures, leading to local adaptation and even initiating the formation of new distinct species. Consequently, **adaptive divergence** among subpopulations of a species represents an important way in which evolution works.

Adaptation does not only operate through species-wide adaptations. At the level of an individual subpopulation, we would see positive directional selection driving adaptation. But stepping back to the level of the entire species, local adaptation would look like spatially varying selection, one of the forms of balancing selection. So, how can we use ideas from molecular population genetics to learn about the way that genomes respond to this spatially structured kind of selection?

A key feature of population-specific selection is that the genetic loci that experience positive selection will have differences between populations. As a result, loci that experience distinct modes of selection in different populations, often referred to as loci subject to **divergent selection**, should show distinctive patterns of genetic variation across the species. In particular, we should expect that those selected loci would have *stronger* differentiation between populations than other loci in the genome. But in section 8.2.1, we learned that K_A/K_S is usually not appropriate for comparing very closely related groups like subpopulations of a species. What to use instead? Remember that we also have in hand F_{ST} as a metric to quantify genetic differentiation between populations (see sections 3.6 and 6.2). As a result, we can build on its use for understanding overall population genetic structure to also understand selective differences between populations.

Exceptionally strong differentiation at some loci would give us **candidate loci** that experienced the effects of divergent selection pressures among populations. We would define F_{ST} values higher than a statistical threshold from the null distribution of values as those loci that have exceptionally strong differentiation. Lower than expected values of F_{ST} would indicate that those loci have less differentiation than the rest of the genome, perhaps reflecting consistent, species-wide selection rather than population-specific directional selection.

Richard Lewontin and Jesse Krakauer worked out how much variation among loci (σ^2) we should expect to see in the values of F_{ST}, which for two populations turns out to be:

$$\sigma^2_{F_{ST}} = 2F^2_{ST}. \qquad (8.2)$$

This equation thus provides a simple neutral prediction for the range of genetic differentiation among loci that we should see in the absence of any local adaptation having driven excessive differentiation. For more than two populations, we just have to divide by the number of populations D, less one (i.e. divide $2F^2_{ST}$ by $D - 1$).

With this information about the shape of the distribution of F_{ST} values, we can simply perform a statistical test based on a χ^2 distribution to see which loci, if any, have values more extreme than we would expect by chance. This approach is called the **Lewontin-Krakauer test**. Remember that we calculate F_{ST} for nucleotide sites that are themselves selectively unconstrained, so this approach relies on hitchhiking with nearby direct targets of selection. Unfortunately, Equation 8.2 holds true only under fairly stringent assumptions about how migration works, and so this test can be misleading if gene flow does not follow the island model of migration.

As an alternative to the null distribution indicated in Equation 8.2, we could use computer simulations to produce a null distribution of F_{ST} values based on a realistic model of population structure and migration for a given organism (Box 8.2). When observed real-world values of F_{ST} are either too big or too small, compared to the distribution of simulated values, then we could conclude that they are significant **outliers** that have unexpectedly strong or unexpectedly weak differentiation between populations. Computer simulations are very flexible by letting us set up any demographic scheme with any design of migration among populations to define a null model.

The limitation of simulations is that we must personally decide what is an appropriate collection of parameters to define the basis of the simulations, in terms of population size, size change, migration paths, and migration rates. If we are wrong about the appropriate demographic scenario, then this will of course bias our interpretations because we would be using an inappropriate reference distribution.

A third approach takes the simplest view to using F_{ST} to identify loci that show exceptional amounts of differentiation. The idea is to simply look at the observed distribution of F_{ST} values and to pick the most extreme values as outliers, usually using a standard statistical criterion such as values that are at least two standard deviations away from the mean of the distribution (± 2 SD). If we have many loci (thousands or millions), then we can plot a very detailed empirical distribution to use as a reference null distribution.

The virtue of using the observed distribution as the reference point is that it inherently incorporates whatever peculiarities truly exist about the demographic history of the populations. The major drawback is that *every* distribution, including a neutral distribution, will have extreme values. So, there is no guarantee that the outliers will actually have been caused by a non-neutral process. This kind of outlier approach also isn't actually a test of an explicit neutral model. A more subtle problem is that the incidence of selection having truly driven genetic differentiation will affect our ability to detect it accurately. For example, if local adaptation drove a large fraction of loci to have especially strong differentiation, then the observed distribution of F_{ST} values will be wider than if there were no divergent selection. This means that the "2 SD" cutoff will lead us to pick *too few* selected loci if adaptive divergence is common and to pick *too many* selected loci if adaptive divergence is rare (Figure 8.7). It is not ideal to have the rate of *false* negatives and false positives depend on the rate of *true* positives in the underlying dataset.

It seems like a very practical approach to just look carefully at those loci with the most extreme F_{ST} values in scans of the genome, but could it sometimes lead us astray? Unfortunately, the answer is "yes." The problem is that loci that happen to have low polymorphism within local populations will also tend to have high F_{ST}, so genomic regions with low polymorphism for reasons unrelated to positive selection would be more likely to produce extreme values of F_{ST}. Background selection in low-recombination regions could produce exactly this effect, and it underlies non-random patterns of genetic differentiation along chromosomes in a number of species (Box 8.6). This issue arises because F_{ST} is a *relative* measure of genetic difference between populations, not an *absolute* measure of genetic divergence (see section 3.6.1). Mixing up relative and absolute measures of genetic difference can confuse molecular evolutionary inferences especially when comparing closely related species (Box 8.6).

The "population branch statistic," abbreviated as **PBS**, provides a more targeted version of the Lewontin-Krakauer test that focuses on differentiation of loci in a single particular subpopulation (Figure 8.7). This approach requires that we have three subpopulations available to us and that we are interested in testing for especially strong differentiation in one of them, generally one of the two most closely related populations of the three. In this way, it is similar in spirit to a relative rate test of divergence (see section 8.2.1). We start by calculating F_{ST} at a locus for each pair of populations, and we will assume that they all have the same population size. Luigi Luca Cavalli-Sforza showed that we can estimate the timing in generations of the onset of restricted gene flow T between populations A and B from F_{ST} as:

$$T_{AB} = -\log(1 - F_{ST}) \cdot 2N_e, \tag{8.3}$$

presuming that the timescale of differentiation is short enough that new mutations have at most a negligible influence. This timing essentially is a measure of the branch length separating the pair of populations, but it uses allele frequency shifts captured with F_{ST} rather than how we might ordinarily summarize divergence from mutational input. We can then look for an exaggerated branch length in our subpopulation of interest, deme A, relative to the other two subpopulations B and C by using the PBS metric:

$$\text{PBS} = (T_{AB} + T_{AC} - T_{BC})/2. \tag{8.4}$$

When a locus has symmetric differentiation in demes A and B relative to C, then PBS = 0. However, higher values of PBS indicate that the locus shows especially strong differentiation specifically within deme A compared to the other two subpopulations (Figure 8.7). A nice feature of PBS is that it is capable of detecting adaptive shifts in allele frequencies that represent only partial sweeps, and so indicates very recent adaptive evolution. This virtue was demonstrated by identifying selection on the gene *EPAS1* in

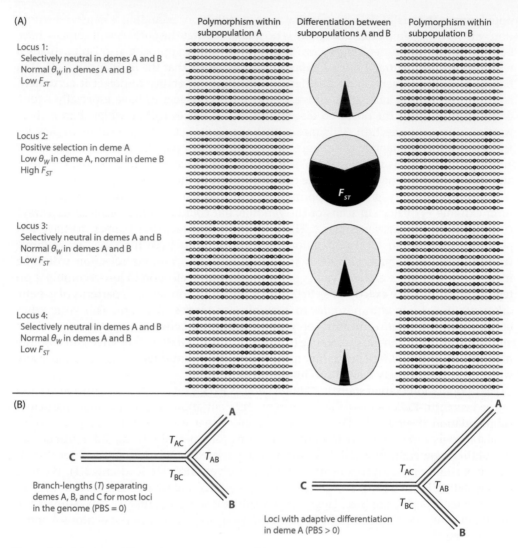

Figure 8.7 Adaptive differentiation among populations.

(A) Most loci will show some degree of genetic differentiation between distinct subpopulations due to genetic drift causing variants to occur at distinct frequencies within each of them (loci 1, 3, 4). However, we expect that loci associated with divergent adaptation, subjected to positive selection in only one subpopulation, will show especially strong genetic differentiation, such as unusually high F_{ST} (locus 2). Using this logic, we can identify candidate genes for deme-specific selective pressures by screening many loci for outlier values of F_{ST} with the Lewontin-Krakauer test or related approaches. (B) The PBS test is a derivative of the Lewontin-Krakauer test that uses genetic differentiation among three demes to infer deme-specific selection from exceptionally strong genetic isolation estimated from the timing of restricted gene flow (pink branch to deme A; pairwise population branch-length times T_{AB}, T_{AC}, T_{BC}) (Yi et al. 2010).

Tibetan humans as a population-specific adaptation to high altitude in the 2010 study by Xin Yi and colleagues that introduced the PBS method. The same approaches to detecting extreme values of F_{ST} also apply to PBS, with the simulation strategy being the best approach to generating a null distribution of values because it allows you to integrate additional biological details.

Box 8.6 LINKED SELECTION AND GENETIC DIFFERENTIATION

Identifying loci with extreme values of F_{ST} seems a logical approach to zoom in on genes likely to be important in local adaptation. However, the chromosomal structure of recombination can drive non-independence among loci. In particular, chromosomal regions with low recombination will experience a stronger influence of linked selection, through both background selection and genetic hitchhiking. As a result, many genes in low-recombination regions may give a signal of extremely high F_{ST}, as well as low θ_{π}. The linkage among the loci, however, means that not all of them are good candidates for the direct effects of divergent selection, giving a false impression of a large fraction of loci subject to divergent selection. This outcome is especially clear in the genomes of crows and other birds, as for the low-recombination region near position 10Mb on chromosome 13 in the crow genome (Figure 5; data redrawn from Vijay et al. 2017). Comparing relative and absolute measures of divergence can help to interpret such patterns in data (see section 3.6.1 and Box 7.2).

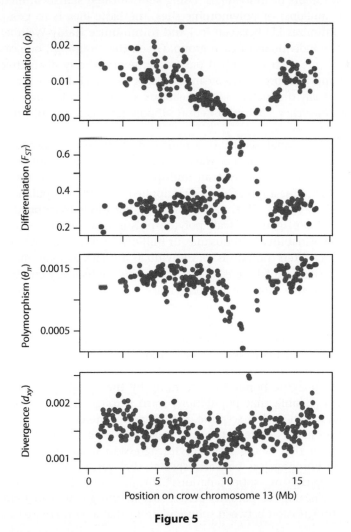

Figure 5

8.2.3 *Characterizing population structure and admixture*

For the purposes of phylogeography, K_A/K_S ratios and F_{ST} outliers are not much help. And sometimes we need to figure out how many subpopulations there are in the first place, how they formed, and what kind of gene flow they have. Figuring out these details of population structure also are important, of course, for developing an appropriate null model for tests of selection. Molecular population genomics, especially for humans, has led to a deluge of sophisticated statistical approaches to include as much information as possible to tackle this challenge. Here we will just give an overview for three of them.

Genetic drift leads different populations to have different allele frequencies at homologous loci, as well as distinct haplotype combinations. In a pooled sample of individuals, this produces linkage disequilibrium (LD) through the two-locus Wahlund effect (see section 6.2). An approach developed by Jonathan Pritchard and his colleagues takes advantage of this effect to help identify how many distinct subpopulations are present within a sample of individuals. Using sophisticated statistical methods applied to hundreds (or millions) of polymorphic sites, the basic idea is to group individuals in a way that minimizes LD between loci and to minimize Hardy-Weinberg deviations within loci for the individuals of each group. Given the groupings, one can then assess how much admixture each individual has from the genetically distinctive groups (see Figure 3.8). This approach provides a means of letting the data inform the researcher about how many genetically distinctive subpopulations are likely to be present in a sample, rather than the researcher having to define subpopulations ahead of time, as with F_{ST} calculations.

Principal component analysis (PCA), a multivariate statistical procedure that is well-known in ecology, provides another way of visualizing and quantifying genetic differences among individuals that may span multiple subpopulations or even non-discrete population structure. This approach was first applied to human population genetic data back in 1978 by Cavalli-Sforza (Menozzi et al. 1978), but was reinvigorated as a technique with the proliferation of genome-scale polymorphism data (see Figure 3.8). The idea is to distill down the variability in thousands or millions of microsatellite or SNP variants, from large numbers (often thousands) of individuals, into a small and manageable set of data dimensions. This data "dimensionality reduction" statistically combines correlated allele frequencies together into so-called principal component axes of variation, which can then be graphed as biplots of just two variables at a time. Distinctive clusters of genotypes in these plots can be used to quantify how many genetically distinct subpopulations are present in the data. Such analysis for human population genetic data shows that the two most important axes often correspond to continuous geographic variation in space (see Figure 3.8).

While we often imagine populations forming by the splitting of an ancestral population, it is also possible that populations form through fusion of pre-existing distinct subpopulations. In such a scenario, we often are interested in better understanding the admixture history. Alternatively, the genome of a population might contain just a small portion that is the result of introgression from another subpopulation. Oftentimes, it is introgression between species that researchers are testing for, but here I use the language of "subpopulations" for generality. For both of these issues, we can look to genealogical patterns across loci in the genome to help learn about admixture and introgression between subpopulations that are otherwise isolated without migration.

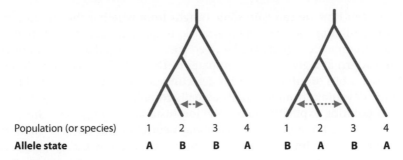

Figure 8.8 The ABBA-BABA test.

In a four-taxon genealogy with population 4 as the outgroup to define the ancestral state of a locus as "A," the ABBA and BABA patterns of variants among the three ingroup taxa are expected to occur with equal abundance. An excess of loci with the ABBA pattern will produce a value of $D_P > 0$, which is consistent with some amount of introgression of sequence between population 3 and population 2; $D_P < 0$ reflects an excess of BABA sites, consistent with introgression between population 3 and population 1.

One way to test for introgression with genealogies is to compare the pattern of data for bi-allelic loci from four subpopulations to the shape of gene trees that would be consistent with introgression, or not, from one of the subpopulations into the other subpopulations. If we have an ancestral allele A and a derived allele B at a locus, then we count up how often we see alleles in each population that are consistent with the "ABBA" pattern versus the "BABA" pattern in the gene tree for three ingroup populations or species (Figure 8.8). Both of these patterns differ from what we should expect from the branching topology of the tree, and so should reflect either introgression or incomplete lineage sorting; incomplete lineage sorting ought to be symmetric for alternative gene trees whereas introgression is likely to be biased toward one gene tree alternative or the other.

With DNA sequence data, we can count up corresponding SNP variants as C_{ABBA} and C_{BABA}, which presumes that we have just a single sequence from each population. From these counts, Nick Patterson and his colleagues (2012) showed how we can get a summary statistic: Patterson's $D_P = (C_{ABBA} - C_{BABA})/(C_{ABBA} + C_{BABA})$. In the absence of admixture or introgression, we expect D_P to equal zero because there should be equal numbers of incomplete lineage sorting events that yield an ABBA pattern and a BABA pattern. A value of $D_P > 0$ implies an overabundance of derived alleles shared between populations 2 and 3, whereas a value $D_P < 0$ implies an overabundance of derived alleles shared between populations 1 and 3 (Figure 8.8); both imply introgression between the populations (or species). Genotypes consistent with a pattern of BBAA, BAAA, ABAA, AABA, or BBBA are ignored for the purposes of this approach, as they match the pattern of population splitting with no introgression. This logic has been applied to data from humans and ancient DNA from Neanderthals, for example, to demonstrate introgression between these two species in our ancestry. Extensions of this **ABBA-BABA test** allow for added complexity, such as including multiple individuals per population, more populations or species, accounting for ascertainment bias, and ancestral state misinference.

8.3 Polymorphism and divergence

So far in this chapter we have detailed how we can use information from polymorphism alone or from divergence alone to test a given set of molecular data against the evolutionary change that we would expect to arise solely from mutation and genetic drift. We have seen how we can gain some insight from whether the amount of genetic variability is too high or too low, whether allele frequencies are skewed, how different classes of sites accumulate differences between species or populations at different rates, and how the pattern at a given locus compares to the rest of the genome. These are all individually useful bits of information for detecting deviations from neutrality. Another way to quantify deviations from neutrality actually combines some of these distinct pieces together, incorporating explicitly both polymorphism and divergence. This integration of information lets us hone our inference about neutral evolution even more finely by focusing on the general aim of detecting adaptive evolution as the specific source of a deviation from neutrality.

8.3.1 *The HKA test*

Let's recap some key points about neutral evolution to frame the ideas behind the HKA test. Mutational input creates polymorphism within populations and genetic drift creates divergence between populations by acting on those polymorphisms. Remember that the rate of input of neutral mutations equals the rate at which they accumulate as substitutions between species, and also remember that the equilibrium amount of neutral polymorphism depends on the mutation rate (see Chapter 4). As a result, those loci that experience a higher mutation rate will, in turn, retain more polymorphism and fix more substitutions, as compared to some other locus that happens to be subject to a lower rate of mutation. These effects of different mutation rates cancel each other out if we look at the **ratio of polymorphism to divergence(r_{pd})** for neutrally evolving sites. Consequently, the amounts of neutral polymorphism and divergence that loci have should be correlated. Furthermore, the ratio of polymorphism to divergence should be constant across loci for neutrally evolving changes. But, if natural selection targets a locus, then that selection will perturb its pattern of polymorphism and alter the value of r_{pd} relative to the value expected under neutrality. A test based on this idea is known as the **HKA test**, named after the three people who first described it in 1987: Richard Hudson, Martin Kreitman, and Montserrat Aguadé.

The HKA test requires two pieces of information: (1) nucleotide polymorphism information for two or more loci in one species and (2) divergence calculated from sequence alignments of homologous loci from a second species. Typically, we choose one locus to represent a "control" region that has changes evolving in a neutral fashion. This neutral region provides the null hypothesis for what value of r_{pd} we should see under neutrality (Figure 8.9). A second locus is a "candidate gene" that we are interested in testing for deviation from neutrality. By calculating divergence at silent sites in the candidate locus, we get the neutral baseline for that locus. The amount of polymorphism relative to that baseline defined by the divergence then gives us a standardized value of r_{pd} for comparison to the control locus (Figure 8.9). Key to the HKA test is that the sites at which we are measuring both polymorphism and divergence are themselves selectively unconstrained. However, the sites in the candidate gene might be linked to other sites that are direct targets of selection.

Positive selection at or near the candidate gene would lower polymorphism, whereas balancing selection would raise it. That is, the HKA test depends on "selection at linked sites" (see section 7.1.1). It turns out that the genetic hitchhiking does not affect divergence for linked but otherwise selectively neutral variants, despite inducing lower polymorphism at those linked sites. This important fact was shown by William Birky and Bruce Walsh in 1988, whereby the rate of divergence at selectively unconstrained sites is unaffected by linkage to direct targets of selection. However, linked selection that occurred

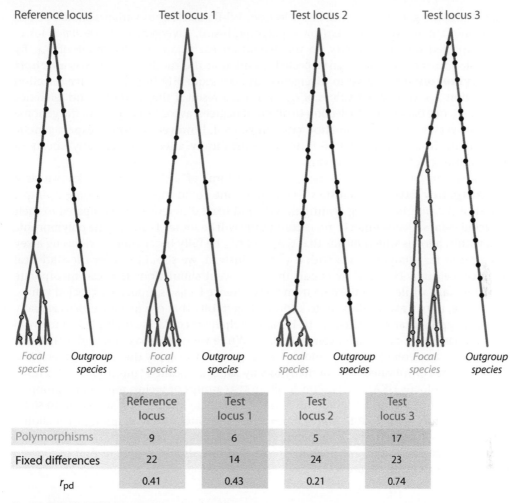

	Reference locus	Test locus 1	Test locus 2	Test locus 3
Polymorphisms	9	6	5	17
Fixed differences	22	14	24	23
r_{pd}	0.41	0.43	0.21	0.74

Figure 8.9 The HKA test.
By comparing the ratio of polymorphism to divergence (r_{pd}) of a candidate locus to reference loci with neutral changes, the HKA test evaluates how consistent evolution at the candidate locus is with neutrality. In these examples, the candidate test locus 1 differs in mutation rate from the reference locus, giving it both lower polymorphism and divergence, but yielding similar r_{pd}. By contrast, test locus 2 has similar divergence as the reference, but reduced polymorphism resulting in much lower r_{pd}, consistent with positive selection acting on sites linked to those used in the analysis of locus 2. Locus 3 shows an excess of polymorphism relative to divergence (higher than neutral r_{pd}), which could result from balancing selection.

in the ancestral population can lead to variation among loci in the amount of sequence divergence observed between descendant species, as showed by Tanya Phung, Christian Huber, and Kirk Lohmueller in 2016 (see Box 7.2).

What does all of this imply for the HKA metric, the value of r_{pd}? Conceptually, if we observe r_{pd} to be greater than expected, such that r_{pd} for the candidate gene is higher than the control locus, then it indicates that the candidate gene has an **excess of polymorphism** relative to the observed divergence. This excess of polymorphism relative to divergence is a signature of balancing selection (see section 7.1.2). This interpretation is appropriate if the sites being analyzed in the HKA test are themselves selectively neutral, though weak purifying selection on the sites could induce a similar pattern. Weakly deleterious mutations can persist for a while as polymorphisms before they eventually get driven extinct by selection. While deleterious variants should be less common than neutral variants (lower polymorphism), divergence will be *much* lower than expected under neutrality, so this would appear as an elevated value of r_{pd}. By contrast, a lower r_{pd} value than expected means that the candidate gene has a **deficit of polymorphism** relative to divergence and consequently implies positive selection (see section 7.1.1). Distinct values of r_{pd} also could result if the candidate and reference loci experience background selection to different degrees, as for loci in parts of the genome with low versus high recombination rates. In general, however, we would expect genetic hitchhiking from positive selection to more drastically affect r_{pd} than any effects of background selection from purifying selection.

Because the data in an HKA test are in the form of a 2×2 matrix known as a **"contingency table"** in statistics (Figure 8.9), one could imagine just doing a simple χ^2 test of the numbers of segregating sites (S) and fixed differences (K) compared to their expected values. Unfortunately, there commonly will be some LD among the polymorphic sites within a locus which means that they will not be fully independent, violating a key assumption for analyzing data with a χ^2 test. Instead, we should test for the statistical significance of an HKA test by conducting coalescent simulations that can incorporate partial linkage while using the data from our reference locus as input for expected values. In practice, it may also be difficult to know with certainty that our reference locus actually is a good representative for neutrality. It might help if we could use multiple loci to avoid the peculiarities of any single reference locus. While we have talked about the HKA test for two loci, a reference and a candidate locus, you can also apply the same approach in a slightly more complicated way for many loci by using a so-called "multi-locus HKA test." For the multi-locus HKA test, you must set up two groups of loci to represent a group of neutral reference loci and a set of one or more candidate loci that are suspected to share a similar history of selection. The r_{pd} also is used in hitchhiking mapping with sliding window scans for selection with population genomic data because neutral divergence acts to standardize the amount of polymorphism that one should expect to find in a given region of the genome (Figure 8.1).

8.3.2 *The MK test*

One goal of molecular population genetic tests of neutrality is to find the genetic targets of ongoing natural selection. Therefore, it would be especially convenient if we could detect adaptive evolution directly at the locus being examined, rather than depending on evolutionary signatures that come from linkage and genetic hitchhiking. The McDonald-Kreitman test does just that. Commonly abbreviated as the **MK test** and named after John McDonald and Martin Kreitman, this influential approach was devised in 1991. The MK test is conceptually related to the HKA test in that it also uses polymorphism information from one species and homologous sequence from a second species to calculate divergence. The difference is that the MK test focuses specifically on silent versus replacement changes at just a single locus, reminiscent of K_A/K_S analyses. The idea is that the Neutral Theory predicts that the ratio of replacement to silent polymorphisms (p_r/p_s) will equal the ratio of replacement to silent substitutions (d_r/d_s) because polymorphism is just a transient phase of the substitution process.

A simple "MK table" shows how the data are summarized: it is just a matrix of counts of the number of sites that are polymorphic (p_r and p_s) or fixed differences (d_r or d_s), summed separately for each of two classes of sites (Table 8.1). The polymorphisms come from the species we are most interested in, with divergence coming from alignment of

Table 8.1 MK table of polymorphism and divergence.

	Polymorphisms	Substitutions	Row Total
Replacement	p_r	d_r	$p_r + d_r$
Silent	p_s	d_s	$p_s + d_s$
Column Total	$p_r + p_s$	$d_r + d_s$	$p_r + p_s + d_r + d_s$

orthologous sequence to an outgroup species. Remember that fixed differences represent a special metric of divergence between species (see section 3.6.1), and d_r is different from K_A (and d_s different from K_S) because we are summing substitutions but not dividing them by the total number of replacement sites. If we are dealing with distantly related species, then we apply a multiple hits correction to the counts.

The MK test of neutrality aims to determine whether the divergence ratio d_r/d_s equals the corresponding polymorphism ratio p_r/p_s, using silent sites as a neutral standard. If $d_r/d_s = p_r/p_s$, then we should accept the null hypothesis of no positive selection on replacement sites (Figure 8.10). However, if we find an excess of replacement substitutions ($d_r/d_s > p_r/p_s$), then it implies that positive selection is driving fixed differences at replacement sites so that they accumulate to a greater extent than one would expect in the absence of positive selection (Figure 8.10). Although I have explained the MK test as the comparison of d_r/d_s to p_r/p_s, which I think is easiest to grasp, some people prefer to think of MK tables as a comparison of p_r/d_r to p_s/d_s (more analogous to how I described r_{pd} for the HKA test in section 8.3.1). This is totally fine to do. You just need to orient your interpretation appropriately. In any case, because the sites that are used to make the MK table are *all* interspersed throughout the *same* locus, we can use a standard χ^2 test to determine the statistical significance of the patterns in the data.

Usually we would have orthologous sequence from a single outgroup species, but that is just the minimum. When two or more outgroup species are used, rather than just one, then we can isolate and focus on just those substitutions that are specific to the branch of the lineage for the species we are most interested in (Figure 8.10). Such a **lineage-specific** analysis would make us more confident that any signature of selection arose in the history of the species we are focusing our analysis on.

The **neutrality index (NI)** provides a simple single metric introduced by David Rand and Lisa Kann in 1996 to summarize the polymorphism and divergence information from an MK table and to help with qualitative interpretation of this information:

$$NI = \frac{p_r}{p_s} \bigg/ \frac{d_r}{d_s} = \frac{d_s p_r}{d_r p_s}. \tag{8.5}$$

Positive selection will induce NI < 1, as a result of an excess of divergence at replacement sites. When, instead, NI > 1, we intuit that the locus contains an overabundance of replacement-site polymorphisms, which can arise from balancing selection or from slightly deleterious polymorphic variants at replacement sites not yet having been eliminated by purifying selection. A drawback to raw NI values is that their range of values is not symmetric around 1: the minimum is 0 and the maximum is infinity. For comparing among loci, then, it is nicer to use $-\log(NI)$ because the log-scale makes it easier to visualize how NI values of, say, 0.1 and 10 both imply a 10-fold magnitude of change (Figures 8.10 and 8.11).

This point about weak purifying selection in interpreting MK tests might seem subtle, but is important. It is important because it provides the tweak of logic required to interpret

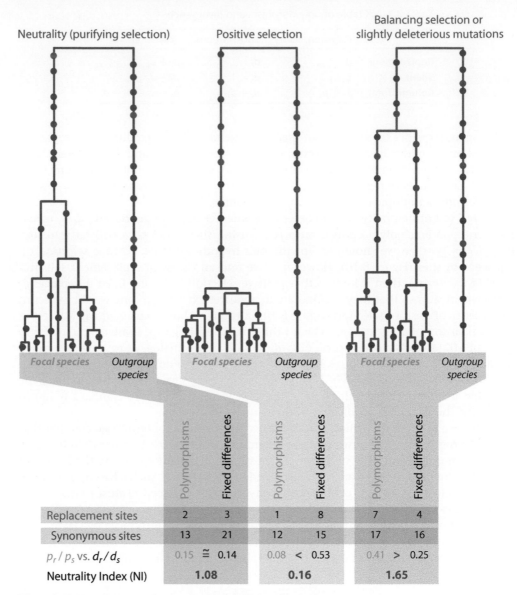

Figure 8.10 Visualization of polymorphism and divergence for an MK test.
An MK test contrasts the ratio of changes of replacement to synonymous sites for polymorphisms of a focal species (p_r/p_s) with fixed differences between species (d_r/d_s). Synonymous sites are used as a neutral reference to test for selection on replacement sites; only p_s and d_s would be used in an HKA test, by comparison. The genealogies in the diagram depict replacement changes as red dots and synonymous changes as purple dots. When p_r/p_s and d_r/d_s have similar values, as for the locus on the left, the neutrality index (NI) will be close to 1, indicating no evidence of positive or balancing selection on the replacement sites in the genealogical history. The locus depicted in the middle genealogy has a very low value of NI, consistent with positive selection causing at least some of the replacement sites to have been driven to fixation. A value of NI > 1, as indicated in the genealogy on the right, could result from balancing selection on the locus that maintains an overabundance of replacement-site polymorphisms, or it could result from the presence of slightly deleterious variants at replacement sites that have not yet been weeded out by purifying selection. By including a second outgroup lineage, it is possible to calculate divergence that is restricted just to the single focal species; otherwise it may be ambiguous which species' history contains an excess of replacement-site divergence.

an MK test in light of the more realistic Nearly Neutral Theory versus the narrower Neutral Theory. Mildly deleterious mutations arise frequently, but can take a long time to ultimately get purged from the population. During that time, deleterious mutations will contribute to calculations of polymorphism (p_r). But because they *do* eventually get eliminated rather than fixed, those mildly deleterious mutations do *not* factor into the value of divergence (d_r). In the metrics of the MK test, slightly deleterious mutations will lead to $p_r/p_s > d_r/d_s$ (NI > 1, Figure 8.11), and so can potentially lead us to underestimate the extent of positive selection on the locus. Fluctuations in population size over time could exacerbate this complication of weakly deleterious mutations. Population size fluctuations can cause the incidence of "effectively" deleterious mutations to differ for polymorphisms and fixed differences, because the proportion of weakly deleterious mutations that get purged by selection would similarly fluctuate over time. This complication of purifying selection on weakly deleterious mutations arises from the fact that the SNM oversimplifies the distribution of fitness effects of mutations as being only neutral or strongly deleterious. Remember that strongly deleterious mutations don't contribute much at all to either polymorphism or divergence because they get weeded out of populations so quickly by selection. As a consequence, MK tests are generally agnostic about strong purifying selection on a locus.

One of the most powerful features of an MK test is that, when it identifies non-neutral evolution, it implies that the sites *at that locus* are direct targets of natural selection. It evaluates two classes of sites that are interspersed within a single locus and it is inconsequential to the test whether or not selection acts at other linked loci (Table 8.2).

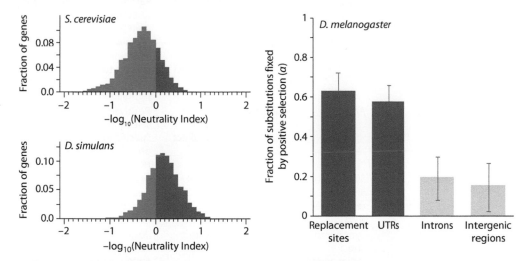

Figure 8.11 Neutrality index and fraction of adaptive substitutions.
(A) The distribution of $-\log_{10}$(NI) among genes from brewer's yeast (*S. cerevisiae*) is shifted toward negative values (i.e. NI > 1), potentially implying either that there are deleterious replacement-site polymorphisms present in most genes or that there is widespread balancing selection. By contrast, a much greater fraction of genes in the fruit fly *D. simulans* are consistent with positive selection having shaped their recent evolution: most genes have values of $-\log_{10}$(NI) > 0 (i.e. NI < 1). (NI data redrawn from Li et al. 2008) (B) In another fruit fly (*D. melanogaster*), estimates of the rate of adaptive substitutions imply that about 60% of nucleotide substitutions were fixed by positive selection, both for replacement sites and for untranslated regions that flank genes (UTRs) and contain important regulatory elements (data redrawn from Andolfatto 2005). Regulatory elements in intronic DNA and intergenic regions also likely are responsible for a lower but significant incidence of adaptive evolution, where synonymous sites were used as a neutral standard.

Table 8.2 Summary of tests of neutrality.

Test	Number of species or populations required	Number of loci required	Genomic context	Sensitivity to non-standard demography	Direct vs. indirect selection detected	Mode of selection best able to detect	Timescale of detection
Non-neutral θ	1	1	Any locus	High	Indirect	Recent selective sweep	Recent
Tajima's D (D_{Taj})	1	1	Any locus	High	Indirect	Recent strong positive selection at a single target	Recent ($<0.25\,N_e$ gen)
Ewens-Watterson	1	1	Any locus	High	Indirect	Recent selective sweep	Recent
HKA	2	2	Selectively neutral and candidate for selection	Moderate	Indirect	Repeated positive selection at multiple sites	Intermediate
MK	2	1	Synonymous and replacement sites	Low	Direct	Repeated positive selection at multiple sites	Intermediate
EHH	1	>100	Any genomic region	Moderate	Indirect	Partial selective sweep	Very recent
F_{ST}	2	1*	Any locus	Moderate	Direct and indirect	Positive selection specific to one population	Recent
PBS	3	1	Any locus	Moderate	Direct and indirect	Positive selection specific to one population	Recent
K_A/K_S	2	1	Synonymous and replacement sites	Low	Direct	Repeated positive selection at multiple sites	Ancient
d_N/d_S	>2	1	Synonymous and replacement sites	Low	Direct	Positive selection at a codon or in a set of lineages	Ancient

By contrast, the HKA test compares the same class of sites at different loci and so the "candidate locus" need not actually be the direct target of selection, but could simply be linked to some other selected locus by genetic hitchhiking or even from the influence of background selection. The interspersion of sites used in the MK test also makes it more robust to recombination within the locus and to non-equilibrium demography, compared to the HKA test. However, the MK test is not without limitations. Like any method that requires multiple species, incomplete lineage sorting of ancestral polymorphism for closely related species will complicate its interpretation (see section 5.3). The MK test also usually can only detect repeated bouts of selection having driven replacement-site substitutions, not just a single selective sweep. A more fundamental limitation is that the MK test is restricted to loci that have two categories of sites, such as coding sequences (replacement vs. synonymous), whereas the HKA test does not have this restriction in the kind of loci analyzed.

8.3.3 Extensions to the "MK framework"

The clever way that the MK test exploits the genetic code to divide up unconstrained sites and selected sites and polymorphisms and divergence turns out to have sparked more applications than just testing for deviation from neutrality. This "**MK framework**" also points to a way of calculating the fraction of replacement-site substitutions that were fixed by positive selection: the fraction of adaptive substitutions. This metric for selection also is popularly referred to as α (pronounced "alpha").

To calculate α using the MK framework (Table 8.1, Figure 8.10), again, the logic starts from the viewpoint of the SNM from Neutral Theory and then we will come back to think about the implications of nearly neutral mutations. We presume that mutations to synonymous sites are all neutral and that mutations to replacement sites will be either neutral, strongly detrimental, or advantageous. Strongly detrimental mutations will get eliminated by selection, so we can presume that they will not contribute to the magnitude of divergence. As a result, mutations that arise and fix to create divergence at replacement sites, d_r, could only have been neutral or beneficial. So, with our next steps, we need to figure out what fraction of them are neutral and subtract it off of the total to reveal the fraction that are advantageous.

Polymorphisms will be neutral no matter what kind of site we are looking at, because we shall make the approximation that selection will quickly drive non-neutral mutations to extinction or to fixation. Consequently, p_r/p_s tells us about the relative incidence of neutral mutation at replacement sites versus synonymous sites. We can then use this ratio to weight divergence to figure out how many neutral substitutions we should expect to see at replacement sites, given the overall accumulation of divergence: $(p_r/p_s) \cdot d_s$. By subtracting these neutral changes from the total number of changes to replacement sites (d_r), we get an estimate of the amount of divergence that accumulated from the only cause of fixation other than genetic drift: positive selection. Excellent. This chain of logic leaves us with a measure of the incidence of adaptive substitutions (d_α):

$$d_\alpha = d_r - (p_r/p_s) \cdot d_s. \tag{8.6}$$

If we then divide d_α by d_r, Nick Smith and Adam Eyre-Walker showed in 2002 that we get the fraction of sequence divergence that resulted from positive selection:

$$\alpha = \frac{d_\alpha}{d_r} = \frac{d_r - \frac{p_r}{p_s}d_s}{d_r} = 1 - \frac{d_s p_r}{d_r p_s}. \tag{8.7}$$

You should also note that the combination of these MK table values that we used to define the neutrality index (NI) also relate directly to α as $\alpha = 1 - \text{NI}$ (section 8.3.2; Equation 8.5).

A practical concern with calculating α is that, given that NI can take on values >1, it is possible that α could be <0, which has no convenient biological interpretation. One potential way this could happen is if balancing selection were prevalent, but this interpretation does not seem to be the case in real data, making it of little practical concern. Another way that we could arrive at negative values for α is if slightly deleterious replacement-site polymorphisms are present in the data: this is a more serious concern, which we will return to in a moment.

So what values of α do we usually see in genomes? The fraction of replacement-site substitutions fixed by positive selection tends to be lower in animal species with small N_e, like primates, with α in humans estimated to be approximately 0.2. Plant species also tend to have very low estimates of α. By contrast, the average value of α observed across animal species is 0.66 with some birds and invertebrates having α estimated to be greater than 0.8 (Galtier 2016). These high values are impressive, as they imply that 80% or more of the differences between species in their coding sequences got there as a result of positive selection.

We can also look at d_α in a slightly different way. If we divide d_α by d_s (using per-site measures), we get a ratio that looks strikingly similar to K_A/K_S: $\omega_a = d_\alpha/d_s$. But there is a powerful distinction between ω_a and K_A/K_S. The K_A/K_S ratio gives us the overall rate of change at replacement sites relative to neutrality whereas ω_a isolates specifically the *adaptive* rate of change at replacement sites relative to neutrality.

My overview of these applications of the MK framework in terms of replacement versus synonymous sites is the most common and simplest approach. But you can apply it to other site types as long as mutations to one group of sites can be presumed selectively neutral and mutations to the other group of sites are potential targets of selection. For example, the interspersion of regulatory elements in regions otherwise free of selective constraint can let us explore the role of adaptive evolution in gene regulation (Figure 8.11).

If these extensions to the MK framework to calculate α and ω_a seem almost too good to be true, it is because I have saved the caveats until now. First of all, by depending on ratios, they are susceptible to problems if either d_r or p_s equals zero. For some questions, we can get around this division-by-zero problem by combining information across many genes; in fact, α is usually calculated as an integrative measure of genome-wide adaptive evolution.

More conceptually, we know from Nearly Neutral Theory that deleterious mutations do not all have strong fitness effects, so what influence do slightly deleterious mutations have for α and ω_a? It takes time to eliminate slightly deleterious mutations, so they will occur more readily as polymorphisms than as divergence and so will cause p_r/p_s to be higher than under the SNM. As a result, Equation 8.7 would underestimate the true value of α; similarly, ω_a would be smaller than the true value. Population bottlenecks in a species' history could also lead to slightly deleterious mutations contributing to the magnitude of d_r, leading to overestimation of α. Because slightly deleterious mutations are a reality of nature, this is an inherent problem. One scheme to ameliorate it is to exclude all rare polymorphisms, because rare variants will be enriched for deleterious mutations. Despite helping to make estimates of α and ω_a less biased, this "frequency cut-off" approach is a somewhat arbitrary, imperfect, and unsatisfying solution. A better solution is to use more sophisticated statistical approaches than the simple calculation shown in Equation 8.7; current practice uses these more technical methods to more accurately quantify the incidence of positive selection on the substitution process, but the basic concept remains the same.

In answering some evolutionary questions, we do not actually need to pinpoint specific instances of adaptive fixation of particular beneficial variants. This is true in answering broad questions such as, how much of the genome is affected by adaptive evolution? We already saw the approach using the α metric that could address this issue. Now we will think about an alternative approach that I will refer to as the analysis of "substitution neighborhoods."

The idea behind the substitution neighborhood method is to look at *all* replacement sites in the genome that have fixed a new mutation in one population relative to a closely related species. We then quantify the amount of neutral polymorphism at synonymous sites as a function of distance away from that fixed replacement-site variant along the chromosome. If positive selection was important in causing the fixation of many of the replacement-site changes, averaged across the genome, then we should see reduced neutral polymorphism in the close vicinity of those substitutions. That is, we expect the substitution neighborhoods of targets of positive selection to show low polymorphism from the influence of selective sweeps. In a plot, this will look like a trough of polymorphism close to the position on the x-axis that represents the location of all the replacement-site substitutions superimposed on one another (Figure 8.12).

With stronger and more pervasive selection driving adaptive substitutions throughout the genome, the troughs of diversity in the substitution neighborhoods should be wider and deeper. This analysis can be repeated to assess neutral polymorphism in the substitution neighborhood of selectively neutral fixed mutations to synonymous sites, as a neutral control. To quantify how much and how strong selection must be to account for the patterns in data, one would then need to perform fairly sophisticated computer simulations of recurrent positive selection that incorporates recombination. Note also that this approach does not tell us anything about any given substitution, just the average effect across all of them.

8.4 Connecting quantitative genetics and molecular population genetics

Most of this book focuses on DNA to the exclusion of phenotypes. But wouldn't it be great to integrate DNA and phenotypes with an understanding of natural selection and other

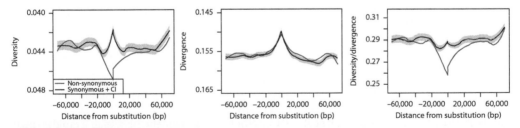

Figure 8.12 Diversity in substitution neighborhoods.
In the genome of *Drosophila simulans*, nucleotide diversity at synonymous sites is lower for sites that are closely linked to fixed replacement-site substitutions (data redrawn from Sattath et al. 2011). The position at 0 on the x-axis corresponds to 26,834 replacement-site substitutions superimposed from throughout the genome (or 66,984 synonymous-site substitutions). Neutral reference patterns indicate that this trough of linked variation is a result of the cumulative effects of positive selection, integrated across the replacement-site substitutions in the genome. In particular, no reduced polymorphism is seen near fixed synonymous-site substitutions, and divergence is not higher near replacement-site substitutions, indicating that heterogeneity in mutation rates cannot explain the pattern. Photo published under the CC-BY license. Sattath et al. (2011). Pervasive adaptive protein evolution apparent in diversity patterns around amino acid substitutions in *Drosophila simulans*. *PLoS Genetics* 7, e1001302. https://doi.org/10.1371/journal.pgen.1001302

evolutionary pressures? Genetic mapping studies help to make this happen, using the tools of quantitative genetics, molecular genetics as well as molecular population genetics. **Quantitative genetics** is the branch of population genetics that aims to understand the selective pressures and genetic basis to phenotypic traits that vary in natural populations.

The usual perspective of quantitative geneticists is that many genes can affect a given trait, which can be identified in a genome through **quantitative trait locus (QTL)** mapping. QTL mapping methods take many forms, including use of controlled cross-breeding experiments and genome-wide association studies (GWAS) (see section 6.2). A permutation of this idea is to map loci by associating allelic variation with a continuous environmental variable across a species' range, rather than an organismal trait, with the logic that allele–environment correlations should reflect the influence of natural selection. These approaches exploit molecular markers and LD to home in on loci that may have a direct connection to adaptive evolution, having let researchers cherry-pick those traits that they suspect to be especially important.

One approach to integrating molecular population genetic approaches with phenotypes is to test genomic regions at QTL for signatures of selection with methods like those outlined previously in this chapter. Another approach is distinct from both genetic mapping and the purely molecular population genetic view of evolution: it combines information from both molecular and phenotypic differentiation. Just as we can think of F_{ST} as telling us about the fraction of the total sequence variation that is due to differences between populations, we can make a similar metric for phenotypic variation. This phenotypic version of F_{ST} is termed Q_{ST}. The idea behind the Q_{ST}-F_{ST} method is that the molecular variation encapsulated in F_{ST} acts as a neutral reference point to use in identifying phenotypes that are more strongly differentiated than expected by chance. It aims to distinguish demographic differentiation in phenotypes caused by chance from the differentiation in phenotypes driven by selection. There are complications in comparing molecular differentiation to phenotypic differentiation because they do not share the same mutational process. Moreover, knowing that a trait was subjected to divergent selection does not tell us what underlying genes were involved and so we cannot directly view the genomic consequences of local adaptation with this approach.

Further reading

Andolfatto, P. (2005). Adaptive evolution of non-coding DNA in *Drosophila*. *Nature* 437, 1149–52.

Andolfatto, P. (2008). Controlling type-I error of the McDonald–Kreitman test in genomewide scans for selection on noncoding DNA. *Genetics* 180, 1767–71.

Birky, C. W. and Walsh, J. B. (1988). Effects of linkage on rates of molecular evolution. *Proceedings of the National Academy of Sciences USA* 85, 6414–18.

Casals, F., Hodgkinson, A., Hussin, J. et al. (2013). Whole-exome sequencing reveals a rapid change in the frequency of rare functional variants in a founding population of humans. *PLoS Genetics* 9, e1003815.

Colautti, R. I. and Barrett, S. C. H. (2013). Rapid adaptation to climate facilitates range expansion of an invasive plant. *Science* 342, 364–6.

Eyre-Walker, A. (2006). The genomic rate of adaptive evolution. *Trends in Ecology & Evolution* 21, 569–75.

Fay, J. C. and Wu, C. I. (2000). Hitchhiking under positive Darwinian selection. *Genetics* 155, 1405–13.

Fu, Y. X. and Li, W. H. (1993). Statistical tests of neutrality of mutations. *Genetics* 133, 693–709.

Galtier, N. (2016). Adaptive protein evolution in animals and the effective population size hypothesis. *PLoS Genetics* 12, e1005774.

Hoban, S., Kelley, J. L., Lotterhos, K. E. et al. (2016). Finding the genomic basis of local adaptation: pitfalls, practical solutions, and future directions. *American Naturalist* 188, 379–97.

Huber, C. D. and Lohmueller, K. E. (2016). Population genetic tests of neutral evolution. In: Kliman, R. M. (ed) *Encyclopedia of Evolutionary Biology*. Academic Press: Oxford. Vol. 3, pp. 112–18.

Hudson, R. R., Kreitman, M. and Aguadé, M. (1987). A test of neutral molecular evolution based on nucleotide data. *Genetics* 116, 153–9.

Kim, Y. and Stephan, W. (2002). Detecting a local signature of genetic hitchhiking along a recombining chromosome. *Genetics* 160, 765–77.

Kreitman, M. and Hudson, R. R. (1991). Inferring the evolutionary histories of the *Adh* and *Adh-Dup* loci in *Drosophila melanogaster* from patterns of polymorphism and divergence. *Genetics* 127, 565–82.

Lamagni, T. L., Keshishian, C., Efstratiou, A. et al. (2013). Emerging trends in the epidemiology of invasive group B streptococcal disease in England and Wales, 1991–2010. *Clinical Infectious Diseases* 57, 682–8.

Li, Y. F., Costello, J. C., Holloway, A. K. and Hahn, M. W. (2008). "Reverse ecology" and the power of population genomics. *Evolution* 62, 2984–94.

McDonald, J. H. and Kreitman, M. (1991). Adaptive protein evolution at the *Adh* locus in *Drosophila*. *Nature* 351, 652–4.

Marques, D. A., Taylor, J. S., Jones, F. C. et al. (2017). Convergent evolution of SWS2 opsin facilitates adaptive radiation of threespine stickleback into different light environments. *PLoS Biology* 15, e2001627.

Melville-Smith, R. and de Lestang, S. (2006). Spatial and temporal variation in the size at maturity of the western rock lobster *Panulirus cygnus* George. *Marine Biology* 150, 183–95.

Menozzi, P., Piazza, A. and Cavalli-Sforza, L. (1978). Synthetic maps of human gene frequencies in Europeans. *Science* 201, 786.

Nachman, M. W. and Payseur, B. A. (2012). Recombination rate variation and speciation: theoretical predictions and empirical results from rabbits and mice. *Philosophical Transactions of the Royal Society B-Biological Sciences* 367, 409–21.

Nielsen, R. (2005). Molecular signatures of natural selection. *Annual Review of Genetics* 39, 197–218.

Nordborg, M., Hu, T. T., Ishino, Y. et al. (2005). The pattern of polymorphism in *Arabidopsis thaliana*. *PLoS Biology* 3, e196.

Patterson, N., Moorjani, P., Luo, Y. et al. (2012). Ancient admixture in human history. *Genetics* 192, 1065–93.

Pease, J. B. and Hahn, M. W. (2015). Detection and polarization of introgression in a five-taxon phylogeny. *Systematic Biology* 64, 651–62.

Phung, T. N., Huber, C. D. and Lohmueller, K. E. (2016). Determining the effect of natural selection on linked neutral divergence across species. *PLoS Genetics* 12, e1006199.

Qiu, Q., Zhang, G. J., Ma, T. et al. (2012). The yak genome and adaptation to life at high altitude. *Nature Genetics* 44, 946–9.

Rand, D. M. and Kann, L. M. (1996). Excess amino acid polymorphism in mitochondrial DNA: contrasts among genes from *Drosophila*, mice, and humans. *Molecular Biology and Evolution* 13, 735–48.

Sattath, S., Elyashiv, E., Kolodny, O. et al. (2011). Pervasive adaptive protein evolution apparent in diversity patterns around amino acid substitutions in *Drosophila simulans*. *PLoS Genetics* 7, e1001302.

Shah, P. and Gilchrist, M. A. (2011). Explaining complex codon usage patterns with selection for translational efficiency, mutation bias, and genetic drift. *Proceedings of the National Academy of Sciences USA* 108, 10231–6.

Smith, N. G. C. and Eyre-Walker, A. (2002). Adaptive protein evolution in *Drosophila*. *Nature* 415, 1022–4.

Tajima, F. (1989). Statistical method for testing the neutral mutation hypothesis by DNA polymorphism. *Genetics* 123, 585–95.

Tavare, S. (1984). Line-of-descent and genealogical processes, and their applications in population-genetics models. *Theoretical Population Biology* 26, 119–64.

van't Hof, A. E., Campagne, P., Rigden, D. J. et al. (2016). The industrial melanism mutation in British peppered moths is a transposable element. *Nature* 534, 102–5.

Vijay, N., Weissensteiner, M., Burri, R. et al. (2017). Genomewide patterns of variation in genetic diversity are shared among populations, species and higher-order taxa. *Molecular Ecology* 26, 4284–95.

Vitti, J. J., Grossman, S. R. and Sabeti, P. C. (2013). Detecting natural selection in genomic data. *Annual Review of Genetics* 47, 97–120.

Wright, S. I. and Charlesworth, B. (2004). The HKA test revisited: a maximum-likelihood-ratio test of the standard neutral model. *Genetics* 168, 1071–6.

Yi, X., Liang, Y., Huerta-Sanchez, E. et al. (2010). Sequencing of 50 human exomes reveals adaptation to high altitude. *Science* 329, 75–8.

CHAPTER 9

Case studies in molecular population genetics

Genotype to phenotype to selection

Collections of DNA for multiple loci give us the raw material for studying evolution. We quantify genetic variation, in the form of SNPs or other mutational differences, and summarize it with polymorphism metrics and the site frequency spectrum (see Chapter 3). We can compare this molecular population genetic data within a species to orthologous DNA sequences from related species in order to calculate divergence as fixed mutational substitutions in the loci (see Chapter 5). The Neutral Theory gives us a baseline reference point for what we should expect to see, all else being equal (see Chapter 4). It details the patterns that we should expect to see for polymorphism and divergence under circumstances in which natural selection does not occur. Only by ruling out the possibility that the observed patterns could be easily explained simply by neutral processes can we infer that natural selection has operated (see Chapter 7). We can do this with tests to infer selection, including the HKA and MK tests that explicitly contrast observed data to a neutral scenario (see Chapter 8). So far, we have primarily discussed the conceptual issues relating to evolutionary forces. We will now fully bring these ideas to life! We will step outside the brief self-contained examples used so far to illustrate the basic principles. In doing so, we will deep-dive into several case studies of exciting real-world organisms that demonstrate the application from A to Z of the concepts we have developed throughout this book.

9.1 Coat color adaptation in mice

When you go for a walk in nature and pay close attention, you can't help but be amazed at the color matching that many animals have with the backdrop of the landscape. It is shocking how well many animals blend in with their environment, so cryptic are they in coloration and shape and even behavior. Extreme examples of camouflage include some leaf-mimicking insects like walking sticks and katydids whose shape mimics leaves so closely that they gently wave their bodies as if caught in a faint breeze and appear to have brownish discoloration and shape malformation on their lush green bodies as if chewed on by herbivores! For vertebrates, it is often the color of the ground that drives cryptic coloration, as in the white coats of snowshoe hares and arctic foxes that let them blend in to snowy backgrounds to either avoid predation or avoid detection by prey.

A Primer of Molecular Population Genetics. Asher D. Cutter, Oxford University Press (2019).
© Asher D. Cutter 2019. DOI: 10.1093/oso/9780198838944.001.0001

Figure 9.1 Coat color adaptation by desert rock pocket mice.

(A) Most of the desert in southwestern North America is covered with sandy-brown rocks inhabited by *Chaetodipus intermedius*, except for locations with ancient volcanic lava flows that created a dark black substrate to the landscape (left vs. right). (B) The fur color of *Chaetodipus* mice usually blends in with the color of the habitat they are found in, with mismatched mice being extremely conspicuous to visual predators like owls and snakes. (C) The Pinacate site in southwestern Arizona was analyzed in detail by Hopi Hoekstra, Michael Nachman, and their colleagues (Nachman et al. 2003, Hoekstra and Nachman 2003, Hoekstra et al. 2004). (photos courtesy M. Nachman and H. Hoekstra, © 2002 National Academy of Sciences).

In the deserts of southwestern North America, small mammals like mice have fur coloration that lets them blend in with the sandy-brown rock substrate. But in the last 500,000–750,000 years, lava flows deposited black rock into a few spots near what is now the USA–Mexico border. All of this region overlaps the geographic range of the rock pocket mouse *Chaetodipus intermedius*. Different populations of *Chaetodipus* live in habitats marked by dramatically different color characteristics (Figure 9.1). Most members of the species live on light-colored rocks in Arizona and New Mexico and northern Mexico. These nocturnal mice have a light pelage (fur) that allows them to blend in to their environment and to avoid getting eaten by owls and other visual predators. However, the mice look different that inhabit those zones covered by the ancient lava flows with their large swaths of dark black rock substrate. The rock pocket mice that live among the dark rocks have dark fur (Figure 9.1). How did this camouflage evolve? Was it driven by natural selection? Can we see signatures of natural selection in the genome at the target of selection? What does molecular population genetic analysis tell us about this case of adaptation of pigmentation?

This striking difference in mouse fur color within the same species, along with the nearly perfect correspondence of dark-furred mice among dark rocks and light-furred mice among light rocks, prompted Hopi Hoekstra, Michael Nachman, and their colleagues to quantify genetic variation more closely. But where to start? It turns out that several genes were known to be involved in melanin production in *laboratory* mice (Figure 9.2). Eumelanin and pheomelanin are pigments produced by skin cells that give skin and hair their color. So, they first sequenced the DNA for one such gene, *Agouti*, to quantify polymorphism and to compare the allelic haplotypes to the pelage phenotypes of the mice. *Agouti* started out as their prime candidate because of its key upstream position in the pathway controlling melanin production (Figure 9.2). Expression of the *Agouti* gene is responsible for making

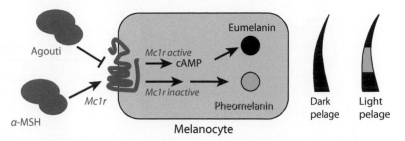

Figure 9.2 Coat color genetics in mammals.
Experiments in laboratory mice have helped establish the pathway leading to pigment production in melanocyte skin cells, which works similarly in other mammals. Depending on the activity of *Agouti* and *Mc1r*, as well as other downstream proteins, either dark eumelanin pigment will get made or red-yellow pheomelanin pigment will get made and deposited into a growing hair to determine its color. The individual hair shafts of most *Chaetodipus* mice first get a burst of eumelanin deposition at the tip, then a band of pheomelanin, and finally eumelanin at the base to give the mouse a mottled sandy appearance. Mice with dark fur, however, have eumelanin pigment deposited along the entire hair shaft (Hoekstra 2006).

hair follicles produce eumelanin (brown or black) or pheomelanin (yellow or red) pigment by acting as a negative regulator of the *melanocortin-1 receptor (Mc1r)*. A pulse of *Agouti* expression during part of the production of a hair results in the typical mouse light fur of most *C. intermedius*, such that eumelanin is present in the base and tip of the hair and pheomelanin is present in most of the middle portion of the hair shaft (Figure 9.2). Some laboratory mouse mutants of *Agouti* produce hairs containing only eumelanin. Much to their chagrin, however, *Agouti* haplotypes in nature showed no association with the coat color of the mice.

Hoekstra and Nachman next examined *Mc1r*, a gene encoding a G-protein coupled receptor, that operates downstream of *Agouti* in the genetic pathway (Figure 9.2). Amazingly, they found that mice from southwestern Arizona (Pinacate, Figure 9.1) with one haplotype of the *Mc1r* gene, containing four amino acid changing substitutions in complete linkage disequilibrium (LD) with one another, corresponded perfectly with the dark coat color phenotype of the mice. They refer to haplotypes containing these four amino acid replacements as *D* "dark" alleles, and other haplotypes as *d*. Individual mice having either *DD* or *Dd* genotypes had dark pelage, whereas all of the *dd* mice were light in color (Figure 9.3). This correlational finding led them to collect additional data to determine whether *Mc1r* was the actual target of natural selection.

They sampled mice along a 35-km transect across the Pinacate region from light- to dark- to light-colored habitat, and found the *D* haplotype of *Mc1r* to occur at high frequency only within the dark lava flow area (Figure 9.3). If simple population structure were responsible, then we should see a similar pattern at other loci because demography should affect all loci in the genome. Instead, they found the frequency of the most common mitochondrial DNA haplotype (which is not associated with coat color) to be *unrelated* to habitat type. This indicates that demography, in the form of population structure, is not likely to be responsible for the pattern observed at *Mc1r*.

In addition, they examined patterns of nucleotide polymorphism in *Mc1r*. Genetic differentiation between the alternative haplotypes of *Mc1r* on light vs. dark rock substrates in the Pinacate location was $F_{ST} = 0.48$, 8 times higher than F_{ST} for reference mitochondrial loci (0.06) and consistent with *Mc1r* as an F_{ST} outlier locus. They found 13 polymorphic

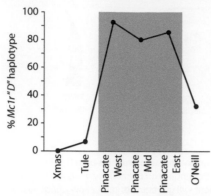

Mc1r haplotype combination	Number of mice with coat color phenotype	
	Dark	**Light**
DD	11	0
Dd	6	0
dd	0	12

Figure 9.3 Haplotype frequencies associated with coat color and geography.
In a transect across the dark rock lava flow at the Pinacate site in Arizona, most mice are dark on dark rock substrates and the stereotypical "*D*" haplotype of those mice is at high frequency. Pooling mice from all locations, all mice with at least one copy of a "*D*" haplotype have dark fur coloration (data redrawn from Hoekstra et al. 2004).

sites among the d haplotypes, and only 1 polymorphic site among the D haplotypes (Figure 9.4). Nucleotide diversity for *Mc1r* was about 20 times higher for d haplotypes: $\theta_\pi = 0.21\%$ per site among d haplotypes versus just 0.01% among D haplotypes. This rarity of polymorphism among D haplotypes is consistent with the idea that one (or more) mutations conferring dark coloration experienced positive selection in the dark rock habitat. However, they found no significant deviation from neutrality when they applied MK and HKA tests to *Mc1r* for these mice sampled from the dark lava rock substrate, suggestive of an incomplete selective sweep.

To see whether the amino acid changes to *Mc1r* were functionally important, they next examined laboratory cell lines that expressed *Mc1r* alleles with different replacement-site substitutions. These experiments found altered cAMP levels resulting from some of the changes, which is a telltale sign of altered production of eumelanin and pheomelanin (Figure 9.2). These results indicate a specific relationship between *Mc1r* haplotypes, biochemical activity, mouse phenotype, and ecological environment mediated by natural selection.

Selection for coat color in rock pocket mice varies over space within the species. Functional differentiation in *Mc1r* is strong, reflecting selection favoring D haplotypes on dark rock substrates, despite extensive gene flow at other loci across the light–dark substrate boundary that the rock pocket mice live on. We see molecular signatures of this selection in the especially high F_{ST} for *Mc1r*, in the strong LD across the *Mc1r* gene, and in the especially low nucleotide polymorphism among D haplotypes. At the scale of the species, we would interpret the case of *Mc1r* to be an example of balancing selection whereby variation is maintained at the locus due to spatially varying selection. At the local scale of a population on the dark rock of a lava flow, however, we would interpret it as a partial selective sweep in which the D haplotypes have reached high frequency but have not yet fixed. Partial sweeps are trickier to detect definitively from molecular population genetic data. Moreover, the high dispersal of the mice creates extensive gene flow in and out of the zone with dark lava rocks and this complicates tests of a simple neutral model. Indeed, despite the clear associations between genotype, phenotype, and environment, HKA and MK tests did not detect a significant deviation from neutrality.

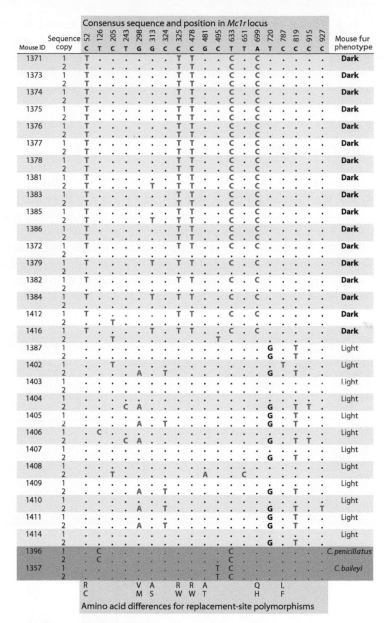

Figure 9.4 Table of polymorphism for *Mc1r*.

The 954-bp-long *Mc1r* coding sequence has 19 polymorphic sites, 8 of which represent replacement-site differences. When the coat color phenotype is mapped onto the haplotypes from each mouse, the four replacement sites at positions 52, 325, 478, and 699 in the sequence are in complete linkage disequilibrium among dark mice and define the "*D*" haplotypes (data redrawn from Nachman et al. 2003).

How general is the result that selection on coat color acts on allelic variants at *Mc1r*? Interestingly, Nachman and Hoekstra found a role for *Mc1r* in fur color differences of rock pocket mice only in the Pinacate region, whereas other lava flows showed no such association. Consequently, dark pelage appears to have evolved convergently by independent genetic mechanisms among mice in different portions of the *C. intermedius* species range. At a broader phylogenetic scale, however, it has turned out that other mutations to *Mc1r* lead to pigmentation adaptation in a diversity of other animals, including other mice like the beach mouse *Peromyscus polionotus*, several species of fence lizards (see Figure 2.3), as well as birds (Hoekstra 2006, 2010). Even humans (and Neanderthals) are polymorphic for *Mc1r*, being responsible for redheads!

9.2 Flight ability in butterflies

The enzyme phosphoglucose isomerase, *Pgi*, functions at a key branch point in the glycolysis pathway, the metabolic process that extracts energy in the form of ATP from glucose sugar molecules. In butterflies, *Pgi* is crucial for supplying energy-demanding flight muscles with ATP. It turns out that the populations for many butterfly species contain a diversity of allozyme isoforms of *Pgi*. For example, in *Colias* orange sulphur butterflies, individuals that are heterozygous for distinct isoforms of this dimeric protein show higher *Pgi* biochemical activity and perform better than homozygous individuals, according to several fitness-related metrics (Figure 9.5). Ward Watt and his colleagues found that

Figure 9.5 Functional effects of distinct allozyme genotypes in *Colias* butterflies.
The biochemical activity (V_{max} / K_m) of the *Pgi* protein dimer increases with temperature in *Colias eurytheme* butterflies, and is highest when different isoforms are together from heterozygous individuals (data redrawn from Watt 1983). Heterozygous individuals also have higher fitness, including through overall survival, male mating success, and female fecundity (data redrawn from Watt 2003).

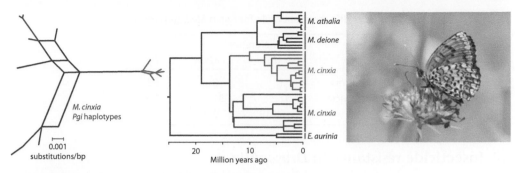

Figure 9.6 Molecular signatures of balancing selection on *Pgi*.
The genealogy for *Pgi* in *Melitaea cinxia*, shown as a haplotype network diagram, clusters haplotypes into distinct groups separated by long branches. Applying a molecular clock to the alternative *Pgi* haplotypes in the context of the *Melitaea* phylogeny shows that the alternate alleles pre-date speciation times separating other *Melitaea* species (*M. athalia* and *M. deione*), suggesting potential for trans-specific polymorphism (data redrawn from Wheat et al. 2010). Glanville Fritillary butterfly © Ivan Marjanovic/Shutterstock.com.

individuals heterozygous for *Pgi* allozyme variants flew more vigorously throughout the day, had higher survival, males had higher mating success, and females had higher fecundity. This classic phenotypic analysis from the 1980s that combined measurements in nature and biochemical analysis pointed to heterozygote advantage at *Pgi*. Might such balancing selection leave a molecular signature in the genome? We don't have the answer for *Colias*, but it turns out that other butterflies also show substantial polymorphism for *Pgi*.

Melitaea cinxia, the Glanville fritillary butterfly found in the Åland Islands of Finland, gives a nice example of molecular population genetic analysis of *Pgi* (Figure 9.6). Nucleotide diversity for the *Pgi* gene in *M. cinxia* is over 10 times higher than for other genes in the genome and with very high values of Tajima's D. Christopher Wheat and his colleagues estimated $\theta_{\pi\_syn} = 0.046$ for *Pgi*, and calculated D_{Taj} at synonymous sites to be +2.48, which lends support to the idea that balancing selection operates on *Pgi*. Remember that we generally predict balancing selection to yield an overabundance of polymorphism and unusually high values of Tajima's D relative to neutral reference loci. Indeed, in contrast to *Pgi*, two reference loci had a much lower average $\theta_{\pi\_syn} = 0.0015$ and $D_{Taj} = 0.94$. LD is strong across the gene, with haplotypes corresponding well to previously known allozyme alleles. A multi-locus HKA test shows an excess of polymorphism in *Pgi* (Table 9.1), though an MK test did not find statistical evidence of deviation from neutrality through repeated replacement-site substitutions. The divergence of alternative haplotypes was estimated to have arisen longer ago than the divergence between *Melitaea* species (Figure 9.6). Coalescent simulations rule out any obvious influence of demographic population size change as being a potential cause of the patterns, although relatively few loci were analyzed to calibrate demographic inference.

Altogether, these molecular data support the idea of long-term balancing selection on *Pgi* in *M. cinxia*. Given that the MK test was not significant, it is likely that one or just a small number of replacement sites represent the target of selection, or potentially that regulatory polymorphisms are the selective targets with linked selection affecting the *Pgi* coding sequence. The fact that allozyme heterozygotes show higher fitness in experiments suggests that it is heterozygote advantage that represents the specific form of balancing selection that is detectable in the molecular population genetic patterns.

Table 9.1 HKA test data table for *Pgi* in *Melitaea cinxia*.

	Reference loci		
	Idh	*Mdh*	*Pgi*
Polymorphisms	1	0	23
Fixed differences	52	31	79
r_{pd}	0.019	0.000	0.291

9.3 Insecticide resistance in *Drosophila*

Both of the case studies for *Mc1r* and *Pgi* used a **candidate gene approach**, meaning that scientists knew something about what genes might underlie variation in a phenotype and accordingly focused their studies on those genes. An alternative approach is to start out agnostic with respect to phenotype and gene function, and to conduct a genome scan to identify sequence regions that exhibit patterns that deviate from what would be expected under neutrality. The idea is to let the genome tell us what parts have been especially important in adaptation most recently, and then to figure out why.

This genome scan perspective is the approach taken by Todd Schlenke and David Begun. They characterized patterns of polymorphism for 28 loci spread across a 3.2-Mb portion of chromosome 2R in *Drosophila simulans* (Figure 9.7). Using this approach, they identified a 100-kb region with exceptionally low polymorphism: they found 0 polymorphic sites among 7092 nucleotides from 8 loci in a collection of flies from California. The average silent-site polymorphism in this species is $\theta_W = 0.0086$, implying that they would have expected to find about 61 differences between any pair of sequences with an equivalent number of silent sites. Needless to say, this genomic region was a desert of polymorphism, which was very peculiar.

The highly unusual pattern of reduced genetic variation prompted Schlenke and Begun to investigate this region more closely. One possible cause of reduced polymorphism is simply a lower mutation rate. To address this possibility, they tested to see if silent-site divergence was low between *D. simulans* and *D. melanogaster* for homologous sequences, because the Neutral Theory predicts that such divergence will accumulate at the rate that it is input by mutation. They found that the per-site substitution rate was actually slightly *higher* than average in this 100-kb interval (Figure 9.7). This result indicated that a reduced mutation rate could not be responsible for the low polymorphism for that stretch of DNA in the *D. simulans* genome.

Unfortunately, the zero polymorphism in the interesting region meant that they could not use variant frequency spectrum methods or HKA tests or MK tests. However, they did find that Tajima's *D* was significantly less than zero for the loci immediately flanking the 100-kb region of zero nucleotide heterozygosity (Figure 9.7). They therefore conjectured that positive selection resulted in a selective sweep within the 100-kb interval on chromosome 2R of *D. simulans*, which also left its imprint on the flanking portion of the chromosome. For such a selective scenario to explain the data, it would have to have occurred very recently in history because the genetic hitchhiking eliminated all polymorphism across such a wide segment of the chromosome that undergoes a normal rate of recombination. In fact, they estimate that a selection coefficient of $s = 2.2\%$ is consistent with the sweep signature, which is a very large value for the strength of selection on a single locus.

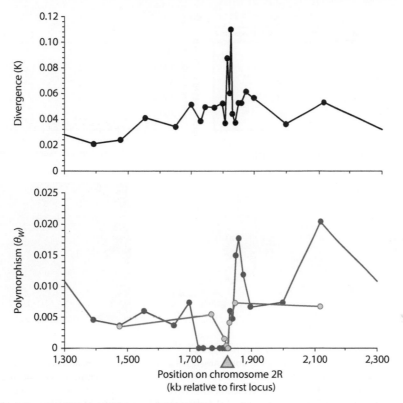

Figure 9.7 Valley of polymorphism around *Cyp6g1*.

In a scan of polymorphism along the right arm of Chromosome 2 in *D. simulans* (red points), Todd Schlenke and David Begun found no polymorphisms at all in a series of 8 loci spanning 100 kb. Silent-site divergence with *D. melanogaster* was not unusual in this region (black points), and the homologous region in *D. melanogaster* near the *Cyp6g1* gene also had exceptionally low polymorphism (orange points; *Cyp6g1* location indicated by the triangle) (data redrawn from Schlenke and Begun 2004).

If the selective sweep was very recent, then perhaps they could use geography to help them understand what happened. What might polymorphism look like in populations from the ancestral region of the world that *D. simulans* came from? It turns out that *D. simulans* has an "out of Africa" history similar to humans. So, they went ahead and measured nucleotide variation for samples of flies from Zimbabwe. These fly populations from Africa had amounts of nucleotide polymorphism in the 100-kb region typical of elsewhere in the genome, with silent-site θ_W averaging 1.76% per site. Now that researchers can sequence entire genomes, the valley of polymorphism in this region of the genome has been recapitulated in multiple non-African populations of *D. simulans* (Figure 9.8). This result suggested that the selective sweep must have occurred following the emigration by *D. simulans* from its ancestral home in Africa.

Not only do humans and *D. simulans* have an "out of Africa" history, so too does *D. melanogaster*, which also colonized the Americas relatively recently. That led Schlenke and Begun to quantify polymorphism for a subset of homologous loci in a California sample of *D. melanogaster* to see whether it might show similar patterns to *D. simulans* in the unusual region. Surprisingly, they found that, in fact, one locus in the homologous genomic region of *D. melanogaster* also exhibited zero nucleotide polymorphism

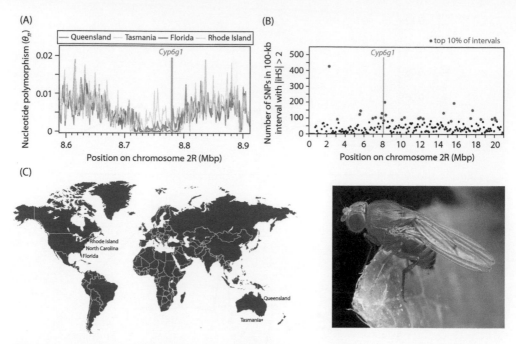

Figure 9.8 Genomic sliding windows around *Cyp6g1*.

(A) Reduced polymorphism around *Cyp6g1* in North America (Florida, Rhode Island) and in Australia (Queensland but not Tasmania) for *D. simulans* based on whole-genome polymorphism data (data redrawn from Sedghifar et al. 2016). (B) The iHS metric for detecting a selective sweep shows a peak in the number of SNPs with high iHS scores near *Cyp6g1* for a collection of *D. melanogaster* from North Carolina (data redrawn from Garud et al. 2015). The slight differences in location of *Cyp6g1* on the x-axis in the two graphs reflect revisions to the reference genome sequence annotation used in each study. (C) The map of the world shows the locations where *D. simulans* populations came from in these studies of *Cyp6g1*. Photo by Dr. Andrew Weeks and reproduced under the CC BY-SA 3.0 license.

(Figure 9.7)! This locus also showed the highest amount of divergence between the species. Subsequent genome-scale analysis of polymorphism confirmed this result in *D. melanogaster*, using the integrated haplotype score (iHS; see section 8.1.2) as a way to detect deviation from neutrality that incorporates LD information (Figure 9.8). These results that are exceedingly unlikely under a neutral evolutionary scenario support the idea that positive selection has occurred at homologous loci independently in both species: independent selective sweeps in each species by different mutations.

The locus that was remarkable in both fly species is named *Cyp6g1*. Upon close inspection of *Cyp6g1*, they found another surprise. They found an insertion of a *Doc* retrotransposable element upstream of *Cyp6g1* in *D. simulans* that was present in all of the California samples but absent from the African collection. Subsequent analysis of a larger collection of wild-caught flies found the *Doc* insertion to occur at a frequency of 98% among Californian *D. simulans*, and that it was a recent insertion. This finding provides a rare example of a transposable element occurring at high frequency in a population. Amazingly, Charles ffrench-Constant and his colleagues reported that an *Accord* retrotransposable element had inserted in the flanking region of the *D. melanogaster Cyp6g1* gene. *Accord* is a different kind of transposable element than *Doc*, but both had inserted in a similar genomic location near *Cyp6g1*, independently in each fly species. Even more surprising in its parallel between the species, the *Accord* element insertion was found in

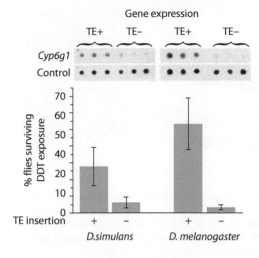

Figure 9.9 Functional effects of mutations to *Cyp6g1*.

The presence of a transposable element upstream of the *Cyp6g1* gene drastically increases gene expression for both *D. simulans* and *D. melanogaster* relative to an experimental control gene (*Gapdh1*), as shown in this dot-blot of mRNA expression (presence TE+; absence TE−). Flies with the transposable element insertion also survive better when exposed to the insecticide DDT (data redrawn from Schlenke and Begun 2004). Photo © 2004 National Academy of Sciences.

98% of wild-caught *D. melanogaster* from California, as well as in many flies found around the globe.

What does *Cyp6g1* do that it seems to have been the target of adaptive evolution independently in both species? *Cyp6g1* encodes a cytochrome P450 protein, a class of proteins that detoxifies noxious compounds. The researchers then performed experimental assays to see if the transposable element insertion near to *Cyp6g1* might affect *Cyp6g1* function in any way. Lo and behold, they showed that the *Doc* and *Accord* insertions cause increased expression of the *Cyp6g1* gene (Figure 9.9). Interestingly, given the molecular function of the *Cyp6g1* cytochrome P450 protein in breaking down nasty chemicals, the higher expression of *Cyp6g1* was associated with increased resistance of flies to the insecticide DDT (Figure 9.9). The *Doc* insertion in *D. simulans* also conferred some resistance to DDT, but it was not as strong as the effect of the *Accord* insertion in *D. melanogaster*. It turns out that even more transposable element mutations have created additional haplotypes of *Cyp6g1* in *D. melanogaster* along with duplication of *Cyp6g1*, implying that pesticide resistance has led to continual evolution of this locus. It is not known whether DDT was the key agent responsible for the parallel selective sweeps in the *Cyp6g1* region for both species of flies, or whether it might have been an alternative insecticide or some other correlated factor. The functional consequences and fitness effects demonstrated for these mutational insertions that were swept to high frequency strongly support the hypothesis that they were the causal mutations in adaptation to handle the new presence of toxins in their environments. Even though the disruptive effects of transposable element insertions usually make them deleterious, when beneficial, they can promote rapid adaptation.

9.4 Body armor in stickleback fish

When the ice sheets receded from the west coast of North America at the end of the last glacial period, they left the land sunken from having been pressed under their huge weight. As the land rebounded, the raised elevation created countless isolated lakes that previously had been connected to the Pacific Ocean. The fish that were caught in these lakes have been prisoners to their new environment for the past 14,000 years, subjected to new selective pressures.

Figure 9.10 Threespine stickleback armoring.
Stickleback fish in marine environments have bony plates running the length of their bodies, as well as a rigid pelvic structure with a pair of bony spines (bony structures dyed red). Fish adapted to freshwater habitats lack this defensive armor (bottom fish in middle and right panels) (Chan et al. 2010). Photo of three-spine stickleback by UW News and reproduced under the CC-BY 2.0 license. Photo of stained whole fish courtesy Nicholas Ellis and Craig Miller; zoom photos of stained fish traits courtesy Frank Chan.

The populations of three-spine stickleback fish, *Gasterosteus aculeatus* in particular, evolved a similar phenotypic response to lake life across the many isolated lakes: they lost their armor. The ancestral population of stickleback from the ocean coast have their bodies covered in bony plates and pelvic spines that help protect them from predators (Figure 9.10). But the freshwater lake fish no longer bear the heavy predator burden of their marine ancestors. They also experience more limited mineral availability, and no longer produce the costly calcified structures that make up the defensive armor. Experiments showed that predation by insects on juvenile fish in the lakes also may induce a selective pressure that favors fish that grow quickly to larger size, so fish grow fast at the expense of not producing much bony armor. The fact that freshwater stickleback evolved this distinct phenotype independently and in parallel across many lakes raised the specter that it represents an adaptation to their new environment. What might the genetic basis be for reduced armor, and does the genome show molecular evidence of a history of directional selection?

David Kingsley, Dolph Schluter, and their collaborators first aimed to pinpoint where in the genome are genes that influence stickleback armoring. This genetic mapping let them home in on two key genes, by analyzing the genotypes and phenotypes of genetically recombinant fish from controlled crosses of fish with extreme phenotypes from lakes and from the ocean. One of the genes is called ectodysplasin (*Eda*) and different allelic haplotypes of *Eda* explain 70% of the variability in armor plating that is seen in nature (Figure 9.10). Given the very recent timescale of adaptation, we will not be able to rely on approaches that depend on inter-species divergence. Instead, we will need to assess deviations from neutrality in terms of polymorphism, differentiation between populations, and haplotype tract lengths.

In a scan of 45,000 SNP variants spread across the stickleback genome by William Cresko, Paul Hohenlohe, and their colleagues, they found that the *Eda* locus was a strong outlier for F_{ST} in comparisons of fish from freshwater versus oceanic collections (Figure 9.11). LD also is high around *Eda*, with extended haplotype tracts in comparisons of oceanic and freshwater fish leading to extreme values of the XP-EHH statistic (see section 8.1.2). It turns out that stickleback are found around continental coasts all over the northern hemisphere, with *Eda* being a genomic F_{ST} outlier and having played a similar evolutionary role in colonization of freshwater in Europe and Asia as well (Figure 9.11).

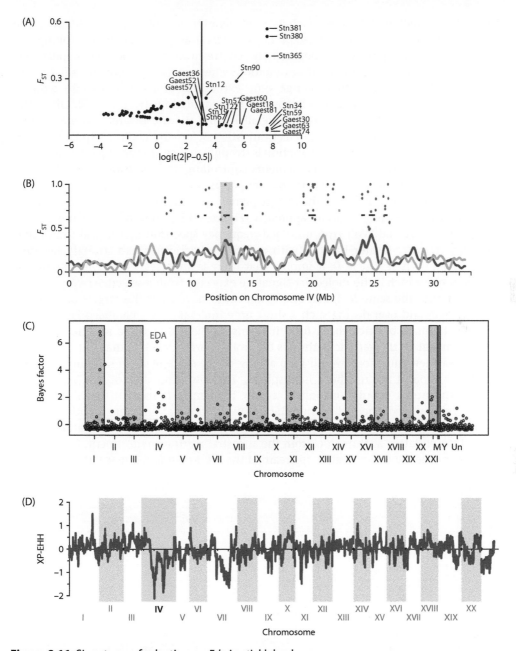

Figure 9.11 Signatures of selection on *Eda* in stickleback.

(A) Microsatellite loci that are closely linked to the *Eda* gene on stickleback chromosome IV are extreme outliers for high F_{ST} in the genome (Mäkinen et al. 2008). Published with permission of the publisher John Wiley and Sons: Mäkinen et al. (2008). Identifying footprints of directional and balancing selection in marine and freshwater three-spined stickleback (*Gasterosteus aculeatus*) populations. *Molecular Ecology* 17 (15), 3565–82. (B) In a scan of F_{ST} along the genome, the *Eda* region (pink region) again sticks out as having higher F_{ST} between marine and freshwater populations (dark blue line) relative to differentiation between different freshwater subpopulations (light blue line) (Hohenlohe et al. 2010), also indicated in (C) by it having a high Bayes Factor statistical score (Jones et al. 2012). Reprinted from *Current Biology* 22, Jones et al., A genome-wide SNP genotyping array reveals patterns of global and repeated species-pair divergence in sticklebacks, pp. 83–90, © 2012 with Permission from Elsevier. Red points in (B) indicate individual SNPs with individually significant differences in F_{ST}. (D) The *Eda* region also is characterized by extended haplotype homozygosity (XP-EHH) on chromosome IV in comparisons of marine versus freshwater samples (Hohenlohe et al. 2012).

Frank Chan and his fellow researchers used genome mapping also to identify the pituitary homeobox transcription factor 1 gene (*Pitx1*) as a key factor controlling the difference in pelvic spine development between freshwater and oceanic populations of stickleback. A spectrum of deletion mutations alter regulatory elements adjacent to the *Pitx1* coding sequence that change gene expression of *Pitx1* and cause reduced production of the bony structure in freshwater fish populations (Figure 9.10). Given the non-coding nature of the function-affecting alleles and that freshwater and marine fish have identical protein sequences of *Pitx1*, an MK test on the coding sequence is not appropriate in this case. Nucleotide variation is strongly reduced in the sequence upstream of *Pitx1* that contains the regulatory elements controlling its expression, and this valley of polymorphism specifically affects freshwater populations (Figure 9.12).

The researchers also quantified the site frequency spectrum to test for deviation from neutrality, using a sliding window approach (Figure 9.12). They polarized mutational changes to use Fay and Wu's H_{FW} unfolded frequency spectrum metric and found that stickleback from freshwater habitats with reduced pelvic armor had strongly negative values of H_{FW}. When $H_{FW} < 0$, it implies that there is an excess of derived variants at high frequency which is the molecular signature expected from a selective sweep.

How is it that the same loci, *Eda* and *Pitx1*, could have been the targets of adaptive evolution over and over during such a short time interval in the parallel adaptation to freshwater habitats by stickleback fish? It seems unlikely that these are the only two genes important in the genetic pathways controlling armor development. It also seems unlikely that new mutations would keep arising in the same place in so many replicate populations around the northern hemisphere. The low-armor haplotypes, in fact, *are* found in the ocean populations, albeit at low frequency. It seems that this standing variation got sampled repeatedly when the freshwater populations were founded, setting the stage for selection to favor increase in frequency of the low-armor alleles and creating genetic differentiation at those loci from the ancestral marine environment. In addition, in the case of the regulatory region upstream of the *Pitx1* coding sequence, that stretch of

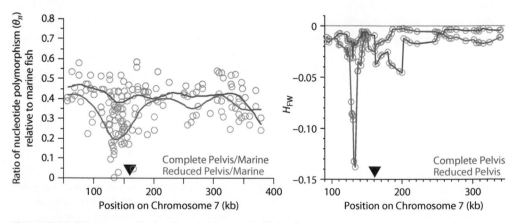

Figure 9.12 Signatures of selection on *Pitx1* in stickleback.

The regulatory region upstream of the *Pitx1* gene that influences its expression has reduced polymorphism among freshwater fish haplotypes with a "reduced pelvis" phenotype compared to fish that have a complete pelvis phenotype collected from both freshwater and marine habitats. In addition, freshwater fish with reduced pelvic spines have a skewed unfolded SFS in this region toward an excess of derived variants at high frequency, as summarized by more negative values of H_{FW} (data redrawn from Chan et al. 2010). Location of *Pitx1* indicated by black triangles, with the upstream regulatory region to the left of it.

DNA is especially fragile, potentially leading to unusually high likelihood of DNA breaks and an elevated indel mutation rate. The marine fish are anadromous, meaning that they swim into freshwater streams to reproduce; flooding of lakes and headwaters also could disperse freshwater fish downstream. This fish movement could facilitate the persistence of low-armor haplotypes in ocean populations, making them available for new rounds of colonization of and adaptation to freshwater habitat.

9.5 Alcohol tolerance in *Drosophila*

Drosophila larvae feed on the yeast that grow on rotting fruit and other decomposing vegetative matter, and a consequence of yeast growth is fermentation and production of alcohol. Alcohol is toxic to most eukaryotes, including flies, and the protein alcohol dehydrogenase is a major enzyme responsible for metabolizing alcohol into non-toxic metabolites. The alcohol dehydrogenase (*Adh*) gene from *Drosophila melanogaster* has a long history of study in evolutionary genetics and now provides a classic case of a gene subject to selection.

Old studies of allozyme variation in the ADH protein identified two forms: a fast (F or ADH^4) and a slow (S or ADH^6) form. The F form of ADH breaks down alcohols more rapidly than the S form, but the S form is more stable at high temperatures (Figure 9.13). Population samples from across the United States by Charles Vigue and Frank Johnson in 1973 showed a cline in allozyme allele frequency: more southerly populations had a higher frequency of the heat-tolerant S form of ADH (Figure 9.13). Moreover, the cline in *Adh* allele frequency was repeated elsewhere in the world, even with the corresponding flip in allele frequencies that one would expect south of the equator in Australia (Figure 9.13). The genotype frequencies in several populations also exhibited deviation from Hardy-Weinberg expectations (Miami and Raleigh had an excess of homozygotes; Portland, Maine had an excess of heterozygotes). This association between protein function and geography in the form of a cline, coupled with non-Hardy-Weinberg frequencies in some populations, suggested that balancing selection might provide an explanation for the unusual pattern of variation in *Adh*. What about the DNA?

With the invention of DNA sequencing technology, in 1983 Martin Kreitman used *Adh* in *D. melanogaster* to demonstrate the first example of DNA sequence variation in a population sample (see Figure 3.1). A later study by Kreitman and Richard Hudson attempted to identify the specific nucleotide region of *Adh* that might be subject to natural selection. By sequencing 11 copies (4,750 nucleotides long in each) of *Adh* and the neighboring locus *Adh-dup*, they showed that the amount of polymorphism varied drastically in different parts of this region of the genome (see Figure 8.1). They observed that silent-site nucleotide polymorphism in most of the region averages $\theta_W = 0.008$ differences per site, whereas the coding region and introns of *Adh* had two to three times higher levels of polymorphism (coding region silent-site $\theta_W = 0.023$, introns 2 + 3 silent-site $\theta_W = 0.018$; Figure 9.14). This result suggests that non-neutral processes might be operating on *Adh*, leading it to retain unusually high genetic variation.

The next goal was to test *Adh* for signals of deviation from neutrality. They quantified the variant frequency spectrum, but none of the regions near *Adh* deviated significantly from the neutral expectation for Tajima's $D = 0$ (the upstream non-coding flanking region $D_{\text{Taj}} = 0.43$, *Adh* $D_{\text{Taj}} = -0.08$, and *Adh-dup* $D_{\text{Taj}} = -1.2$). However, tests of neutrality based on only the frequency spectrum have low power to detect a real effect, especially given the small sample size they had at the time. So, they ended up inventing and using

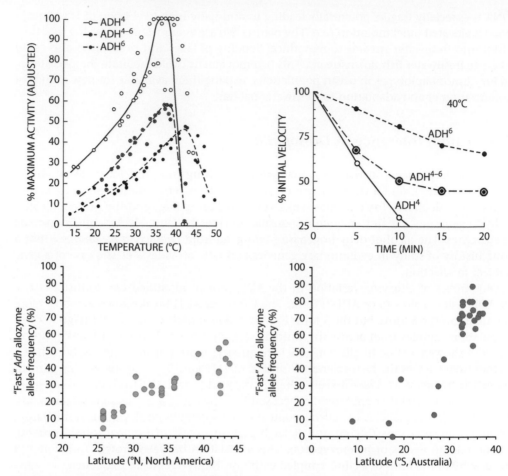

Figure 9.13 Functional activity and geographic cline of *Adh* allozyme variants.

The biochemical activity of the "Fast" enzyme isoform (*Adh*[4]) is reduced at high temperatures compared to the "Slow" isoform (*Adh*[6]) (Vigue and Johnson 1973). Allele frequencies for the two *Adh* isoforms show latitudinal clines in North America and Australia such that the "Fast" isoform is relatively more common at cooler latitudes nearer to the poles of the earth (data redrawn from Vigue and Johnson 1973 and Oakeshott et al. 1982). Biochemical activity graphs reprinted by permission from Springer Nature: Vigue, C. L. and Johnson, F. M. (1973). Isozyme variability in species of the genus *Drosophila*. VI. Frequency-property-environment relationships of allelic alcohol dehydrogenases in *D. melanogaster*. *Biochemical Genetics* 9 (3), 213–27.

a new test of neutrality for the first time: the now-classic HKA test. Homologous sequence from *D. simulans* let them calculate divergence, and they used the 5′ flanking non-coding sequence as a neutral reference locus to compare to the *Adh* coding sequence as the test locus. Amazingly, they found a significant deviation from neutrality! The pattern at *Adh* showed an excess of polymorphism relative to the observed divergence: an especially high value for r_{pd} in the *Adh* coding sequence (Figure 9.14). Balancing selection on alternative haplotypes of *Adh* could create just such a pattern. Complementary HKA tests found a deviation between *Adh* and *Adh-dup*, but not between the flanking region and *Adh-dup* (Figure 9.14), lending further support for balancing selection acting specifically on the *Adh* coding sequence.

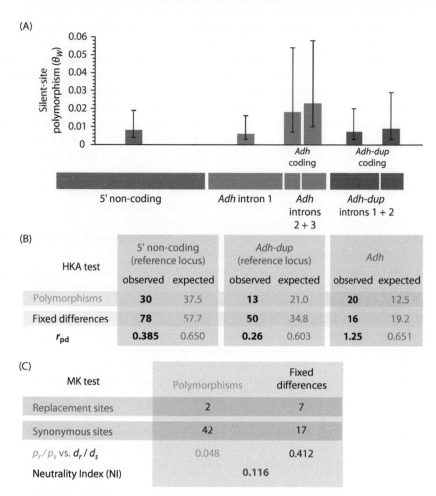

Figure 9.14 Nucleotide polymorphism and tests of neutrality for the *Adh* locus.
(A) Silent-site polymorphism is elevated within the *Adh* coding sequence of *D. melanogaster*. Horizontal colored bars indicate the relative length of sequence in each region assessed for polymorphisms. (B) HKA tests show a significant excess of polymorphism compared to reference sequence. (C) An MK test finds an excess of replacement-site substitutions for *Adh* using *D. yakuba* as the outgroup for calculating divergence (data redrawn from Kreitman and Hudson 1991).

To try to narrow down the region within the *Adh* gene that might be experiencing balancing selection, they used a **sliding window analysis** to visualize polymorphism across the 4,750-bp region. In this case, they calculated polymorphism with θ_π within subsets of 100 silent sites incrementally along the length of the region. A peak of polymorphism stands out near nucleotide position 1,500 (see Figure 8.1), much higher than would be predicted given the observed divergence. It turns out that the replacement nucleotide polymorphism responsible for the Fast and Slow allozyme variants occurs at nucleotide position 1,490, providing a correspondence between genotype, phenotype, patterns of polymorphism, and selective pressures.

Does *Adh* represent a simple case of long-term balancing selection on the "Fast" and "Slow" protein forms related to heat stability of the protein? Perhaps not. Trying to understand how selection influences the evolution of *Adh* at the nucleotide level also led to the invention and first application of the MK test (Figure 9.14). John McDonald and Kreitman found a significant excess of replacement-site substitutions in the coding sequence of *Adh*, showing a Neutrality Index value of NI = 0.12. This MK result implies a history of positive directional selection, and is consistent with the idea that adaptive evolution has played an important role in the recent history of the *Adh* gene. What is going on?

A more recent MK test analysis that looked for lineage-specific effects found that the signature of positive selection appears to come not from the *D. melanogaster* lineage but from *D. yakuba*, the species used as the outgroup in calculating divergence in the original MK analysis of *Adh*. Biochemical studies of the "Slow" form found that it has no obvious functional difference compared to proteins that were synthesized to represent sequences from the common ancestor of the gene for *D. melanogaster* and other species. So it would seem that we can now rule out repeated positive directional selection on *Adh* in the *D. melanogaster* lineage. That means we still have some form of balancing selection as a key part of understanding the unusual molecular population genetic signatures in *Adh*, but does it fit with *long-term* balancing selection?

It turns out that most of the high synonymous-site polymorphism in *Adh* is associated with the "Slow" allozyme haplotypes (see Figure 3.1). This observation implies that the "Fast" haplotypes have a more recent origin and increased quickly to intermediate frequency, especially at latitudes farther from the equator. *Adh* haplotypes found in the ancestral African range of *D. melanogaster* all tend to be of the "Slow" form. This rules out very long-term balancing selection, indicating that the persistence of the distinct functional isoforms of the *Adh* protein must reflect more recent selective pressures. The allozyme clines also correlate better with local rainfall conditions than with temperature, at least in some studies, suggesting that heat tolerance might not be the key factor. Recent studies also show that ethanol can act to "medicate" flies and their larvae to stave off parasitoid infections. Whether the alternative protein forms of *Adh* influence this immune response differently in a way that relates to clinal variation remains to be understood, despite nearly half a century of study on alcohol dehydrogenase in *D. melanogaster*.

9.6 Advanced issues in molecular evolutionary signatures

In this "primer" of a book, I have laid out what I see as the introductory essentials to what is a fairly advanced subdiscipline within biology: molecular population genetics. There remain many interesting and sophisticated topics and approaches that I have neglected in order to focus as closely as possible on the primary aims of conveying these essentials, their rationale, and an appreciation of them. Modern molecular population genetics has embraced advances to DNA sequencing, and the genomic scale of polymorphism and divergence that it brings. As a result, increasingly sophisticated statistical and computational tools are needed to parse all that raw data, and to integrate it into more and more detailed mathematical models of evolution that can best incorporate the subtleties of the evolutionary process at the molecular level. But most all of such analysis is still built on the core ideas and approaches that we have worked through in this book. To give you a sense of further topics that you can dive into more deeply, now that you have this firm backing, here I will mention a few areas of especially active research.

Selection on standing variation: In our discussion of selective sweeps of beneficial variants, we have generally assumed that this occurs on a new mutation that arises that is then immediately subject to positive selection. Fixation of such a mutation is termed a hard selective sweep. In contrast, a soft selective sweep occurs when multiple haplotypes with the same beneficial variant are present and selected in the population simultaneously (Hermisson and Pennings 2017). This can happen when there is standing genetic variation that was neutral (or even slightly deleterious) until a new selective regime caused a variant to be favored (e.g. change in environmental conditions). This means that the newly beneficial variant will be linked to multiple distinct haplotypes because of recombination during the past, prior to when it experienced positive selection. The molecular population genetic signatures of soft sweeps are not as dramatic as for hard sweeps, and this is an area of research that currently is receiving vigorous attention in the scientific literature to understand the theoretical and empirical consequences. For example, the **H12** metric combines information about the frequencies of the two most common haplotypes in a linkage block in a calculation of expected homozygosity (Garud et al. 2015). The idea is that a soft sweep ought to bring multiple haplotypes to high frequency because the fixed beneficial variant will be linked to more than one genetic background. When selection acts on standing variation for a polygenic trait, then in the short term it may just shift allele frequencies at many loci resulting in partial sweeps. Again, the molecular signatures of partial sweeps and polygenic selection will be more subtle than for hard sweeps, making genome-scale data analysis a necessity in deciphering their role in adaptation in nature. Soft sweeps, partial sweeps, and polygenic selection all have been proposed to be very important in recent human adaptation, in particular. Understanding the relative importance of these different ways that selection affects genomes may also help to resolve Lewontin's "paradox of variation."

Population genomics and computational methods: By sequencing the genomes of multiple individuals from a population, we must bring molecular population genetic ideas to the genome scale: population genomics. Rather than pulling out one or a few loci to use in tests of neutrality, population genomics lets researchers conduct scans for selection across all the loci in a genome at once to learn about selection on DNA in a more holistic way. This can help identify particular loci at which selection has acted in interesting ways, and also can address general questions about how much evolution is influenced by linkage so that the efficacy of natural selection in one part of the genome gets limited by natural selection that targets somewhere else in the genome. Population genomics also lets us more easily incorporate more complicated features of molecular evolution besides just SNPs, including duplications and copy number variants (CNVs), indels, and non-coding genes like smallRNA genes. The challenge with population genomics is in dealing with all the data. New computational methods get devised all the time to help manage and understand it all. They range from handling technical details of artifacts or limitations of the DNA sequencing technologies to efficient and clever analysis software development. Approaches like approximate Bayesian computation (ABC) that allow estimation of model parameters by exploring broad ranges of parameter space, hidden Markov models (HMM), Markov chain Monte Carlo (MCMC) simulation, machine learning, and many others all find a place in this area.

Coalescent models and simulating evolution: Throughout this book, I have emphasized the importance of generating predicted distributions from null models to compare to observed data. Coalescent theory has been instrumental in providing a concrete way to simplify evolution and in making robust tests of neutrality in a computationally feasible manner. The basic coalescent model has been extended in many ways to explore a broad range of details of organisms. For example, some models allow multiple mergers in the coalescent genealogy rather than just bifurcations, which is likely important for some species like oysters that can have extreme variability in reproductive output among individuals.

Other models help understand dispersal, both to think about the effects of sampling of haplotypes across subpopulations in different ways and to think about what the implications are for different modes of migration; or what happens when migration involves a group of individuals rather than just single individuals. Researchers interested in macroevolution and the accurate construction of phylogenies of species are actively incorporating coalescent theory into "phylogenomic" approaches to better account for genetic variation within species and their common ancestors. Coalescent theory has also been applied to analyze genome-scale data, as with the pairwise-sequential Markovian coalescent (PSMC) approach that considers heterozygosity in a pair of chromosome copies and the distribution of haplotype blocks to determine a time series back in time of effective population size. Given the details of genome structure and their large size coupled with recombination, classic coalescent simulation approaches actually are becoming computationally inefficient to address some questions at that scale. Coalescent simulations mimic selection by changing N_e, but this is not totally accurate, and becomes more of a problem with large datasets. Genome-scale data also presume a small sample size relative to the total population, but this assumption will not hold for very large collections of DNA. This issue is especially true for some rare species, but also applies to humans since there is the real possibility that someday soon every human will have its genome sequenced at birth. As a result, for some questions, researchers are turning to new and efficient "forward-time" simulations of the evolutionary process in contrast to the "reverse-time" simulations of the coalescent approach.

Human molecular anthropology and human-influenced evolution: I have intentionally tried to use a diversity of organisms as examples throughout this book, at the expense of the fascinating studies of molecular population genetics in our own species. Many of the technical advances in molecular population genetics owe a debt of gratitude to this study of humans, which has a large contingent of researchers working to illuminate the intricate details of our past from our DNA. This includes figuring out complex patterns of migration, our genetic connections with extinct human species like Neanderthals, pinpointing the identity, strength, and consequences of selection among human populations, and deciphering selective targets that distinguish us from other primates. And, how do these differences influence health and sociologically interesting traits? Some of the most fascinating molecular anthropology stories include the genomic signatures of human adaptation to high altitudes in Tibet, or use of food sources like milk and omega-3 fatty acids. When you order a personal genome analysis kit, you get a summary read-out that is based in large part on analysis from molecular population genetic principles. Molecular population genetics also has proved crucial in understanding the history and evolution of many of the domesticated organisms that we know best and depend on: dogs, maize, rice. As experimenters, researchers use lab organisms to monitor evolution in real time, using population genomic analysis to test theories about how evolution proceeds in viruses and bacteria and yeast and animals and plants. There is now a steady stream of fascinating new molecular stories that we are learning about ourselves and the species we interact with on a daily basis.

Further reading

Chan, Y. F., Marks, M. E., Jones, F. C. et al. (2010). Adaptive evolution of pelvic reduction in sticklebacks by recurrent deletion of a *Pitx1* enhancer. *Science* 327, 302–5.

Daborn, P. J., Yen, J. L., Bogwitz, M. R. et al. (2002). A single P450 allele associated with insecticide resistance in *Drosophila*. *Science* 297, 2253–6.

Garud, N. R., Messer, P. W., Buzbas, E. O. and Petrov, D. A. (2015). Recent selective sweeps in North American *Drosophila melanogaster* show signatures of soft sweeps. *PLoS Genetics* 11, e1005004.

Haasl, R. J. and Payseur, B. A. (2016). Fifteen years of genomewide scans for selection: trends, lessons and unaddressed genetic sources of complication. *Molecular Ecology* 25, 5–23.

Hermisson, J. and Pennings, P. S. (2017). Soft sweeps and beyond: understanding the patterns and probabilities of selection footprints under rapid adaptation. *Methods in Ecology and Evolution* 8, 700–16.

Hoekstra, H. E. (2006). Genetics, development and evolution of adaptive pigmentation in vertebrates. *Heredity* 97, 222–34.

Hoekstra, H. E. (2010). From Darwin to DNA: the genetic basis of color adaptations. In: Losos, J. B. (ed) *In the Light of Evolution*. Roberts and Co.: Greenwood Village, CO, pp. 277–96.

Hoekstra, H. E., Drumm, K. E. and Nachman, M. W. (2004). Ecological genetics of adaptive color polymorphism in pocket mice: geographic variation in selected and neutral genes. *Evolution* 58, 1329–41.

Hoekstra, H. E. and Nachman, M. W. (2003). Different genes underlie adaptive melanism in different populations of rock pocket mice. *Molecular Ecology* 12, 1185–94.

Hohenlohe, P. A., Bassham, S., Currey, M. and Cresko, W. A. (2012). Extensive linkage disequilibrium and parallel adaptive divergence across threespine stickleback genomes. *Philosophical Transactions of the Royal Society B-Biological Sciences* 367, 395–408.

Hohenlohe, P. A., Bassham, S., Etter, P. D. et al. (2010). Population genomics of parallel adaptation in threespine stickleback using sequenced RAD tags. *PLoS Genetics* 6, e1000862.

Jones, F. C., Chan, Y. F., Schmutz, J. et al. (2012). A genome-wide SNP genotyping array reveals patterns of global and repeated species-pair divergence in sticklebacks. *Current Biology* 22, 83–90.

Kreitman, M. (1983). Nucleotide polymorphism at the alcohol-dehydrogenase locus of *Drosophila melanogaster*. *Nature* 304, 412–17.

Kreitman, M. and Hudson, R. R. (1991). Inferring the evolutionary histories of the *Adh* and *Adh-Dup* loci in *Drosophila melanogaster* from patterns of polymorphism and divergence. *Genetics* 127, 565–82.

Mäkinen, H. S., Cano, M. and Merila, J. (2008). Identifying footprints of directional and balancing selection in marine and freshwater three-spined stickleback (*Gasterosteus aculeatus*) populations. *Molecular Ecology* 17, 3565–82.

Milan, N. F., Kacsoh, B. Z. and Schlenke, T. A. (2012). Alcohol consumption as self-medication against blood-borne parasites in the fruit fly. *Current Biology* 22, 488–93.

Nachman, M. W. (2005). The genetic basis of adaptation: lessons from concealing coloration in pocket mice. *Genetica* 123, 125–36.

Nachman, M. W., Hoekstra, H. E. and D'Agostino, S. L. (2003). The genetic basis of adaptive melanism in pocket mice. *Proceedings of the National Academy of Sciences USA* 100, 5268–73.

Oakeshott, J. G., Gibson, J. B., Anderson, P. R. et al. (1982). Alcohol-dehydrogenase and Glycerol-3-phosphate dehydrogenase clines in *Drosophila melanogaster* on different continents. *Evolution* 36, 86–96.

Schlenke, T. A. and Begun, D. J. (2004). Strong selective sweep associated with a transposon insertion in *Drosophila simulans*. *Proceedings of the National Academy of Sciences USA* 101, 1626–31.

Schrider, D. R. and Kern, A. D. (2018). Supervised machine learning for population genetics: a new paradigm. *Trends in Genetics* 34, 301–12.

Sedghifar, A., Saelao, P. and Begun, D. J. (2016). Genomic patterns of geographic differentiation in *Drosophila simulans*. *Genetics* 202, 1229–40.

Vigue, C. L. and Johnson, F. M. (1973). Isoenzyme variability in species of the genus *Drosophila*. VI. Frequency-property-environment relationships of allelic alcohol dehydrogenases in *D. melanogaster*. *Biochemical Genetics* 9, 213–27.

Watt, W. B. (1983). Adaptation at specific loci. II. Demographic and biochemical-elements in the maintenance of the *Colias Pgi* polymorphism. *Genetics* 103, 691–724.

Watt, W. B. (2003). Mechanistic studies of butterfly adaptations. In: Boggs, C. L., Watt, W. B. and Ehrlich, P. R. (eds) *Butterflies: Ecology and Evolution Taking Flight*. University of Chicago Press: Chicago, IL, pp. 319–52.

Wheat, C. W., Hagg, C. R., Marden, J. H. et al. (2010). Nucleotide polymorphism at a gene (*Pgi*) under balancing selection in a butterfly metapopulation. *Molecular Biology and Evolution* 27, 267–81.

Zhu, J. and Fry, J. D. (2015). Preference for ethanol in feeding and oviposition in temperate and tropical populations of *Drosophila melanogaster*. *Entomologia Experimentalis et Applicata* 155, 64–70.

Glossary of units, metrics, acronyms, and abbreviations

α (**"alpha"**): the fraction of non-synonymous substitutions fixed by positive selection; also can refer to the P-value significance threshold in a statistical test

θ (**"thayta"**): the population mutation rate, equal to $4N_e\mu$ in a diploid Wright-Fisher population

θ_π (**"thayta pie"**): a measure of nucleotide diversity calculated from the average number of pairwise differences among haplotype copies of a locus, which estimates the population parameter θ

θ_W (**"thayta w"**): a measure of nucleotide diversity calculated from the number of segregating sites (S), also referred to as "Watterson's θ," estimates the population parameter θ

μ (**"mew"**): mutation rate, usually refers to the point mutation rate per nucleotide site per generation in molecular population genetics

π (**"pie"**): common notation for θ_π, the average number of pairwise nucleotide differences among haplotype copies

ρ (**"rho"**): the population recombination rate, equal to $4N_e c$ in a diploid Wright-Fisher population

σ^2 (**"sigma squared"**): a common variable name to refer to the statistical quantity of variance, also often referred to as V; the square-root of σ^2 (σ) is known as the standard deviation

ω (**"omega"**): common notation for d_N/d_S, a codon-based analog of K_A/K_S usually applied to multi-species phylogenies to summarize the rate of protein evolution relative to the neutral expectation

ω_a (**"omega a"**): the subset of d_N/d_S or K_A/K_S that represents the rate of non-synonymous substitution caused specifically by fixation of positively selected variants

A: the abbreviation for the DNA nucleotide adenine; also the single-letter abbreviation for the amino acid alanine; also a common notation for an arbitrary allele name ("big A")

ABC: acronym for approximate Bayesian computation, a statistical approach to estimating parameter values in complex data, used commonly in molecular population genetics for estimating parameters in demographic models

AFLP: acronym for amplified fragment length polymorphism, a classic molecular biology technique to identify and measure genetic variation; RADseq is a modern incarnation of AFLP analysis

bp, kb, Mb: unit measures for the length of DNA in single nucleotide basepairs (bp), kilobasepairs (1,000 bp), or megabasepairs (1,000,000 bp)

c: recombination rate parameter, usually referring to crossover recombination measured in genetic distance units of Morgans with a range from 0 to 0.5

C: abbreviation for the DNA nucleotide cytosine; also the single-letter abbreviation for the amino acid cysteine; also sometimes refers to the amount of sequence constraint

CLR: the composite-likelihood ratio test of neutrality

cM: a common unit of genetic distance and recombination (centiMorgans)

CNV: acronym for copy number variation, a form of molecular genetic variability that includes indels, duplications, and rearrangement polymorphisms that are usually large in size

D: the disequilibrium coefficient is a measure of non-random association of alleles at two loci to summarize the amount of linkage disequilibrium; also commonly used as a variable name to define the number of demes or subpopulations; also sometimes refers to Tajima's D statistic; also sometimes refers to "divergence" or "distance" metrics

D' ("dee prime"): a version of the disequilibrium coefficient that is normalized to range between -1 and $+1$

d_α: substitutions at replacement sites fixed by positive selection, used in calculations of the fraction of adaptive substitutions for replacement sites (α)

d_a: the net sequence distance between subpopulations is a relative measure intended to capture the split time of the populations, derived from d_{xy} by subtracting off ancestral polymorphism

d_f: the number of fixed differences between subpopulations, a relative measure of divergence

DFE: acronym for the distribution of fitness effects of new mutations, often limited to summarizing non-beneficial mutational effects

D_{FL}: Fu and Li's D statistic summarizes the skew in the site frequency spectrum as a test of neutrality similarly to Tajima's D, except that D_{FL} focuses on singleton variants

d_N: non-synonymous-site divergence per non-synonymous site, analogous to K_A but usually calculated using codon-based methods, often reported as branch-specific values for a phylogeny with ≥ 3 species

DNA: deoxyribonucleic acid, the genetic material that provides the basis for heredity

d_N/d_S: a codon-based analog of K_A/K_S to summarize the rate of protein evolution relative to the neutral expectation as the ratio of the substitution rates for non-synonymous and synonymous sites, commonly denoted ω

d_S: synonymous-site divergence per synonymous site, analogous to K_S but usually calculated using codon-based methods, often reported as branch-specific values for a phylogeny with ≥ 3 species

DSB: double-strand breaks of DNA, a kind of DNA damage

D_{Taj}: Tajima's D statistic summarizes the skew in the site frequency spectrum as a test of neutrality, with negative values implying an excess of low-frequency variants and positive values indicating a deficit of low-frequency variants

d_{xy}: a measure of "absolute" genetic distance between subpopulations as the average sequence distance between haplotypes

EHH: the extended haplotype homozygosity statistic that provides a test of neutrality to infer recent positive selection in genomes

ENC: the effective number of codons measure of codon usage bias for coding genes, a value of 61 indicates uniform usage of alternative codons with lower values indicating non-random codon usage

F: the inbreeding coefficient as defined by Wright, also termed the fixation index, and equivalent to the probability of identity by descent

f_0: a variable often used to denote the fraction of new mutations that are selectively neutral

FDR: false discovery rate, a statistical multiple-test correction procedure to reduce the false positive rate when conducting many tests of significance

F_{IS}: Wright's hierarchical fixation index for structured populations that is equivalent to F as applied to individual subpopulations

F_{IT}: Wright's hierarchical fixation index that summarizes the species-wide deviation from Hardy-Weinberg genotype frequencies due to the combined effects of population subdivision, inbreeding within subpopulations, and any other factors

f_{op}: the frequency of optimal codons measure of codon usage bias, ranging from 0 to 1 with higher values indicating non-random usage of synonymous codons in a gene

F_{ST}: the most widely used measure of genetic differentiation among subpopulations, ranging from 0 to 1 it describes the fraction of the species' genetic variability that is due to differences between subpopulations

G: homozygosity $(1-H)$, the fraction of diploid individuals with identical alleles at a locus; also the abbreviation for the nucleotide guanine; also the single-letter abbreviation for the amino acid glycine

G_{exp}: expected homozygosity

G_{obs}: observed homozygosity

G_{ST}: a measure of genetic differentiation among subpopulations, essentially equivalent to F_{ST}

GWAS ("gee-woss"): genome-wide association mapping study, a statistical approach to explain phenotypic variation with genetic differences using molecular markers that correlate with trait values but are not necessarily themselves the alleles that causally explain the trait differences

h^2, H^2: heritability, the predictability of offspring phenotype given parental phenotypes; narrow-sense heritability h^2 includes just the additive genetic effects whereas broad-sense heritability H^2 also includes non-additive effects like dominance and epistasis

H: heterozygosity, the fraction of diploid individuals with different alleles at a locus $(1 - G)$

H12: a test of neutrality metric that incorporates information about the two most common haplotypes to help distinguish soft sweeps from hard sweeps

H_{exp}: expected heterozygosity

H_{FW}: Fay and Wu's H is a statistical summary of the unfolded SFS used as a test of neutrality to identify positive selection, analogous to Tajima's D

H_I: observed heterozygosity, calculated within each subpopulation separately and then averaged across the subpopulations

HKA: the Hudson-Kreitman-Aguade test of neutrality that uses both polymorphism and divergence information from each of two loci

HKY: the Hasegawa-Kishino-Yano mutational substitution matrix used to correct molecular divergence for multiple hits

H_{obs}: observed heterozygosity

H_S: expected heterozygosity, calculated for each subpopulation separately and then averaged across the subpopulations

H_T: expected heterozygosity, calculated for the total combined set of individuals for the pooled sample across all subpopulations

iHS: integrated haplotype score metric for deviation from neutrality, a version of EHH that incorporates information about ancestral versus derived mutations

indel: abbreviation for a mutation or polymorphism involving an insertion or a deletion

JC: the Jukes-Cantor mutational model to correct molecular divergence for multiple hits

k: an arbitrary variable name, often used to represent the number of unique haplotypes

K: the substitution rate per site between species for all sites at a locus or for any given class of sites defined in an ad hoc way

K2P: the Kimura 2-parameter substitution model used to correct molecular divergence for multiple hits, incorporates different rates for transitions and transversions

K_A: non-synonymous-site divergence per non-synonymous site, analogous to d_N, which measures the amount of protein sequence change between species

K_A / K_S: the rate of substitution at non-synonymous sites relative to substitution at synonymous sites as a neutral reference, analogous to d_N/d_S, measures the rate of protein evolution relative to that expected by mutation and genetic drift

K_{neu}: neutral divergence per site between species, may include inter-genic, intronic, and synonymous sites

K_S: synonymous-site divergence per synonymous site between species, analogous to d_S, which measures the accumulation of mutations by genetic drift when changes to synonymous sites are selectively neutral

LD: linkage disequilibrium, the non-random association of alleles at different loci

m: a common variable name used in population genetics to refer to the rate of migration

MA: mutation accumulation, often referring to the experimental evolution paradigm of bottle-necking a population to minimize selection and allow new mutations to fix so that the mutation rate and distribution of fitness effects may be inferred

MAF: minor allele frequency, often used when the ancestral state of an allele is unavailable to polarize alternative alleles as ancestral versus derived, used to construct the folded site frequency spectrum

MK: the McDonald-Kreitman test of neutrality that uses both polymorphism and divergence for different classes of sites at a single locus

MRCA: most recent common ancestor of a set of alleles within a species or a set of homologous gene copies in different species

Mya: million years ago

n: a common variable name used to refer to sample size

N: census size of a population

N_{anc}: ancestral effective size of a population

N_e: effective population size, the size of an idealized population that would experience the same amount of genetic drift as the actual population being considered

NI: the Neutrality Index, a single number summary of polymorphism and divergence information from an MK table

PBS: the population branch statistic, used to test for population-specific selection

PCA: principal components analysis, a multivariate statistical approach for reducing the dimensionality of large datasets into a smaller number of statistically independent factors

PCR: polymerase chain reaction, a molecular biology technique for amplifying the number of copies of DNA sequences experimentally

P_s: the proportion of nucleotide sites that are polymorphic in a locus

PSMC: pairwise-sequential Markovian coalescent, a technique for estimating population size through time based on analysis of tracts of heterozygosity in a single diploid genome

QTL: quantitative trait locus, a polymorphic locus in a genome with alleles that correlate with phenotypic variation in a population

Q_{ST}: a measure of trait differentiation between populations, the phenotypic analog of F_{ST}

r^2: the linkage disequilibrium metric that is based on the squared correlation coefficient for allelic state between loci

RADseq, RADtag: restriction associated digest tag-sequencing, a modern version of AFLP that uses high-throughput DNA sequencing to obtain genetic polymorphisms from random segments of DNA in a genome

RFLP: restriction fragment length polymorphism, a classic molecular biology technique for measuring molecular variation based on the presence or absence of endonuclease cut sites within a segment of DNA that is amplified by PCR and visualized by gel electrophoresis

RNA: ribonucleic acid, the product of transcription of DNA by an RNA polymerase; common examples include messenger RNA (mRNA), ribosomal RNA (rRNA), transfer RNA (tRNA), and small interfering RNAs (siRNA) like micro-RNA (miRNA) and piwi-interacting RNA (piRNA), among others

r_{pd}: ratio of polymorphism to divergence, the summary metric used for data in an HKA test

R_{ST}: a measure of genetic differentiation between populations based on microsatellite loci, analogous to F_{ST}

s: a variable name usually used in population genetics to refer to the selection coefficient

S: the number of segregating sites; also can refer to the selection differential in quantitative genetics; also can refer to the rate of self-fertilization, especially in plant reproductive biology; also the one-letter abbreviation for the amino acid serine

SD: the standard deviation of the mean, the square-root of the variance of a statistical distribution

SFS: the site frequency spectrum, a distribution that summarizes how many variants at a locus are found at a given population frequency; also referred to as the allele frequency spectrum or variant frequency spectrum or the distribution of variant frequencies

SNM: single nucleotide mutation, also known as a point mutation; also can refer to standard neutral model

SNP ("snip"): single nucleotide polymorphism, the genetic variation found at an individual nucleotide site that was created by a point mutation; in human genetics, usually understood to have a population frequency $\geq 1\%$

SNV: single nucleotide variant, includes SNPs regardless of population frequency

SSR: simple sequence repeat, another name for a microsatellite locus

STR: short tandem repeat, another name for a microsatellite locus

T: time, often scaled in generations in population genetic models

TE: transposable element

T_{MRCA}: time to the most recent common ancestor of alleles within a population or homologous gene copies between species

TMRCA: time to the most recent common ancestor

T_S: time to speciation, the time since a set of lineages ceased gene flow

UTR: untranslated region, the upstream (5′) and downstream (3′) DNA segments of a gene sequence that get transcribed but not translated

V: a common variable name to refer to the statistical quantity of variance, also often referred to as σ^2; also the single letter abbreviation for the amino acid valine

VNTR: variable number of tandem repeats, another name for a microsatellite locus but usually with a longer repeat motif

w: a variable name usually used in population genetics to refer to fitness

XP-EHH: the cross-population extended haplotype homozygosity test of neutrality, a version of EHH that incorporates information for multiple populations

Glossary of terms and jargon

ABBA-BABA test: a test for genetic introgression between populations or species, based on branching patterns expected when admixture has or has not taken place (Ch. 8)

absolute fitness: a measure of reproductive success that integrates survival from birth to adulthood and the total number of offspring produced (Ch. 2)

adaptationist: the view that biological features evolved as a result of adaptation by natural selection with little influence of other evolutionary forces (Ch. 4)

adaptive divergence: evolution of distinct traits and underlying DNA sequence in different subpopulations or species that confers increased reproductive success in their respective environments (Ch. 8)

additive: the allelic effect on a trait or on fitness that corresponds to equal effects for each allele in diploid individuals, indicated by a dominance coefficient of $h = 0.5$ (Ch. 4)

admixture: the merger of two or more populations, or the establishment of gene flow between populations (Ch. 6)

alignment: linear arrangement of multiple copies of homologous DNA, RNA, or protein sequences from distinct individuals or species, with the aim of identifying conserved versus diverged positions (Ch. 4)

allele: an alternative version of a segment of DNA in a population (Ch. 1)

allele frequency spectrum (AFS): the abundance distribution for variants that occur at distinct allelic frequencies in a population sample (site frequency spectrum) (Ch. 3)

allelic gene conversion: an outcome of non-crossover recombination, in which a short segment of DNA from one chromosome copies over the DNA of the homologous region on its sister chromatid (Ch. 6)

allozyme: a protein molecular marker with allelic protein isoforms distinguishable by gel electrophoresis (Ch. 3)

amplified fragment length polymorphism (AFLP): a type of molecular marker similar to RFLP, except usually applied to many anonymous loci simultaneously (Ch. 3)

analogous: traits that are similar in state but distinct in evolutionary origin (Ch. 5)

ancestral polymorphism: genetic variation in the population that was in the common ancestor of a set of present-day populations being examined, which can influence interpretation of patterns of polymorphism and divergence in genomes (Ch. 5)

ancestral recombination graph: coalescent theory and graphical visualization of genealogies that incorporates recombination (Ch. 5)

ancestral state: the combination of variants in a DNA sequence that were present in the common ancestor of a set of observed sequences, which must be estimated and therefore is not known with perfect accuracy (Ch. 5)

ancestral variant: the allelic state that was present in the past that mutated to give rise to a derived variant (Ch. 3)

ancient DNA: DNA samples isolated from archaeological or subfossil material (Ch. 7)

anonymous loci: molecular markers that are assayed for polymorphism in the absence of knowledge about their sequences or location in the genome, as for AFLPs (Ch. 3)

approximate Bayesian computation (ABC): a statistical approach to estimating parameter values in complex data, used commonly in molecular population genetics for estimating parameters in demographic models (Ch. 7)

ascertainment bias: the statistical sampling bias inherent to genotyping loci already known to be polymorphic because they are a non-random sample of loci, which skews estimates of polymorphism and the site frequency spectrum in predictable ways (Ch. 3)

association mapping: application of linkage disequilibrium to correlate allelic states of a segment of a genome with phenotypic states in a sample of individuals from a population (linkage disequilibrium mapping, genome-wide association study) (Ch. 6)

associative overdominance: linkage of neutral variants to a locus subject to heterozygote advantage in a balancing selection version of genetic hitchhiking (Ch. 7)

background selection: decrease in allele frequency at a locus that is caused by its linkage to a deleterious allele at another locus, resulting in lower neutral polymorphism near targets of purifying selection (Hill-Robertson effect) (Ch. 7)

balance theory: the idea that genetic variation is common, with the polymorphism actively maintained by selection, a popular view prior to empirical and theoretical molecular population genetics (Ch. 4)

balancer chromosome: chromosomes with a large inversion allele used in experimental genetic cross-breeding (Ch. 2)

balancing selection: natural selection that favors the maintenance of multiple distinct alleles in a species (Ch. 2)

base pair (bp): a single nucleotide position in a DNA sequence composed of the paired bases in the DNA double helix (Ch. 3)

Bayesian inference: a statistical approach to estimating the parameters of a model that aims to identify the probability that the observed data support a particular statistical hypothesis, often applied to gene tree inference from DNA sequence data in phylogenetics (Ch. 5)

beneficial mutation: a mutation that confers a fitness advantage relative to the ancestral allelic state (Ch. 2)

bi-allelic: a locus that has exactly two allele variants (Ch. 2)

bifurcate: the splitting of one branch on a gene tree into two descendant branches (Ch. 4)

binomial distribution: a probability distribution that describes systems with just two possible outcomes per event, even if each possible outcome is not equally likely in any given event (Ch. 3)

biological species concept: the idea that different species represent groups of individuals that do not interbreed with one another (Ch. 5)

Bonferroni correction: a statistical multiple-test correction procedure in which the significance threshold is reduced in proportion to the number of tests being performed (Ch. 8)

branch: the lines that connect nodes in a gene tree, often scaled in length to reflect the amount of time or accumulation of sequence change (edge) (Ch. 5)

candidate gene approach: a perspective on characterizing the relationships between genotype, phenotype, and selection in nature by focusing on those loci for which pre-existing information provides a hint that they might be biologically important (Ch. 9)

candidate loci: sequence regions identified to be unusual in a biologically interesting way warranting further study (Ch. 8)

case and control samples: a common experimental paradigm in medical genetics to compare a set of individuals that have a focal characteristic, such as a disease state (case sample), to another set of individuals who lack that feature (control sample) (Ch. 6)

census population size: the actual count of the number of adult individuals in a population (Ch. 3)

centiMorgan (cM): unit of measure for the distance between loci based on the likelihood of recombination between them (Ch. 6)

centromere: region of a chromosome that acts as the site of kinetochore assembly, where microtubules bind to separate sister chromatids in mitosis and meiosis, which usually contains highly repetitive DNA, low recombination, and low density of genes (Ch. 6)

character: a feature used to construct a gene tree or phylogenetic tree, which could include a single nucleotide difference at an orthologous position or a phenotypic trait (Ch. 5)

chromatin: the complex of molecules around chromosomes comprised of DNA, protein, and RNA that influences the accessibility of the DNA to transcription (Ch. 2)

cluster of orthologous groups: a set of genes from at least three species that are all pairwise orthologs of one another (Ch. 4)

coadapted gene complex: a set of linked loci that interact non-additively to influence fitness (supergene) (Ch. 6)

coalescence: the merging of two haplotypes into their common ancestor from the reverse-time perspective of genealogical history (Ch. 5)

coalescent process: the reverse-time genealogical view of common ancestry of a sample of sequence haplotypes (Ch. 5)

coalescent time: the expected time to the most recent common ancestor from coalescent theory, $4N_e$ generations for a diploid population (Ch. 5)

coding sequence (CDS): the portion of a gene sequence that encodes a protein, which excludes introns and untranslated regions of a gene (Ch. 2)

codon: a set of three nucleotides in a coding gene, defined by the genetic code to encode a particular amino acid when the gene gets translated into a protein (Ch. 2)

codon usage bias: the non-random usage of alternative synonymous codons in a coding sequence, which can be caused by mutational and nucleotide composition skews in the genome or by selection actually distinguishing fitness effects of alternative synonymous codons in highly expressed genes (Ch. 4)

compensatory mutation: mutational change that offsets a fitness cost of another mutation (Ch. 2)

complete linkage: a pair of loci with zero crossover recombination between them, corresponding to a recombination fraction of $c = 0$ (perfect linkage) (Ch. 6)

concerted evolution: the homogenization of paralogous gene copies within the genome of a species as a result of non-allelic gene conversion events (Ch. 6)

conservation: the preservation of nucleotide sequence as a result of purifying selection against new deleterious mutations (Ch. 7)

conservative test: a statistical analysis that is biased against detecting a significant effect if it exists, thus providing greater confidence in the detection of true positives at the expense of a high false negative rate (Ch. 8)

contingency table analysis: a statistical approach to test for unequal representation of different features among groups, as for a 2×2 data matrix in a χ^2 test (Ch. 8)

continuum model: a model of migration that does not use discrete subpopulations to characterize the deviation from panmixia in a species with a contiguous range across space (Ch. 3)

convergent evolution: the independent evolutionary origins of similar features in different species (Ch. 5)

copy number variant (CNV): the allelic state of a locus that is polymorphic for an insertion, deletion, or duplication mutation (Ch. 2)

coupling phase: two-locus haplotype combinations each comprised of the most common alleles at each locus, or the least common alleles at each locus used in calculation of the disequilibrium coefficient D (Ch. 6)

crossover recombination: reciprocal exchange of genetic material between sister chromatids in meiosis (Ch. 6)

daughter lineage: the descendant haplotype(s) or species from a given ancestral haplotype or species, which could include internal portions of a tree in addition to a terminal node (Ch. 5)

deficit of polymorphism: an underabundance of nucleotide variation relative to the amount expected from what Neutral Theory would predict from the amount of sequence divergence as an indicator of mutation rate (Ch. 8)

deficit of rare variants: a pattern observed in a site frequency spectrum of an underabundance of singleton variants and other low-frequency variants relative to the neutral expectation for the site frequency spectrum (Ch. 7)

degenerate: when a codon has a synonymous alternate codon that encodes the same amino acid (Ch. 2)

deme: a subset of individuals in a species that may experience restricted gene flow and genetic distinctiveness from other subpopulations (subpopulation, local population) (Ch. 3)

demographic history: the past events of a species that describe its population dynamics, including expansions, contractions, bottlenecks, population subdivision, migration, and admixture (Ch. 1)

demography: the population dynamic events of a species that describe its size and distribution, including expansions, contractions, bottlenecks, population subdivision, migration, and admixture (Ch. 7)

derived variant: the variant allele that arose most recently by mutation (Ch. 3)

diffusion theory: mathematical models originally developed by physicists to describe the dynamics of particle diffusion, subsequently applied to population genetics to describe allele frequency changes (Ch. 7)

dinucleotide repeat: a kind of microsatellite with a pair of nucleotide bases repeated many times in a row, as for the CG motif in CGCGCGCGCGCGCG (Ch. 2)

direct selection: selection that induces allele frequency change at a locus due to differential fitness effects of the alleles (Ch. 7)

disruptive selection: phenotypic mode of selection that favors extreme trait values or disfavors intermediate trait values (Ch. 7)

distance method: means of constructing genealogies or phylogenies based on the similarity of aligned orthologous DNA or protein sequences among species (Ch. 5)

distribution of fitness effects (DFE): the probability distribution describing the spectrum of how new mutations affect relative fitness (Ch. 2)

divergence: mutational differences between sequences, usually fixed differences between species or distinct populations (Ch. 2)

divergent selection: selection that favors different trait values or genotypes in the environments experienced by different subpopulations (Ch. 8)

DNA transposon: a transposable element that excises itself before inserting elsewhere in a genome (Ch. 2)

dominance coefficient (h): the relative influence of an allele on the fitness of heterozygous individuals, with a fully recessive allele having $h = 0$, an additive effect $h = 0.5$, and a fully dominant allele having $h = 1$ (Ch. 2)

drift barrier: the threshold of $2N_e s$ that serves as a heuristic for the point below which allele frequency changes will be controlled primarily by genetic drift with little influence of selection (Ch. 4)

duplication-degeneration-loss model: the conceptual description of the evolutionary path taken by gene duplicates in a genome such that one or the other paralogous sequence copy will acquire a disruptive mutation that turns it into a pseudogene that will eventually accumulate sufficient mutations to be lost from the genome (Ch. 2)

ectopic recombination: crossover between non-homologous loci, which can lead to chromosome rearrangement (Ch. 2)

edge: the lines that connect nodes in a gene tree, often scaled in length to reflect the amount of time or accumulation of sequence change (branch) (Ch. 5)

effective population size: the size of an idealized Wright-Fisher population in which genetic drift would operate at the same rate as in the actual population (Ch. 3)

effectively neutral mutation: a mutation that experiences changes in frequency that are primarily controlled by genetic drift ($4N_ehs<1$, $2N_es<1$ assuming additivity) (Ch. 2)

endonuclease: a protein restriction enzyme that cleaves DNA at particular sequence motif cut sites (Ch. 3)

epigenetics: the study of heritable change across generations that does not involve changes to DNA, as for the influence of histone marks, DNA methylation state, and smallRNA molecules transmitted from parent to zygote (Ch. 2)

epimutation: a change to the state of an epigenetically inherited factor, for example the methylation state of a segment of DNA (Ch. 2)

epistasis: the dependence of one allelic effect on the identity of an allele at another locus (Ch. 7)

epistatic selection: the strength or direction of natural selection on a locus that depends on the allelic context at other loci (Ch. 6)

equilibrium: a stable state of a population or mathematical model with predictable properties (Ch. 4)

estimate: a value, usually a mean or median, calculated from data intended to infer the true population parameter value (Ch. 3)

estimator: a metric based on empirical data that aims to measure a population parameter from a mathematical or statistical model (Ch. 3)

Ewens-Watterson test: a test of neutrality that evaluates observed versus expected homozygosity, based on the infinite alleles model of mutation (Ch. 8)

excess of polymorphism: an overabundance of nucleotide variation relative to the amount expected from what Neutral Theory would predict from the amount of sequence divergence as an indicator of mutation rate (Ch. 8)

excess of rare variants: a pattern observed in a site frequency spectrum of an overabundance of singleton variants and other low-frequency variants relative to the neutral expectation for the site frequency spectrum (Ch. 7)

exome: the portion of a genome that encodes exons for all protein-coding sequences in the genome (Ch. 3)

expectation: a value to be calculated from data, usually the mean value, intended to estimate the true population parameter value (Ch. 3)

expected heterozygosity (H_e): proportion of pairs of gene copies that are predicted to have different alleles when sampled randomly from a population (Ch. 3)

experimental evolution: the use of controlled laboratory or field experiments to monitor evolutionary change across generations (Ch. 4)

extended haplotype: a haplotype of a length much longer than expected in the absence of a partial selective sweep (Ch. 8)

fail to reject neutrality: test of a neutral null model that is unable to exclude the null model as an explanation for observed data, either because the null model appropriately explains the data, or due to insufficient statistical power to exclude the null model, or due to overlap in empirical prediction of the null model and other explanatory processes (Ch. 4)

false discovery rate (FDR): a statistical multiple-test correction procedure to control the false positive rate when conducting many tests of significance (Ch. 8)

F_{IS}: Wright's fixation index that summarizes the amount of non-random mating that takes place within populations, based on the difference between observed and expected heterozygosity of subpopulations (Ch. 3)

Fisher's exact test: a test of statistical significance for count data in contingency tables (Ch. 8)

F_{IT}: Wright's fixation index that summarizes the overall deviation from Hardy-Weinberg genotype frequencies, based on the difference between observed heterozygosity of individuals and expected heterozygosity for the total pooled sample across populations (Ch. 3)

fitness (w): a quantitative representation of the ability of individuals with a particular trait or genotype to survive and reproduce, usually meant in terms of relative fitness rather than absolute fitness in population genetics (Ch. 2)

fitness component: a measurement of one factor that contributes to overall fitness, such as egg to adult survival or female fecundity (Ch. 2)

fixation indices: a set of statistical metrics devised by Sewall Wright to summarize the degree of difference from an idealized randomly mating population, including F_{IS}, F_{IT}, and F_{ST} (Wright's hierarchical F-statistics) (Ch. 3)

fixed: a variant that has risen in frequency to become the only allele present in the population at that locus (Ch. 1)

fixed difference: a locus that is fixed for one allelic type in one population and for another allelic type in another population (Ch. 2)

folded site frequency spectrum: version of the site frequency spectrum based on minor variant frequencies only (Ch. 3)

forensics: the study of science for investigation in a court of law (Ch. 2)

four-fold degenerate: synonymous-site nucleotide positions in codons of amino acids for which any change to that position yields a synonymous codon (Ch. 4)

fragment length polymorphism: a type of molecular marker that can be genotyped by separating alleles of different sequence lengths by gel electrophoresis, including RFLP and AFLP (Ch. 3)

frame-shift mutation: an insertion or deletion mutation of a length that is not a multiple of three nucleotides and thus alters a coding sequence to change the codon frame of translation (Ch. 2)

free recombination: a pair of loci with alleles that segregate independently, as for loci on different chromosomes, corresponding to a recombination fraction of $c = 0.5$ (unlinked, independent assortment) (Ch. 6)

F_{ST}: Wright's fixation index that summarizes how much genetic variation in a collection of subpopulations is due to differences among them, based on the difference in expected heterozygosity within subpopulations compared to the total pooled sample across populations (Ch. 3)

gametic phase disequilibrium: the non-random association in a population of allelic variants from different loci leading to biased haplotype combinations (linkage disequilibrium) (Ch. 6)

GC-biased gene conversion: the phenomenon in which the sister chromatid with higher GC nucleotide content is used as the donor sequence in non-reciprocal exchange recombination (Ch. 6)

gel electrophoresis: method for separating DNA, RNA, or protein molecules by applying an electric charge to a sample of them in an agar or agarose gel matrix (Ch. 3)

gene: a segment of DNA that gets transcribed to exhibit biological activity in a way that can affect an organism's phenotype and fitness (Ch. 1)

gene conversion: the non-reciprocal exchange of genetic material between sister chromosomes in meiosis, which involves copying of short tracts of DNA from one chromosome to the other during both crossover and non-crossover recombination; sometimes inaccurately used as shorthand to refer to non-crossover recombination in general, but these terms are not synonyms (Ch. 6)

gene flow: the exchange of genetic material between individuals from different subpopulations through migration (Ch. 3)

gene tree: branching diagram that summarizes the descent with modification of segments of homologous DNA copies from different individuals in a population or from different species (Ch. 5)

gene tree heterogeneity: differences in the topology of gene trees among loci, reflecting incomplete lineage sorting in a multi-species context (Ch. 5)

gene tree–species tree discordance: mismatch between gene trees and the species tree, often a result of ancestral polymorphism resolving as different gene tree topologies for different genes when speciation events occur in rapid succession and when ancestral population sizes were large (incomplete lineage sorting) (Ch. 5)

generation time effect: the pattern in gene trees of longer branch lengths in lineages that have faster turnover of generations, reflecting a faster molecular clock per year despite potentially similar mutation rates per generation (Ch. 5)

genetic architecture: the make-up of the genetic loci and allelic effects responsible for variation in a trait (genotype-phenotype map) (Ch. 7)

genetic background: the set of allele combinations for a given individual at a biologically important locus or across the genome more generally (Ch. 6)

genetic code: the mapping of DNA codon triplet nucleotides to the amino acid that they encode (Ch. 2)

genetic differentiation: a relative measure of the degree of genetic distinctiveness of different subpopulations due to unequal allele frequencies between them, as quantified with metrics like F_{ST} (Ch. 3)

genetic distance: depending on context, an absolute measure of sequence divergence between pairs of subpopulations or species, as quantified with metrics like K_S, or, the spacing between loci measured in centiMorgans, based on the likelihood of recombination between them (recombination distance) (Chs 3, 6)

genetic draft: fluctuations in allele frequency induced by positive selection at linked loci that can superficially mimic the effects of genetic drift (Ch. 7)

genetic drift: the stochastic change in allele frequency across generations due to statistical sampling effects in finite populations (Ch. 3)

genetic hitchhiking: increase in allele frequency at a locus that is caused by its linkage to a beneficial allele at another locus, resulting in lower neutral polymorphism near targets of positive selection (Hill-Robertson effect) (Ch. 7)

genetic map units: distances between loci measured in centiMorgans, based on the likelihood of recombination between them (Ch. 6)

genome scan: statistical analysis of all or most of a genome sequence for signatures of selection (scan for selection) (Ch. 8)

genome-wide association study (GWAS): application of linkage disequilibrium to correlate allelic states of a segment of a genome with phenotypic states in a sample of individuals from a population (association mapping, linkage disequilibrium mapping) (Ch. 6)

genotype-phenotype map: the relationship between genetic loci and their allelic effects with variation in a trait (genetic architecture) (Ch. 7)

genotyping: the determination of a particular genotypic state for a locus in a given individual (Ch. 3)

genotyping chip: a generic term for high-throughput molecular methods used to determine the variants present in a genome for a set of loci known to be polymorphic (Ch. 3)

haplotype: a unique sequence of linked DNA (Ch. 2)

haplotype block: a segment of DNA with a distinctive pattern of linked variants having boundaries that reflect historical recombination events, common in human genetics due to hotspots of recombination (Ch. 6)

haplotype network: branching diagram that does not require unidirectional bifurcation, but can allow visualization of reticulation (Ch. 5)

hard selective sweep: the fixation of a particular allelic variant at a locus that was subject to positive selection following its origin as a new beneficial mutation (Ch. 7)

Hardy-Weinberg principle: the effect that a single generation of random mating will produce predictable genotype frequencies, with the stereotyped ratio of $p^2{:}2pq{:}q^2$ for a bi-allelic locus with allele frequencies p and q (Ch. 3)

harmonic mean: a kind of average that is especially sensitive to low values, calculated as the reciprocal of the mean of reciprocal values for a set of observations (Ch. 3)

hemizygous: a chromosome found in only one copy in an organism that is otherwise diploid, as for X and Y chromosomes in male mammals (Ch. 7)

heterogametic sex: the sex of a species that receives different sex chromosomes from each parent, for example, male mammals and female birds (Ch. 7)

heterozygosity (*H*): proportion of individuals that are heterozygous at a given locus (Ch. 3)

heterozygote advantage: fitness function at a locus defined by highest fitness for heterozygous individuals relative to either of the homozygous genotypes, a form of balancing selection (overdominance) (Ch. 7)

heterozygote disadvantage: fitness function at a locus defined by lowest fitness for heterozygous individuals relative to either of the homozygous genotypes (Ch. 7)

Hill-Robertson effect: the phenomenon of selection on one locus influencing the allele frequency changes at other linked loci that are not direct targets of that selection pressure, including genetic hitchhiking, background selection, and selective interference (linked selection, indirect selection) (Ch. 6)

hitchhiking mapping: an approach that exploits linkage and genetic hitchhiking to find targets of selective sweeps by first identifying signatures of selection in linked loci, often used in genome scans for selection (Ch. 8)

homologous: segments of DNA sequence that derive from a common ancestral sequence (Ch. 3)

homoplasy: convergent evolution in distinct lineages to the same observed state (Ch. 2)

homozygosity (*G*): proportion of individuals that are homozygous at a given locus (Ch. 3)

Hudson-Kreitman-Aguadé test: a test of neutrality that uses both polymorphism and divergence information from each of two loci (HKA test) (Ch. 8)

hybridization: the genetic merger of individuals from differentiated populations or distinct species (Ch. 2)

identical by descent: the observation of two copies of a locus that share their equivalent form due to shared common ancestry (Ch. 2)

identical by state: the observation of two copies of a locus that are indistinguishable irrespective of whether the state originated convergently or from common ancestry (Ch. 2)

inbreeding coefficient (*F*): deviation between observed and expected heterozygosity (fixation index) (Ch. 3)

incomplete lineage sorting: mismatch between gene trees and the species tree, often a result of ancestral polymorphism resolving as different gene tree topologies for different genes when speciation events occur in rapid succession and when ancestral population sizes were large (gene tree–species tree discordance) (Ch. 5)

indel: abbreviation for a mutation or polymorphism involving an insertion or a deletion (Ch. 2)

indel-associated mutation: a mutational process in which heterozygous indels increase the point mutation rate at nearby nucleotides (Ch. 2)

independent assortment: a pair of loci with alleles that segregate independently, as for loci on different chromosomes, corresponding to a recombination fraction of $c = 0.5$ (unlinked, free recombination) (Ch. 6)

indirect selection: selection that induces allele frequency change on a locus due to its linkage to a direct target of selection, through background selection or genetic hitchhiking (linked selection, selection at linked sites) (Ch. 7)

ingroup: the main set of haplotypes, populations, or species under investigation in a gene tree or phylogeny (Ch. 5)

internal node: a point of merger in the interior of a gene tree that corresponds to the hypothesized ancestral sequence of all sequence copies descending from it (Ch. 5)

inversion: mutation that reverses the ordering of loci in a DNA sequence (Ch. 2)

island model: an idealized mode of migration in which gene flow occurs between all pairs of subpopulations with equal likelihood (Ch. 3)

isolation by distance: observation of greater genetic differentiation between subpopulations that are farther apart (Ch. 3)

joint site frequency spectrum: a multidimensional site frequency spectrum that summarizes variant frequency abundances from two or more subpopulations (Ch. 3)

jumping gene: colloquial name for a transposable element (Ch. 2)

length variant: a type of allele that differs in sequence length from other allelic variants, as for indel polymorphisms and microsatellites (Ch. 2)

levels of selection: hierarchy of selection pressures on a genetic locus that operate separately due to competitive factors within a genome, among individuals, and among populations (Ch. 2)

Lewontin-Krakauer test: a test of neutrality based on exceptional genetic differentiation between populations, aimed to detect locus-specific and population-specific adaptive evolution (Ch. 8)

lineage: a temporal trajectory of shared ancestry for a haplotype, population, or species (Ch. 5)

lineage-specific: use of three or more species to determine sequence changes that are unique to a single branch between nodes on the gene tree (Ch. 8)

linkage disequilibrium (LD): the non-random association in a population of allelic variants from different loci leading to biased haplotype combinations (gametic phase disequilibrium) (Ch. 6)

linkage disequilibrium mapping: application of linkage disequilibrium to correlate allelic states of a segment of a genome with phenotypic states in a sample of individuals from a population (association mapping, genome-wide association study) (Ch. 6)

linkage equilibrium: combinations of allelic variants from different loci occurring in the frequencies expected from independent assortment of the loci (Ch. 6)

linked selection: selection that induces allele frequency change on a locus due to its linkage to a direct target of selection, through background selection or genetic hitchhiking (indirect selection, selection at linked sites) (Ch. 7)

local adaptation: evolution of higher fitness of a subpopulation to its distinct environment relative to the genotypes of individuals found in other subpopulations, when they experience the same conditions (Ch. 7)

local population: a subset of individuals in a species that may experience restricted gene flow and genetic distinctiveness from other subpopulations (deme, subpopulation) (Ch. 3)

local sample: a collection of individuals from a single subpopulation (Ch. 3)

locus ("low-kuss"; plural = loci, "low-sigh"): a genetic position on a chromosome, which corresponds to a segment of DNA of arbitrary length, often representing a single nucleotide or a coding gene (Ch. 2)

macroevolution: evolutionary change at the scale of divergence between species, usually involving comparison of many species that may share very distant common ancestors (Ch. 1)

major variant: the more common allelic variant at a polymorphic site in a sample (Ch. 3)

Markov chain: a statistical modeling approach that propagates events through time, where the probability of each event only depends on the state of the system in the preceding timestep (Ch. 3)

Markov property: the feature of the Markov statistical process in which the state of a system at the next timestep depends only on its current state, not the particular sequence of events that led to its current state (Ch. 4)

massively parallel DNA sequencing: a generic term for high-throughput DNA sequencing technologies (Ch. 3)

maximum likelihood: a statistical approach to estimating the parameters of a model that aims to identify the model parameter values that best predict the observed data, often applied to gene tree inference from DNA sequence data in phylogenetics (Ch. 5)

McDonald-Kreitman framework: analysis of molecular evolution by partitioning DNA sequences from two species into fixed versus polymorphic differences for each of two categories of sites (usually replacement sites vs. synonymous sites), with one site category presumed to contain only neutral differences; forms the basis of the McDonald-Kreitman test of neutrality and of calculations of the proportion of sites fixed by positive selection, α (MK framework) (Ch. 8)

Mendel's laws: the genetic Law of Segregation holds that allele copies segregate randomly into the haploid gametes; the Law of Independent Assortment holds that alleles of distinct loci segregate independently of one another during gamete formation (Ch. 1)

metapopulation: a set of multiple subpopulations connected by migration (Ch. 7)

metapopulation dynamics: the recurrent process of subpopulation extinction and subsequent re-establishment through migration across a set of multiple subpopulations (Ch. 7)

microevolution: evolutionary change at the scale of allele frequency changes within a species or divergence between closely related species (Ch. 1)

microsatellite: a segment of DNA with a repeated short sequence motif (variable number of tandem repeat locus, simple sequence repeat, short tandem repeat locus) (Ch. 2)

migration: the transfer of genetic material among populations through movement of individuals or gametes (Ch. 2)

migration-drift equilibrium: the distribution of allele frequencies that occur in a subpopulation at the mathematical balance between new allelic input by migration and loss of alleles by genetic drift (Ch. 3)

minor allele frequency (MAF): relative abundance of the rarer variant at a polymorphic site in a sample, used in describing the folded site frequency spectrum (Ch. 3)

minor variant: the rarer variant allele at a polymorphic site in a sample (Ch. 3)

mismatch distribution: the frequency distribution of pairwise differences between haplotypes from a population sample (Ch. 7)

missense mutation: change to a replacement site that yields a different amino acid being encoded (Ch. 2)

model: a representation of reality, usually using simplifying assumptions and often based on mathematic or statistical principles to make quantitative descriptions or predictions (Ch. 2)

molecular clock: the idea from the Neutral Theory of Molecular Evolution that divergence between species accumulates at the rate of mutational input, so that measures of sequence divergence measure the time since common ancestry of species, given constant mutation rates (Ch. 4)

molecular marker: a locus with allelic differences that are measurable with direct or indirect molecular biology techniques (Ch. 3)

mononucleotide repeat: a kind of microsatellite with a single nucleotide base repeated many times in a row, as for TTTTTTTTTT (Ch. 2)

Moran model: an idealized model of a population similar to the Wright-Fisher model except that it better approximates populations with overlapping generations and genetic drift operates twice as fast (Ch. 3)

most recent common ancestor (MRCA): the ancestral haplotype or historical population most proximate in time that gave rise to a given set of descendants (last common ancestor) (Ch. 5)

multifurcate: the splitting of one branch on a gene tree into three or more descendant branches (Ch. 4)

multi-gene family: a set of paralogous loci in a genome that arose through one or more rounds of duplication (Ch. 2)

multiple hit: the phenomenon of >1 mutation having occurred at a given nucleotide position, violating the infinite sites mutational model (Ch. 5)

multiple sequence alignment: linear arrangement of three or more copies of homologous DNA, RNA, or protein sequences from distinct individuals or species, with the aim of identifying conserved versus diverged positions (Ch. 8)

multiple-test correction: a statistical procedure applied to analyses that include many tests of significance in order to control the false positive rate, such as the Bonferroni correction or false discovery rate (FDR) procedure (Ch. 8)

multi-species coalescent: application of coalescent theory to infer the species tree from a collection of many gene trees that include sequences for multiple individuals from each of multiple species (Ch. 5)

mutation: any change to DNA in a cell, with germline mutations that get passed through gametes to the next generation being most relevant to evolution (Ch. 2)

mutation rate heterogeneity: differences in the rate of mutation at different loci in the genome (Ch. 5)

mutational target size: the length of sequence or number of loci that mutation could alter to influence a given trait or fitness (Ch. 2)

mutation-drift equilibrium: the distribution of allele frequencies that occur at the mathematical balance between new mutational input and loss of alleles by genetic drift (Ch. 2)

mutator: an allele that induces a higher rate of mutation in the genome (Ch. 2)

nearly neutral mutation: a mutation that experiences changes in frequency that are sensitive to both genetic drift and natural selection ($4N_e hs$ close to 1) (Ch. 2)

Nearly Neutral Theory: modified view of the Neutral Theory of Molecular Evolution that emphasizes the combined effect of selection and population size in determining whether allele frequency changes will be affected mostly by genetic drift, selection, or both (Ch. 4)

negative frequency-dependent selection: a form of balancing selection in which the rarer alleles at a locus experience the strongest selection, favoring their increase in frequency (Ch. 7)

negative selection: natural selection that favors the elimination of new deleterious mutations that enter a population (purifying selection) (Ch. 2)

neo-functionalization: the phenomenon of one gene copy in a set of gene duplicates acquiring novel functional activity while another copy retains the ancestral functionality (Ch. 2)

neo-sex chromosome: a sex chromosome that recently acquired its role in genetic sex determination in the evolutionary history of the organism (Ch. 7)

neutral locus: jargon term to be avoided; used to refer to selectively unconstrained loci at which any mutational change would not alter fitness relative to other possible allelic states (Ch. 4)

neutral mutation: change to a DNA sequence that does not affect fitness relative to the ancestral allelic state (Ch. 2)

neutral polymorphism: the amount of sequence variation that involves selectively neutral variants at each variant site (Ch. 4)

neutral site: jargon term to be avoided; used to refer to selectively unconstrained nucleotide positions at which any mutational change would not alter fitness relative to other possible allelic states (Ch. 2)

neutral site frequency spectrum: the site frequency spectrum predicted for neutral variants from the Neutral Theory of Molecular Evolution, usually based on the standard neutral model (Ch. 4)

Neutral Theory of Molecular Evolution: a body of mathematical models first introduced by Kimura to describe the interaction between mutation and genetic drift to produce molecular polymorphism within populations and divergence between species (Ch. 4)

non-additive effects: epistatic or dominance effects of an allele on a trait or on fitness (Ch. 6)

non-allelic gene conversion: an outcome of non-crossover recombination, in which a short segment of DNA from one chromosome copies over the DNA of a paralogous region in the genome (Ch. 6)

non-autonomous element: a transposable element that does not encode its own transposition machinery, instead parasitizing the proteins produced by other transposable elements (Ch. 2)

non-crossover recombination: a type of recombination event involving the non-reciprocal exchange of genetic material between sister chromosomes in meiosis, including the copying of short tracts of DNA from one chromosome to the other; non-crossover recombination involves a gene conversion step but these terms are not synonyms (Ch. 6)

non-reciprocal exchange: the outcome of non-crossover recombination and gene conversion, in which a short segment of DNA from one chromosome copies over the DNA on another chromosome (Ch. 6)

nonsense mutation: a mutation that changes a codon from encoding an amino acid to encoding a translation termination signal, leading to premature truncation of the encoded protein (premature stop) (Ch. 2)

non-synonymous site: nucleotide position in a codon that, if changed, would also change the encoded amino acid (replacement site) (Ch. 2)

non-synonymous site substitution rate (K_A): the fraction of replacement sites with substitutions between a pair of species, usually adjusted to correct for multiple mutational hits, intended to summarize divergence in protein sequence evolution (Ch. 5)

normalization: a standardization procedure applied to measurements so that the distribution of values conforms to a convenient form, such as a range from 0 to 1 (Ch. 6)

nucleotide diversity: a measure of genetic variation based on the average number of pairwise differences between DNA sequences in a sample, equivalent to expected heterozygosity at the nucleotide level (θ_π) (Ch. 3)

nucleotide polymorphism: a measure of genetic variation based on the count of sites in a DNA sequence that are variable in a sample of sequences (θ_W) (Ch. 3)

null model: a model used as a reference point to compare against more complicated models, based on fit of observed data to predictions of the models (Ch. 4)

number of segregating sites (S): the count of nucleotide positions that are polymorphic across all individuals in the sample of sequences (Ch. 3)

observed heterozygosity (H_{obs}): proportion of individuals observed to have different alleles at a given diploid locus (Ch. 3)

operational sex ratio: the relative abundance of males and females among breeding individuals (Ch. 7)

orthologous: homologous copies of DNA sequence in different species that descended from the same ancestral sequence, orthologous genes are commonly referred to as "orthologs" (Ch. 4)

outgroup: sequence information from a population or species that is known from other information to be more distantly related to the main set of populations or species under investigation in a gene tree or phylogeny (Ch. 5)

outlier loci: loci that have extreme values of a given statistical metric (Ch. 8)

overdominance: fitness function at a locus defined by highest fitness for heterozygous individuals relative to either of the homozygous genotypes, a form of balancing selection (heterozygote advantage) (Ch. 7)

pairwise-sequential Markovian coalescent (PSMC): an analysis technique for estimating population size through time based on tracts of heterozygosity in a single diploid genome (Ch. 7)

panmixia: a population or species that experiences random mating across its range (Ch. 3)

paradox of variation: the observation that the differences in population genetic diversity among species span a smaller range than one might expect given the differences in population size among species, first pointed out by Richard Lewontin (Ch. 4)

paralogs: homologous copies of DNA sequence at distinct locations in a genome that arose from duplication (Ch. 2)

parameter: an idealized characteristic quantity in a mathematical or statistical model (Ch. 3)

parsimony: the idea that simpler solutions are more likely to be correct; often applied as a means of constructing genealogies or phylogenies based on shared derived characters rather than absolute similarity by minimizing the number of changes required to construct the tree, usually applied to phenotypes rather than DNA sequences (Chs 4, 5)

partial selective sweep: the increase of a beneficial variant by positive selection up to a high frequency but not fixation, potentially due to a sweep that is in progress but as yet incomplete (Ch. 7)

p-distance: the proportion of nucleotide sites that differ between two orthologous sequences, uncorrected for the possibility of multiple hits (Ch. 5)

percent sequence identity: the proportion of nucleotide sites that are the same between two orthologous sequences, the converse of p-distance (Ch. 5)

perfect linkage: a pair of loci with zero crossover recombination between them, corresponding to a recombination fraction of $c = 0$ (complete linkage) (Ch. 6)

phase: the variant combinations along a segment of DNA that define haplotypes, as opposed to unphased diploid genotype information for each variant site in the sequences (Ch. 6)

phylogenetic tree: the evolutionary branching diagram representing the history of speciation and common ancestry among a group of species (phylogeny, species tree) (Ch. 5)

phylogeny: the evolutionary tree branching diagram representing the history of speciation and common ancestry among a group of species (phylogenetic tree, species tree) (Ch. 5)

phylogeography: the application of population genetics to studying the historical factors responsible for present-day geographic distributions of species (Ch. 3)

physical map units: distances between loci measured in number of nucleotides along the DNA sequence (Ch. 6)

plastid: an organelle that contains a genome separate from the cell nucleus, including mito-chondria and chloroplasts (Ch. 7)

point mutation: change of one DNA base nucleotide to another nucleotide (single nucleotide mutation) (Ch. 2)

polarize: to determine the ancestral versus derived state of polymorphic variants or fixed differences (Ch. 8)

polymerase chain reaction (PCR): a technique of molecular biology to amplify small amounts of a DNA segment by repeatedly synthesizing new strands of DNA after separating the two strands of the DNA double helix with heat (Ch. 3)

polyploid: an organism that has more than two copies of each chromosome due to historical whole-genome duplication or failure of meiotic reduction to haploidy (Ch. 2)

pooled sample: a collection of individuals from each of two or more subpopulations (Ch. 3)

population bottleneck: a transient decrease in the size of a population that then rebounds, usually assumed to recover to its pre-bottleneck population size (Ch. 7)

population contraction: decrease in the size of a population (Ch. 7)

population expansion: increase in the size of a population (Ch. 7)

population genetics: the branch of science devoted to the mathematical theory and empirical analysis of changes to gene frequencies over time (Ch. 1)

population mutation rate (θ): a compound population parameter that integrates mutational input into a finite population ($4N_e\mu$ in diploids), equal to nucleotide polymorphism at equilibrium (scaled mutation rate) (Ch. 3)

population parameter: an idealized characteristic quantity in a statistical model of a population (Ch. 3)

population recombination rate (ρ): a compound parameter that captures the overall influence of recombination on allele combinations between partially linked loci, $4N_ec$ at equilibrium in a population (Ch. 6)

population structure: the non-random distribution of individuals in a species that is subdivided into groups (subdivided population) (Ch. 3)

positive selection: natural selection that favors the increase in frequency of a given allele (Ch. 2)

preferred codon: a codon that experiences a fitness advantage over alternative synonymous codon variants (Ch. 8)

premature stop: a mutation that changes a codon from encoding an amino acid to encoding a translation termination signal, leading to premature truncation of the encoded protein (nonsense mutation) (Ch. 2)

private variant: an allelic variant that is unique to one subpopulation (Ch. 8)

probability of fixation: the expected likelihood that an allele will spread to become fixed in a population, given the population size and the allele's frequency, dominance, and selection coefficient (Ch. 4)

probability of identity by descent: probability that two gene copies share their state by virtue of common ancestry (Ch. 3)

product rule: the combined joint probability of a set of independent events is equal to multiplying together their individual probabilities of occurrence (Ch. 6)

protein electrophoresis: method for separating proteins by applying an electric charge to a protein sample in a gel matrix (Ch. 3)

pseudogene: a copy of a gene that experiences relaxed selection because it acquired a disruptive mutation that makes subsequent changes to it selectively neutral (Ch. 2)

purifying selection: natural selection that favors the elimination of new deleterious mutations that enter a population (negative selection) (Ch. 2)

QTL mapping: a set of methods that aim to determine the loci and allelic effects that contribute to phenotypic variation, including use of genome-wide association studies (GWAS) (Ch. 8)

quantitative genetics: the branch of population genetics that studies the genetic basis and evolutionary forces affecting phenotypic variation (Ch. 8)

quantitative trait locus (QTL): a polymorphic locus with alleles that correlate with phenotypic differences in a population (Ch. 8)

random sample: a subset of individuals or gene copies collected at random with respect to those features intended for study (Ch. 3)

rearrangement: movement of a segment of DNA from one location on a chromosome to another location on the same chromosome (Ch. 2)

reciprocal monophyly: given two sets of taxa or haplotypes, the members of each group are more closely related to one another than they are to any members of the other group, reflected in the pattern of common ancestry of the gene tree or species tree (Ch. 5)

reciprocal translocation: exchange of segments of DNA between two non-homologous chromosomes, as from ectopic recombination (Ch. 2)

recombination distance: spacing between loci measured in centiMorgans, based on the likelihood of recombination between them (genetic distance) (Ch. 6)

recombination fraction (c): the proportion of gametes that inherit different combinations of alleles for a pair of heterozygous loci, ranging from 0 up to 0.5 for independent assortment in units of Morgans (Ch. 6)

recurrent selective sweeps: the repeated fixation of beneficial mutations in a genome or segment of DNA by hard selective sweeps (Ch. 7)

recursion: the repeated application of a mathematical function to itself after updating the values from the current iteration (Ch. 5)

relative fitness: the reproductive success of individuals with a given genotype compared to individuals having a different genotype; usually the relative fitness of the genotype with greatest reproductive success is defined to have a value of one (Ch. 2)

relative rate test: phylogenetic comparison requiring at least three species to test for different rates of sequence divergence between lineages (Ch. 8)

relaxed selection: a reduction in the strength of purifying selection on a segment of DNA, as occurs in pseudogenes (Ch. 8)

replacement site: nucleotide position in a codon that, if changed, would also change the encoded amino acid (non-synonymous site) (Ch. 2)

reproductive isolation: the outcome of genetic and environmental factors that keep populations from interbreeding with one another and which thus represent distinct species under the biological species concept (Ch. 5)

repulsion phase: two-locus haplotype combinations with one locus having the most common allele at the locus and the other locus having its least common allele (Ch. 6)

restriction fragment length polymorphism (RFLP): a type of molecular marker that produces distinguishable alleles of different sequence lengths by the presence or absence of endonuclease cut sites, which can be genotyped with gel electrophoresis (Ch. 3)

restriction-site associated digest tag marker (RADtag): a type of molecular marker based on massively parallel sequencing of DNA fragments isolated from endonuclease digestion of a genome (Ch. 3)

reticulation: recombination or hybridization in a gene tree network that results in a given pair of lineages having more immediate ancestors rather than fewer (Ch. 5)

retroelement: a transposable element that uses an RNA intermediate to make a copy of itself for insertion elsewhere in a genome (Ch. 2)

reverse mutation: the ability of mutation to change a derived allele back to its ancestral state (Ch. 2)

reversion mutation: mutation of a derived allele back to its ancestral state (Ch. 2)

root: the most ancestral node in a gene tree or phylogeny, inferred with an outgroup (Ch. 5)

sample: a subset of individuals or gene copies, usually meant to be a representative random collection (Ch. 3)

sample with replacement: statistical sampling process that assumes a random draw of a value from a population will not change the probability of drawing subsequent values (Ch. 5)

sampling error: the statistical noise due to estimating features of the entire population from just a small subset of individuals (Ch. 3)

sampling scheme: an experimental design of collecting individuals from multiple populations, including local samples, pooled samples, and scattered samples (Ch. 7)

saturation: the phenomenon of >1 substitution having occurred at each site in a sequence on average when comparing diverged orthologous sequence, as can happen for synonymous sites in coding sequence alignments between distantly related species (Ch. 5)

scaled mutation rate (θ): a compound population parameter that integrates mutational input into a finite population ($4N_e\mu$ in diploids), equal to nucleotide polymorphism at equilibrium (population mutation rate) (Ch. 3)

scan for selection: statistical analysis of all or most of a genome sequence for signatures of selection (genome scan) (Ch. 8)

scattered sample: a collection of individuals with one representative from each of many subpopulations (Ch. 7)

segmental duplication: repeat of a long chromosomal stretch of DNA sequence, usually >1 Mb in length (Ch. 2)

segregating site: nucleotide position that has two (or more) alternative allelic variants present within a population (Ch. 3)

selection at linked sites: selection that induces allele frequency change on a locus due to its linkage to a direct target of selection through background selection or genetic hitchhiking (linked selection, indirect selection) (Ch. 7)

selection coefficient (s): difference in relative fitness between individuals homozygous for different alleles (Ch. 2)

selective constraint: purifying selection against new deleterious mutations on a segment of DNA that preserves its nucleotide sequence (Ch. 7)

selective interference: natural selection that favors an allele at one locus that inhibits natural selection against a deleterious allele at a linked locus (Ch. 6)

selective sweep: the fixation of a particular allelic variant at a locus (Ch. 7)

selfish genetic element: a DNA sequence that facilitates its own spread within a genome, irrespective of any contribution to organismal fitness, including transposable elements (Ch. 2)

sequence alignment: linear arrangement of multiple copies of homologous DNA, RNA, or protein sequences from distinct individuals or species, with the aim of identifying conserved versus diverged positions (Ch. 4)

sequence divergence: the accumulation of fixed derived mutations in a DNA sequence along a lineage (Ch. 4)

sex chromosome: a chromosome that is found with different ploidy in different sexes (Ch. 7)

sex-linked locus: a genetic locus that is encoded on a sex chromosome (Ch. 7)

sexual conflict: a mode of sexual selection in which the sexes have different optimal fitness strategies over reproduction (Ch. 7)

sexual selection: a form of natural selection in which members of one sex compete for reproductive access to the other sex, or in which members of one sex choose mates of the other sex (Ch. 7)

shared derived character: traits that originated in a common ancestor that are the same in two or more descendant taxa (Ch. 5)

signatures of selection: population genetic features of a segment of DNA that are unusual for the genome overall and consistent with selection having caused their presence, usually identified with tests of neutrality (Ch. 8)

silent site: nucleotide sites in introns, non-coding DNA, or synonymous sites that are often assumed to only experience neutral mutations (Ch. 2)

silent substitution rate: the expected fraction of sites with selectively neutral substitutions between a pair of species, often including synonymous, intronic, and inter-genic sequence; usually adjusted to correct for multiple mutational hits; intended to summarize the accumulation of neutral mutations as a measure of mutation rate and the molecular clock (Ch. 5)

single nucleotide mutation: change of one DNA base nucleotide to another nucleotide (point mutation) (Ch. 2)

single nucleotide polymorphism (SNP): a nucleotide locus with alternative DNA bases found in different copies of the locus in a population (Ch. 2)

singleton: a variant at a polymorphic site that is found in just a single copy in the sample (Ch. 3)

site frequency spectrum (SFS): the abundance distribution for variants that occur at distinct variant allele frequencies in a population sample (allele frequency spectrum, variant frequency spectrum) (Ch. 3)

sliding window analysis: a type of scan for selection that visualizes a population genetic summary statistic as a function of chromosomal position, usually using bins or windows of constant width in nucleotides to calculate the metric from one end of the chromosome to the other (Ch. 9)

soft selective sweep: the fixation of a particular allelic variant at a locus that was subject to positive selection only after having been segregating at moderate frequency in the population, or when the same allelic variant arose by multiple independent mutational events on distinct genetic backgrounds (Ch. 7)

somatic mutation: a change to DNA sequence that occurs in a non-germ cell that will not be transmitted to gametes (Ch. 2)

spatially varying selection: a form of balancing selection in which different alleles are favored in different locations within a species' range, often with different subpopulations experiencing distinct selection pressures (Ch. 7)

species concept: one of dozens of ways of conceiving how to define what is a species, with the biological species concept being most commonly applied (Ch. 5)

species split time: the time at which gene flow ceased between the ancestral populations that gave rise to descendant species (Ch. 5)

species tree: the phylogeny inferred from a collection of multiple gene trees considered in a way that incorporates the partial independence of the histories of different loci (phylogeny, phylogenetic tree) (Ch. 5)

stabilizing selection: natural selection at the phenotypic level that favors intermediate trait values (Ch. 2)

standard neutral model (SNM): the model of molecular evolution stemming from the simplest set of assumptions about a population, an idealized Wright-Fisher population of randomly mating hermaphrodites with non-overlapping generations that is at demographic equilibrium (Ch. 4)

standing variation: the sequence polymorphisms that are currently present in a population (Ch. 4)

start codon: the codon that signals the point of initiation of translation in an mRNA transcript of a coding gene sequence, usually ATG in the standard genetic code that also encodes the amino acid methionine (Ch. 2)

stepping-stone model: an idealized mode of migration in which gene flow occurs only between adjacent subpopulations (Ch. 3)

subdivided population: the non-randomly distributed grouping of individuals in a species (population structure) (Ch. 3)

sub-functionalization: the phenomenon of duplicate gene copies each acquiring just a subset of the ancestral single copy gene's functional activity (Ch. 2)

subpopulation: a subset of individuals in a species that may experience restricted gene flow and genetic distinctiveness from other subpopulations (deme, local population) (Ch. 3)

substitution: a mutation that has fixed to have become the only allele in a population (Ch. 2)

substitution matrix: a molecular evolutionary model of divergence for substitutions between the four nucleotides, such as the Jukes-Cantor or Kimura 2-parameter rate matrices (Ch. 5)

substitution rate: the amount of accumulated sequence change along a lineage, usually measured on a per-site basis with a metric like K_A or K_S (Ch. 4)

summary statistic: metrics that estimate population genetic parameters or features to aid in biological interpretation, such as θ_W, r^2, and Tajima's D (Ch. 3)

supergene: a set of linked loci that interact non-additively to influence fitness (coadapted gene complex) (Ch. 6)

synonymous site: nucleotide position in a codon that, if changed, would not alter the encoded amino acid (Ch. 2)

synonymous site substitution rate (K_S): the fraction of synonymous sites with substitutions between a pair of species, usually adjusted to correct for multiple mutational hits, intended to summarize the accumulation of neutral mutations as a measure of mutation rate and the molecular clock (Ch. 5)

table of polymorphism: a data matrix summarizing the variants among sequence copies in a sample at polymorphic sites only and excluding monomorphic positions (Ch. 3)

tandem duplicate: paralogous copies of DNA that are directly adjacent along a chromosome (Ch. 2)

target of selection: a particular locus with allelic differences that confer different fitness effects (Ch. 7)

taxon: a group of individuals that comprise a biological unit, usually a species (Ch. 5)

temporally varying selection: a form of balancing selection in which selection favors distinct alleles at different points in time across generations (Ch. 7)

test of neutrality: a quantitative, statistical comparison of a feature of observed molecular data to theoretical predictions based on a null model of evolution, usually involving a summary statistic (e.g. D_{Taj}) and the Neutral Theory of Molecular Evolution (Ch. 1)

time to first coalescence: the expected duration to coalescence in a common ancestor for a random pair of haplotypes using coalescent theory (Ch. 5)

tip: a terminal node on a gene tree, corresponding to an observed molecular sequence (Ch. 5)

topology: the pattern of branching relationships among nodes in a gene tree or phylogeny (Ch. 5)

transcriptome: the subset of the genome that gets transcribed in a given set of cells, usually referring to mRNAs but can also include small-RNAs that are identified by reverse-transcription and DNA sequencing (Ch. 3)

transition mutation: a point mutation from a purine nucleotide to another purine (adenine, A; guanine, G), or from a pyrimidine nucleotide to another pyrimidine (cytosine, C; thymine, T) (Ch. 2)

translocation: movement of a segment of DNA from one location on a chromosome to a different non-homologous chromosome (Ch. 2)

transposable element: a DNA sequence that facilitates its own copy number spread within a genome, irrespective of any contribution to organismal fitness (Ch. 2)

trans-specific polymorphism: allelic variation that was present in the common ancestor of different species and persisted as polymorphisms in their descendants to the present day (Ch. 7)

transversion mutation: a point mutation from a purine to a pyrimidine nucleotide or vice versa (Ch. 2)

tree length: total length of all branches in a gene tree, where branches are scaled by number of derived mutations (Ch. 5)

under neutrality: jargon term to refer to a situation in which we draw on the assumptions of the Neutral Theory of Molecular Evolution (Ch. 8)

unfolded site frequency spectrum: version of the site frequency spectrum that includes variant frequencies from near 0 up to 1 by distinguishing derived and ancestral variants (Ch. 3)

unlinked: a pair of loci with alleles that segregate independently, as for loci on different chromosomes, corresponding to a recombination fraction of $c = 0.5$ (free recombination, independent assortment) (Ch. 6)

valley of polymorphism: a segment of DNA in the genome that has reduced population genetic variation, usually identified in a sliding window analysis along a chromosome (Ch. 8)

variable number of tandem repeat locus (VNTR): a segment of DNA with a repeated sequence motif (microsatellite, simple sequence repeat, short tandem repeat locus) (Ch. 2)

variant: an allele as applied to a locus defined as a single nucleotide position (Ch. 2)

variant frequency spectrum: the abundance distribution for variants that occur at distinct variant allele frequencies in a population sample (site frequency spectrum, allele frequency spectrum) (Ch. 3)

variant site: the single nucleotide position in a DNA sequence that has alternative variant allelic types in a population (Ch. 2)

Wahlund effect: the production of a deficit of heterozygotes in a pooled sample of multiple subpopulations relative to Hardy-Weinberg expectations for a single population, with the pooling also yielding an excess of linkage disequilibrium termed the two-locus Wahlund effect (Ch. 7)

Wright-Fisher population: a mathematically convenient idealized model of a population, in which N diploid hermaphrodite individuals mate at random with non-overlapping generations and no changes to population size (Ch. 3)

Wright's hierarchical F statistics: a set of statistical metrics devised by Sewall Wright to summarize the degree of difference from an idealized randomly mating population, including F_{IS}, F_{IT}, and F_{ST} (fixation indices) (Ch. 3)

Index

Boxes are indicated by an italic *b* following the page number. Entries in **bold** can be found in the glossaries.